Pornographic Archaeology

To Lynn and Pep,

this

Pornographic Archaeology

Medicine, Medievalism, and the Invention of the French Nation

not yet rated!

Zrinka Stahuljak

With great admiration and love,

PENN

UNIVERSITY OF PENNSYLVANIA PRESS

PHILADELPHIA

Published by
University of Pennsylvania Press
Philadelphia, Pennsylvania 19104-4112
www.upenn.edu/pennpress

Printed in the United States of America on acid-free paper

10 9 8 7 6 5 4 3 2 1

A Cataloging-in-Publication record is available from the Library of Congress

ISBN 978-0-8122-4447-2

Pour Laure, soleil de ma vie

Contents

Note on Translations

The modern French texts quoted in this book have been translated by Alisa Belanger, with my assistance, except for the Introduction and portions of Chapters 4, 5, and 6, where I translated with her assistance. Whenever possible, I quote from published English translations. In cases where published English translations are unavailable, out of date or not easily accessible, new translation is provided and the original modern French appears throughout in the endnotes. In cases where the official published translation is used, page numbers in square brackets in-text refer to the text of original French editions, but the original text is not provided. All other translations are my own, unless otherwise noted.

The translation of the French "mœurs" deserves particular mention, since it covers a wide range of meanings in English depending on context. Most often "mœurs" is rendered as "mores," "customs and morals," and "culture," and less frequently as "manners," "traditions," "lifestyle," and "conduct." "Histoire des mœurs" is translated as "history of customs and morals" and "history of culture," the latter term taken from the *Encyclopedia Britannica* (11th edition, 1910–1911) or left in the original. The term "historien des mœurs" is rendered as "historian of customs and morals" or "historian of culture."

Introduction

Sex and Nation

In 1896, Marc André Raffalovich (1864–1934), author of the famous *Uranisme et unisexualité*, expressed his hopes and testified to the promise that medieval literature in general, and courtly literature in particular, will contribute to the making of the history of sexuality: "The sustained study of romances of chivalry and literature, known only to scholars and not even touched upon by amateurs, would lead to many an interesting discovery. Let us hope that this fertile field will not remain neglected for too long; as the number of scholars increases, our demands on them also grow and new ones emerge. And the history of unisexuality calls for such study."[1] Almost forty years later, in 1934, André Gide posed a series of questions to Georges Hérelle, translator and author of an unpublished "Nouvelles études sur l'amour grec."[2] In his letter to Hérelle, Gide critiqued the exclusive focus on literature among scholars that, according to him, should be set aside because the state of knowledge in 1934 of medieval sexual practices is poor, incorrect, and possibly misguided, unable to break through the dissimulation of literature:

> What in reality were medieval mores? Does literature inform us enough about the mores of that period? This literature, so tender and delightful, is quite idealist and quite artificial; but does it tell us the truth? What is hiding behind this charming decor? . . . Where and how should we find such information? For example, what was the role of the pages, young companions to knights? The latter professed mystic love: we only hear of that love, it is the only one brought to the fore. But how may we suppose that all these vigorous men remained chaste? And toward whom were directed their erotic desires? Knowing to what extent reality today differs from appearance and how deceptive the dissimulation of mores proves to be, it is reasonable to believe that this deception is not specific

> to our times and that it was even more opaque in the times when opinion, being harsher, required more dissimulation. This problem should be studied in secret chronicles, in criminal trials, in monastic documents, without taking literature into account.[3]

Thus in the forty years that separate Raffalovich's remarks from Gide's, the promise that the study of medieval literature would illuminate the history of sexuality was not fulfilled; to the contrary, the promotion of medieval "mystic love," known to us as courtly and chivalric love, only added to the obscurity of medieval sexual culture, as it eclipsed other source documents, chronicles, trial records, monastic penitentials, and manuals. The 1934 date of Gide's letter is of interest here, since it was the foundation in 1929 of the Annales School by Marc Bloch and Lucien Febvre that opened the doors to the study of the medieval body.[4] It was precisely in 1934 that Marcel Mauss began to speak of "the techniques of the body," "the ways in which men know how to use their body."[5] The body has a history and, starting in the 1930s, it will no longer be understood as a natural given—it is to that history of the body that Gide appealed. But it was only in the 1970s and 1980s that sex and sexuality were included more broadly in medieval historical studies in France, under the impetus of cultural history, and in the 1990s in the United States, with the advancement and legitimation of gender and queer studies.

If Marc André Raffalovich was sure that the medieval period held such promise for the history of sexuality, it was because the nineteenth century had already been filled with the explorations of medieval sexuality. André Gide's letter to Georges Hérelle likewise shows that sources overshadowed by literary study of mystic love were numerous and varied. Indeed, the Middle Ages, a central battlefield in the ideological, political, and aesthetic confrontations over the status of history and the construction of the present for the French nation, were of great interest to antiquarians, archaeologists, writers, doctors, historians, bibliophiles, men of religion, and lawyers, who explored them through the lens of sexuality. In particular, nineteenth-century general medicine (both physiology and psychiatry) and later sexology were deeply interested in the past, and especially the Middle Ages. It is the aim of *Pornographic Archaeology* to restore this rich medical body of work to the larger discussions around the French nation and medieval studies. The book attempts to get at its subject from three different angles, reflected in its tripartite division: sex and blood; sex and race; and sex and love. Each section has a three-fold purpose, namely to show how a particular medieval concept acted upon the nineteenth-century medical opinions on social issues, as well as upon medical theories and hypotheses; how a certain idea of the Middle Ages was constructed around medical notions in the nineteenth century; and how this medical construction of the Middle Ages—medical medievalism—invariably shaped a certain way of thinking about the Middle Ages until the very end of the twentieth century and, in some cases, until today.

Pornographic Archaeology is a cultural history of medical medievalism. At its core is the notion of the double transference of ideas between French medievalists (roughly divided between historians, writers, and philologists) and their professional counterparts practicing medicine. The nineteenth-century milieu and intellectual climate influenced medievalists and conditioned their thinking about the Middle Ages; at the same time, the opinions and ideologies professed and promoted by nineteenth-century doctors were formed by historical knowledge (*le médiéval*) as much as by fictions and fantasies (*le moyenâgeux*) of the Middle Ages.[6] My first objective is to elucidate the impact of these historical discussions concerning the French Middle Ages on medical scientific categories and theories which, moreover, underpin national and colonial thinking; namely, the ways in which nineteenth-century constructions of medieval sexuality are implicated in the contemporary construction of nation and empire, and in the definitions of race, hygiene, marriage, and family. The second objective is to analyze the medical constructions of medieval sexuality in nineteenth-century France and their impact on the conceptualization of the "French Middle Ages." In the margins of these two lines of inquiry, the book also pursues the impact of medical medievalism on the formation of the academic discipline of medieval studies and its professionals. It questions why and how the discourses on the nation and its sexuality were cleft at the foundation of academic medieval studies: Why did philology and sexology embark on parallel but separate tracks, even though medicine valued the medieval past as evidence of its theories? How did the French academic discipline of medieval studies, though scientific and positivistic in its practice of philology, exclude *scientia sexualis*, the study of sexuality that was at the height of its expansion precisely at the time of the institutional formation of medieval studies in France in the 1880s? While the inquiry as to *why* sexuality was occluded in the study of the earliest national literature bears out a fairly obvious answer—all disciplines born in that period and their historiographies have excluded sexuality and some continue to do so—*how* academic medieval studies did this is an important question to answer because of the modalities of exclusion of sexuality particular to academic medieval studies and its effects on disciplinary assumptions.

"Sexuality" is here a set of discourses about sex and the effects that sexual practices and relations have, following Michel Foucault's notion of the apparatus of sexuality (*le dispositif de sexualité*), which regulated, disciplined, and controlled the individual and national body through various pedagogical, medical, and economic institutions: body hygiene, marital hygiene, fertility, birthrate, longevity, public health, housing, migration, and the improvement and expansion of the human race were all discourses of sexuality.[7] In order to encompass this large epistemic space, sexuality is analyzed as genealogy and heredity, race and miscegenation, venereal disease, vice, sodomy, virginity, prostitution, adultery, and, finally, courtly love and marriage.[8] The focus placed on the pathological is conditioned by nineteenth-century medical thought. According to Doctor Cabanès (1862–1928), a great early twentieth-century popularizer of nineteenth-century medical theories, only crisis can inform because crisis delineates the norm by exploring its obverse: "It is easier . . . to study different natures in their crises than in their normal state: their normal state lets us see only the surface and conceals innermost recesses; morbid disorders bring out more clearly what health is."[9]

In order to paint the cultural sphere of medical medievalism, this book encompasses popularizing and scientific discourses. It takes a broad view of the medical-historical scientific field and includes popularizers of scientific knowledge (e.g., Paul Lacroix, Pierre Larousse, Auguste Brachet, Dr. Cabanès, Dr. Garnier, Dr. Fauconney) alongside disciplinary and academic heavyweights (e.g., Jules Michelet, Gaston Paris, Benedict-Augustin Morel, Émile Littré, Richard von Krafft-Ebing, Julien Chevalier). Both groups are equally important to the book's argument because they represent the wide reach of scientific thought, in erudite circles and in the French population, since these scientific passions and interests did not remain limited to a specialized audience but spread among the general public, manifest in "the infatuation [*engouement*] that the public shows today for the *medical point of view*."[10] The book covers the period from the early nineteenth to the early twentieth century. In 1805, François Raynouard's play *Les Templiers* was performed to great critical acclaim and public success, and it marked the entry of sexual exoticism of the Middle Ages into the public sphere. The publication of Raynouard's play contained a sort of a primer on the history of the Order of the Templars and of their trial on charges of heresy and sodomy, as well as Raynouard's plea for their rehabilitation in national history (Chapter 3).[11] The last period surveyed in the book, 1905–1910, was the time when marriage manuals and books popularizing medical knowledge proliferated among the bourgeois and working classes, importing and generalizing a very particular, non-academic vision of love borrowed from the Middle Ages (Chapter 6).

This book takes its title from Paul Lacroix (1806–1884), a major popularizer and historian of customs and morals (*histoire des mœurs*). Lacroix became an important source for doctors but has been completely excluded by the late nineteenth-century academic medievalists and modern historiographers of French medieval studies (Chapter 5). Lacroix described his six-volume *Histoire de la prostitution, chez tous les peuples du monde depuis l'antiquité la plus reculée jusqu'à nos jours* (History of prostitution among all of the peoples of the world from the earliest antiquity to the current day) (1851–1853) as "pornographic archaeology" (*archéologie pornographique*).[12] Both terms went against the grain of an accepted and acceptable wisdom for its time. Etymologically, "pornographic" referred "above all to the books on prostitution," that is historical, moral, and hygienic studies of prostitution.[13] But by the 1850s, when Lacroix was writing, "pornography" in the meaning of "prostitution" was all but outdated; the connotation was increasingly one of explicit sexual arousal by text and/or image. Lacroix thus played on the ambiguity the term had in the mid-nineteenth century in order to attract readership while simultaneously claiming erudition as a warrant against certain side effects of reading (Epilogue). According to the definition in the *Grand dictionnaire universel du XIXe siècle*, Pierre Larousse's "Encyclopedia of the Nineteenth Century,"[14] Lacroix's other term, archaeology, is a discipline that "makes us penetrate the inside, into the depths of that life [of societies]; it teaches us the mores, traditions, customs, and beliefs of the ancients."[15] But archaeology that dug "into the depths of that life" belonged to the history of culture (*histoire des mœurs*) and was thus opposed to history proper—institutional, political history, the history of events. Events were the central preoccupation of the philosophy of history in the nineteenth century, "*the century of history*."[16] History was concerned with "general facts regarding politics, war, revolutions and . . . remarkable events that broadly outline the profile of peoples—in a word, the dramatic and external life of societies."[17] The opposition of event versus *le quotidien* (everyday life), to use Michel de Certeau's term,[18] sums up the opposition of history to archaeology. Archaeology studied the practice of everyday life, including sexual practices and customs. By contrast, history proper did not treat the life of the people from within, on the "inside," but saw events determine human life. By choosing to write a "pornographic archaeology," an "*histoire des mœurs*" (Lacroix, 6:90), Lacroix wrote a counter-historical narrative on the cusp of an epistemic shift (in the meaning of pornography) and disciplinary lesions (history vs. archaeology). These are the cultural and disciplinary tensions that *Pornographic Archaeology* sets out to explore and elucidate.

Philology and *Scientia Sexualis*

It is a well-known fact that the nineteenth century rediscovered the Middle Ages. The French academic discipline of medieval studies was itself born out of the slow and fragile establishment of the French Republic and the nineteenth-century search for national origins in the Middle Ages, a search reaffirmed and mythologized when in December 1870 French philologists grouped around Gaston Paris adopted the French epic *La Chanson de Roland* as the linguistic and literary origin of the nation, in the wake of the devastating Franco-Prussian War (which the French lost) and the declaration of the Third French Republic on September 4, 1870.[19] The double transference between the nineteenth-century preoccupations with the nation and the Middle Ages led not only to the establishment of academic medieval studies, but also to the construction and reification of French national identity through "realms of memory," to employ the term of Pierre Nora's *Les lieux de mémoire*, in medieval figures and sites such as Charlemagne, Joan of Arc, and Notre-Dame of Paris.

In the 1990s a number of volumes and articles were published in the United States with the objective of understanding these nineteenth-century disciplinary origins and the nationalist desires and fantasies at the foundation of academic medieval studies. What these studies had in common was the predisposition "to explore and interrogate presuppositions underlying current philological practices"[20] in order to "reformulate assumptions about the discipline of medieval studies broadly conceived."[21] Following the lead of Paul Zumthor, who advocated already in 1980 "a mistrust toward the very materials of these works, toward the language manifested in these texts, refusing to accept it as existing in nature [*comme nature*],"[22] the North American medievalists criticized the philological status of text as history and language as transparency, both of which denied the theoretical condition of possibility for the text's interpretation and "contributed to the theory of the 'natural text' . . . 'the work which speaks for itself' . . . the fantasy of a language of beginnings, a poetic vehicle that remains to its object invisible."[23] In attempting to understand the fallacies of philology, these new histories of medieval studies wanted to look more broadly at how the discipline was "determined by the specific ideological or local, nationalistic, or religious, political or personal interests" of the fathers of the academic discipline of medieval studies.[24] An archaeology of medieval studies became a legitimate part of a medievalist's work, aiming to interrogate on a permanent basis "mental habits inherited from the nineteenth century" (Zumthor, p. 16 [p. 23]), although the above list of interests that formed the discipline did not cover sexual or imperial interests. This challenge to nationalist nineteenth-century philology has led directly to the rapid growth of postcolonial medieval studies, in the analysis of the formation of national medieval literatures (e.g., after the Norman conquest) and in the opening of national medieval literatures and their scholarship to analysis of extra-metropolitan, "peripheral" influences, exchanges, and interconnectivity (e.g., the travels of Marco Polo).[25]

Sexual interests were also explored in the same time period but under separate cover. As medievalists were questioning their "mental habits" inherited from the nationalist nineteenth century, there was a parallel and seemingly unrelated movement in medieval studies: the advent of the history of medieval sexuality and queer studies in the wake of Michel Foucault's *History of Sexuality*.[26] The Foucauldian insight that truth is an effect of relations of power and that truth is thus contingent on discourses of its time—making truth nothing more than a truth claim—opened new areas of inquiry for medievalists. Medievalists were finally poised to ask what this construct of "sexuality" and its truth claims were in the Middle Ages. However, while we have gained much knowledge of the history of medieval sexuality, medievalists have not questioned as thoroughly the subtle underlying assumptions about medieval sexuality that were imagined by the nineteenth century and that continue to influence medieval studies more broadly, beyond the immediate issues of history of medieval sexuality. Already in 1998, Kathleen Biddick took "new medievalists" to task because their solution was "to distance medieval studies, not from the techniques, but from the 'fathers,'" which in her view changed little as to the methods that continue to prop up the academic discipline of medieval studies to this day.[27] I agree with Biddick and stake the claim throughout this book that it is impossible to write a philological history of nation and empire without the history of sexuality. Both the medieval discourse on sexuality and the nineteenth-century discourse on medieval sexuality are effects of relations of power; each period constructed an object of knowledge whose truth was grounded in contingency, and each period therefore deserves a careful exploration of discourses that constructed it. Moreover, these two simultaneous revisions, philological-national and sexual, of our disciplinary knowledge of the medieval period have remained parallel but separate, uninterrogated at the source of their cleavage, even though they were constructed over 100 years ago by the same nineteenth-century discourse: positivism. Thus, if in the 1990s we interrogated the assumptions about the transparency of language that subtended the national-philological practice, a parallel inquiry into assumptions of scientificity and historicity that play a part in medical medievalism remains to be conducted.

These recent explorations and revisions of the Middle Ages have remained inscribed in the nineteenth-century division of philology from sexology, the division that this book aims to understand.[28] Previous work on the relationship of the nineteenth century and the Middle Ages focused exclusively on the uses of medieval history in the construction of nation, but it neglected the role that medicine played in the larger intersection of sex, nation, and race in the nineteenth-century French nationalist, capitalist, and colonial project. Ann Laura Stoler has argued that we must understand "how a history of empire affects the history of European sexuality"; in her view, it is impossible to write a history of the nation and empire without a history of sexuality.[29] *Pornographic Archaeology* explores how a history of nation and empire, constructed by the medical discourses that return repeatedly and obsessively to its national—medieval—past, has affected the history of European sexuality. The book thus argues that the metropole and the colony triangulate with the past, and that the relationship between the present and the past passes through the colony rather than just the metropole. Furthermore, by proposing that academic medieval studies and *scientia sexualis* were formative of each other and by exploring their confluence, the book fills, on the one hand, a lacuna created in the exclusive focus on the nation as formative of medieval studies and, on the other, a lacuna in the history of sexuality and queer studies of the medieval period. By the same token, this is not a book on the representations of medieval sexuality or, more generally, on representations of the Middle Ages in the culture and literature of the nineteenth century.[30] It is a book on the dialogical convergence of medievalism and medicine in the creation of "medical medievalism," which both shaped the social fabric of the nineteenth century and is the foundation on which medieval studies continues to rest.

Medicine and the Middle Ages

It is difficult to overstate the social importance of medicine and its practitioners in nineteenth-century France. After July 1789 and the fall of the *ancien régime*, doctors entered public life and became engaged citizens, participating in the political life and nation-building, guided by their individual political convictions. Medical space began to coincide with social space. Two preoccupations of equal importance emerged for medicine: the care of the body and the care of the mind, hence medical authority in the physiological and psychological domains.[31] After the Constitutional Assembly abolished the Church in 1793, mental care (conceived as the care of the soul under the *ancien régime*) fell to the newly formed corps of alienists.[32] The next major step in the medical domination of social space was broached with the professionalization of schools of medicine in the opening years of the nineteenth century. Philippe Pinel's "moral treatment" was consecrated when Napoleon granted the medical corps monopoly in 1803. In the same year, the Écoles de santé became Écoles de médecine, changing their name once again in 1808 to Facultés de médecine, and the practitioners of medicine were now licensed by the state. The law of 1838 made mental health the prerogative of the state, in the portfolio of the Ministry of Interior, formalizing the procedure of certification and commission of individuals to asylums.[33] Medicine established itself as a clinical science as much as a social science at the intersection of sociopolitical, cultural, and professional circumstances and interests. After 1793, medicine gradually displaced the Church in its social role and, in particular, as authority on sexuality. This meant that medicine was permeated by the same tensions that inhabited the nineteenth century, namely the continuing clash over the role of religion and the Catholic Church in the nation-state, which lasted from the Revolution until the law on the "Separation of the Churches and the State" (*Séparation des Églises et de l'État*) in 1905. It is important to note that not all doctors were atheists or agnostics and not all Catholics were against science and medicine. Neither were all Republicans secularists nor all monarchists ultra-conservative. Later in the nineteenth century, doctors' political convictions shaped imperial medicine. Medicine was integral to the Western civilizing mission in the colonies, taming diseases and practices, improving the health of the natives while protecting the colonizers and the metropole from imported epidemics and habits. Imperial medicine "became a demonstration of the superior political, technical and military power of the West, and hence a celebration of imperialism itself."[34]

Nineteenth-century doctors practiced a historicist medicine that traced and outlined genealogies of diseases, vices, and sexual practices from the Middle Ages to the nineteenth century. This tradition of historicist medicine was established early in the nineteenth century, when psychiatrists, in trying to understand the devastating consequences of the French Revolution on the human mind, looked for the past examples of madness and identified mystical exaltation of the senses, a common medieval phenomenon, as the probable cause of insanity. This historical sense of medicine was part of the larger reconstitution of French national history in the wake of the traumatic revolutionary rupture: "To understand the Revolution is to glean all of its positive consequences in order to avert its repetition. And to understand it, one has to return to its origins."[35] Doctors thus shared the same impulse as the early historians of France writing prior to the revolutions of 1848, when "the Middle Ages guaranteed . . . continuity with the past and promised a possible future" (Boureau, p. 70 [p. 84]). The purpose of the doctors' historical research was multiple: at once, to look for the etiology of a disease and provide its genealogy, to furnish historical evidence for their medical theories, and, finally, to identify retrospectively diseases whose terminology did not exist before and thus to exercise a "retrospective medicine" (*médecine rétrospective*).[36]

Retrospective medicine saw the Middle Ages as its cradle, the origin of a continuous civilization that led to the present of the nineteenth century. This continuity made it distinct from antiquity, from which a definitive rupture was felt: "In following the chronological order of epochs, we reach the great period of the Middle Ages, which, closer to us, will give more precision to our ideas and furnish information of many kinds."[37] But this national medieval origin is also from the beginning an imperial history, with the foundational moment in medical histories always returning to the first four Crusades of the eleventh to thirteenth century (1096–1204). The uprooting of the West in the pursuit of its eastern conquests caused an enormous upheaval during the Crusades: "We witnessed . . . the West tear itself away from its foundations in order to charge at the East."[38] The loss of the moral, sexual, and civilizational compass is summarized in a well-known article by Alexandre Brierre de Boismont:

> Such exaltation was singularly favorable to the development of mental disorders. The annals of the Crusades are therefore replete with visions of angels and saints, divine revelations, marvelous feats. Hallucinations of sight and hearing are very common; the religious and warring madness belong above all to the era of the first two Crusades. In the third crusade, the warrior customs of the Franks abandon their eminently distinct epic character to take on the character of romance. It is the reign of troubadours and knights, who, turning the imagination toward love and glory, lead amorous and chivalric madness to flourish.

> Erotomania, nymphomania, hysteria in all its vagaries, heroic mania, are all distinctive traits of this period, represented by two main figures, Roland and King Arthur.[39]

The cultural and sexual contacts established in the Muslim East produced "erotic melancholia" that was imported into the European West: "The chivalric spirit in the wake of the Crusades multiplied the cases of erotic melancholia."[40] Often, these medical genealogies considered diseases and sexual practices to be immutable and laid claims to universal, transhistorical essences and natures, reaffirming the Middle Ages as the origin of nineteenth-century French race and sexuality. Thus the medical obsession with finding the seat of the illness or infection in the individual's diseased body found its analog in the search for the origins of diseases in the national past.

Sex and Nation

Medical-historical discourses on sexuality were central to creating a network of correlations between two major nineteenth-century ideological constructs—"the Middle Ages" and "the French nation." Alain Boureau's 1995 *Le droit de cuissage* showed, albeit indirectly, that there is no history of the nation without the history of its medieval sexuality and sexual practices. The debate over the *droit de cuissage* spanned three decades, from 1854 to 1882. *Ius primae noctis*, or the lord's first night, was the putative medieval custom allowing feudal lords to spend the first night after marriage with the newly married brides of their subjects. This physical act was believed to be at the origin of a symbolic tax levied by the lord on his subjects who married, but also, more scandalously, by the clergymen. Instead of focusing on sexuality, Boureau's book traced the political uses and abuses of this fragment of medieval history in the battle over the construction of the French nation. The practice of *cuissage* crystallized what the Middle Ages were about in the wake of the two 1848 revolutions: on the one hand, the enlightened Middle Ages led by "a severe but just Church, . . . the great pacifier" that moderates the abuse of the feudal powers of the lords and, on the other, "a contrary image of a Middle Ages of obscurantism and fanaticism," born from an ancient oppressive alliance of feudalism and the Church (Boureau, p. 72 [p. 86]). The debate transformed the historical narratives of the French nation. Conservative historians created a history of "anachronistic comparisons" (Boureau, p. 83 [p. 98]), a cyclical history that made the Middle

Ages a mirror of the present and in which the Church and the monarchy "protected subjects from the encroachment of secular powers" (Boureau, p. 83 [p. 97]), that is, against the abusive feudal power of overlords. Their view contradicted liberal historians' notion of evolutionary history that saw the Middle Ages as the origin of human rights, a "continuity emerging from Frankish freedoms and urban communes of the twelfth century" (Boureau, p. 75 [p. 89]). According to this Republican version of history, the Middle Ages led to the Revolution, in which the third estate (*le tiers état*) was opposed to the bond between the Church and "a feudal despotism backed by the monarchy" (Boureau, p. 69 [p. 83]). Partisans of each viewpoint used the putative custom of the lord's first night to shape their vision of the Middle Ages that reflected exactly the social, political, and religious battles in post-Revolutionary France between conservatives and progressives, "the Middle Ages as the matrix of the order of their own century" (Boureau, p. 70 [p. 85]). In both views, the Middle Ages were an ideological construct, whether they served as a model or as a foil. The debate that raged was not entirely about the historical existence of the usage itself; rather, the pitched battle between the two political currents concerned whether or not the Church supported, and even practiced, the levy imposed by the feudal lords on its subjects. The battle was bitter and relentless because its outcome would determine which version of medieval, and thus national, history would be written for future generations: a history that would proclaim "a dreamy political nostalgia, or rational and social rejection" for the Middle Ages (Boureau, p. 40 [p. 51]). The debate over the *droit de cuissage* set a blueprint for the organization of knowledge about the Middle Ages, in which they emerged as always different from the present. And although this debate shaped a vision of the medieval period for generations of historians, it was not about the knowledge of the medieval period for its own sake, but precisely about the political role of the historical Middle Ages in the French post-Revolutionary period.

Boureau thus noted indirectly, and without exploring it, the role that medieval sexuality, whether as actual practice or as "a pure effect of discourse" (Boureau, p. 224 [p. 251]), played in the constitution of the national(ist) nineteenth-century present. A political debate about the origins of the French nation and its nation-building that was articulated around the question of whether or not there existed a sexual right and a sexual practice shows that sex and sexuality were implicated in the foundation of the nation and that sex was a major stake for the nation. The present book elaborates the idea that it was not just the debates among historians about sexual practices in the past

(Boureau's focus) that constituted the French nation and the Middle Ages; in addition, medical debates on sexual practices in the past also constituted the French nation and the Middle Ages, respectively. Furthermore, the public debate over the lord's first night shows how arguments over sexual practices translated, on the one hand, into contemporary nation-building, and, on the other hand, into a construct of the remote past whereby the Middle Ages were portrayed as the nation's essential, albeit radically other, point of origin. When various nineteenth-century discourses (historical, legal, medical) were locating the origin of France in the Middle Ages, they were inventing a notion of the Middle Ages from debates as much about the nation as about sexuality. That is, if the Middle Ages were invented by nineteenth-century nationalist desires, they were also invented by nineteenth-century opinions, fears, and fantasies about medieval sexuality.

The history of the nation could not be, and indeed was not, divorced from the history of its own sexuality. The genealogy of the nation and French race depended on its medieval sexuality, which was perceived to dictate the relations of blood and nation: healthy blood and healthy heredity in the past made for a healthy nation in the present. If nineteenth-century France (and European civilization) was perceived to be in decline, was it not because of morbidity introduced in the Middle Ages? The future of the French race was contingent on the classification and etiology of pathologies and on the normalization and regulation of sexual practices: what could be learned from the past about pathologies and morbid heredity? For instance, sexual perversions—a pathological behavior—were seen to be a manifestation of morbid heredity and civilizational degeneration, and lessons from history instructed doctors to push for the regulation of "deviance" in contemporary France. Likewise, venereal diseases had their epidemiological and etiological histories in the medieval "colonies" that could help determine the need for the sanitary cordon around the metropole and could influence the course of colonial public health policies. In short, the stakes for the genealogy of the French "race" lay as much in the present as in the past.

Sex and Blood

In Part I, "Sex and Blood," the history of the Middle Ages is put into relation with the favorite scientific perspective of the nineteenth century: theories of hereditarianism and hereditary degeneration (Chapter 1). Heredity refers to the inheritance of normal and morbid physical and mental characteristics and dispositions over generations of blood descendants. Heredity was a historical concept: medicine could formulate it as an autonomous epistemic space only out of the *longue durée* analysis of generations of descendants and, further, medicine relied on history and past examples for proof of its theoretical tenets. In short, this meant that the dominant nineteenth-century medical theory was impracticable without history. Heredity was in many ways a scientific transformation of the aristocratic concept of genealogy, since it looked for the earliest blood antecedent, just as genealogy looked for the founder of a lineage. The kinship of heredity and genealogy is especially apparent in the debate over the role of consanguinity in heredity. Doctors on both sides of the divide inevitably resorted to the aristocratic Middle Ages and its dynastic genealogies to prove either their anti-consanguinist or consanguinist position. Although the bulk of the debate took place between 1856 and 1866, for several decades consanguinity continued to be a flashpoint of scientific discussions on pathology and heredity, as well as the legislation of marriages, and it reflects the contemporary imbrication of nation, race, and class. We can read the consanguinity debate as a reflection on the nation's history of institutions and class, as much as a disciplinary technology of population.

There was also a converse effect of the historicization of medicine: the medicalization of history. Doctors argued that the course of historical events and institutional and political history could no longer be understood without the medical explanation of psychology and physiology of history's main actors, without looking at how "the health of a given historical figure had exerted considerable influence on the ideas, the doctrines of an age."[41] Doctors had to perform a "pathologic archaeology" (*archéologie pathologique*)[42] in order to understand, for instance, the erratic behavior of Francis I and Louis XIV, after they had both suffered from abscesses. In medical terms, their reigns could be defined as "before and after the fistula."[43] Medicine and history were thus engaged in a dialectic: history taught medicine (*histoire de la médecine*) but medicine alone could understand history (*médecine de l'histoire*) and medicine alone could claim the status of the philosophy of history: "the philosophy of history, the medicine of history."[44] Medicine was always already historical and history was always already a medical category.

That heredity was a discourse of sexuality was first highlighted by Michel Foucault:

> The bourgeoisie's "blood" was its sex. And this is more than a play on words; many of the themes characteristic of the caste manners of the nobility reappeared in the nineteenth-century bourgeoisie, but in the guise of biological, medical, or eugenic precepts. The concern with genealogy became a preoccupation with heredity; but included in bourgeois marriages were not only economic imperatives and rules of social homogeneity, not only the promises of inheritance, but the menaces of heredity; families wore and concealed a sort of reversed and somber escutcheon whose defamatory quarters were the diseases or defects of the group of relatives. (Foucault, *History of Sexuality*, pp. 124–25 [pp. 164–65])

The fortunes of the nineteenth-century family thus depended entirely on biological heredity, both for its symbolic capital (in cases of normal heredity) and the transmission of its economic capital (inheritance). The family was under twice the pressure of heredity: to preserve an inheritance within the family and to keep the lineage free from morbidity. By extension, the fate of the entire French nation rested on family heredities: "The analysis of heredity was placing sex (sexual relations, venereal diseases, matrimonial alliances, perversions) in a position of 'biological responsibility' with regard to the species" (Foucault, *History of Sexuality*, p. 118 [p. 156]). Foucault traces the transformation of genealogy, an old technique revealing the sovereign power to give death, into a technique of sex exhibiting disciplinary power over life (bio-power), and sees it as an overlap of sex and blood: "The family is the interchange of sexuality and alliance."[45] In the apparatus of sexuality, the discourse of genealogy was transformed from a vertical hereditary line of privilege and class distinction into heredity, an egalitarian, pervasive, homogeneous, and horizontal technique of bio-power. An individual's blood, his/her heredity, became the support on which were built the health and prosperity of the nation and species.

Chapter 2 analyzes the overlap of blood and sex, of sanguinity and sexuality, in the foundations of French medieval studies by tracing the transformation of the (medieval) psychology of genealogy into the (nineteenth-century) biology of heredity. It shows how theorists of heredity and consanguinity used medieval dynastic genealogies as historical evidence for their claims of hereditary degeneration and subsequent effort to regulate sexuality. The immediate result of the success of such scientific claims of heredity was the medicalization of the very concept of genealogy that was now grounded in blood (paternity of the biological father) rather than law (matrimonial law, whereby the father is designated as the husband of the biological mother, regardless of actual paternity). In short, while blood had been a psychological (and not physiological) concept in the Middle Ages, nineteenth-century theory of heredity made the semantic field of "blood" exclusively medical-biological; heredity was thus a much narrower concept than medieval genealogy, which did not depend only on physical blood ties but also psychological ones. This in turn means that our modern understanding of medieval genealogies and their deployment is grounded in the nineteenth-century biological understanding of genealogy as bloodline, rather than in medieval notions of genealogy as psychology. Finally, a revision of medieval genealogy and its restitution as a juridical-psychological category, from the nineteenth-century medical-biological one, allows for a reappraisal of Michel Foucault's genealogy as critique, with its emphasis on discontinuity, as a valuable tool of critical analysis for medievalists, historians, and literary historians alike.

Sex and Race

The medical history of the nation is double: on the one hand, the Middle Ages play a significant role in the construction of nineteenth-century discourses about sexuality and, on the other, the nineteenth century shapes the constructions of medieval sexuality. Put differently, nineteenth-century attitudes about sexuality are grounded in the historical research on the Middle Ages; simultaneously, the Middle Ages are invented through the historical research that grounds these modern discourses about sexuality. But medicine also maintains a relationship to fantasy: a fantasy of origins, in other words, an imaginary medical genealogy of the French race and its hereditary diseases and vices. While medical genealogy stimulated fantasy in the search for origins, medicine gave a stamp of science that has legitimated and reified an official history that is to a certain extent made of errors and fantasies. The task in Part II, "Sex and Race," is to reconstitute the objects of medical study in their relation to a phantasmagoric transmission of diseases and vices, and to understand how certain fantasies of racial purity and fears of national degeneration could produce an accepted version of history.

Part II is concerned with two different notions of the French word "race": its modern racial meaning, familiar to English speakers, and race as family, lineage, and, by extension, nation. Race was an integral part of the vast apparatus of sexuality. Michel Foucault pointed to the shift from a historic discourse on the war of races (*la guerre des races*) to a biological discourse of race, or "the struggle for life" (*la lutte pour la vie*): "The theme of the purity of the race replaces that of the struggle between races."[46] Foucault was interested in the supersession of the mythical races by the biological race, that is, the singularization of plural races struggling against each other into one race that had to be defended and controlled with "a whole politics of settlement [*peuplement*], family, marriage, education, social hierarchization, and property, accompanied by a long series of permanent interventions at the level of the body, conduct, health, and everyday life" (Foucault, *History of Sexuality*, p. 149 [p. 197]). Ann Laura Stoler has offered a sustained corrective to Foucault's treatment of race and racism. Foucault thought it in terms of metropolitan France, as "racism in its modern, 'biologizing,' statist form" (Foucault, *History of Sexuality*, p. 149 [p. 197]). Organized by the state to the benefit of the French nation, in Foucault's thinking, statist racism was exported to the colonies only as secondary effect. Neglecting the colonial, racial history meant that Foucault analyzed racism from an entirely Eurocentric perspective in which "racism wasn't initially a political ideology. It was a scientific ideology which manifested itself everywhere."[47] But this universalizing argument of "everywhere," making colonial racism derivative of metropolitan, statist racism, turned Foucault's argument into a "selective (northern) Europe-bound genealogy" (Stoler, *Carnal Knowledge*, p. 151). The symbolics of blood "that pervades nineteenth-century racism may be traced, as Foucault does, from an aristocratic preoccupation with legitimacy, pure blood, and descent, but not through it alone. It was equally dependent on an imperial politics of exclusion that was worked out earlier and reworked later on colonial ground."[48] That is, racism was not merely a product of the deployment of sexuality that was exported to the colonies from the metropole, but rather race *and* sexuality produced together the French nation. The making of the French race, both as genealogy and as heredity, as racism and as eugenics, contributed to the larger project of nation- and empire-building.

Building on Foucault's and Stoler's work, Part II thus argues that it is not enough to pursue the dialectic of present/past or metropole/colony in the production of scientific discourses, and it situates them in the triangulation between the metropole, the colony, and the past. Chapter 3 focuses on the racial meaning of "race," bringing the Middle Ages into the colonial/imperial sphere, while Chapter 4 explores the notion of degeneration and regeneration of race (family; species) in national/metropolitan terms. In 1806, a well-known Austrian archaeologist, Joseph von Hammer-Purgstall, hypothesized that the Templars were accused of worshipping idols called

Baphomet (Chapter 3). This new interpretation of the confessions in the trial record of the Templars verged on an erudite hoax. Hammer-Purgstall's error was not to understand that the transcription of "Baphomet" was either a scribal error or a corrupted pronunciation of "Mahomet." Between the dates of the Napoleonic Egyptian expedition in 1799–1801 and the colonization of Algeria in 1830, Baphomet remained this erudite fantasy of an Oriental idol. But as contemporary colonial contact grew, and with it the fear of contamination, the Templars and their contact during the Crusades with the other races, namely "Mohammedans" (or Muslims), began to be considered the origin of a quintessential nineteenth-century disease (syphilis) and vice (sodomy). The decline of medieval chivalry and civilization was dated to the Crusades and contact with foreigners, and this historical lesson was used over and over again to prepare and control contact with the colonized of the French nineteenth-century colonies. This fantasy of imperial dimensions had its counterpart in the national sphere, since the borders of the nation-empire were carefully policed externally and internally. Chapter 4 thus follows the literary hoax that originated with Paul Lacroix and led to the medical interpretation of Gilles de Rais as a model for sadism, matching marquis de Sade's accomplishments in the domain. At the same time, doctors developed the theory of hereditary degeneration and considered Gilles's criminal tendencies as degenerate as Joan of Arc's hallucinations. Two extremes on the same spectrum, Gilles was a degenerate criminal and Joan of Arc was declared a degenerate genius. Thus the virgin maid, a national symbol, stood out against the background of the degenerate criminal, but one whose medical diagnosis was built on a literary hoax.

These examples suggest that fantasy and error were at work in what became the official history that medieval studies have inherited. This is, then, the other objective of this project: What in the Middle Ages triggered the construction of a certain representation or interpretation by nineteenth-century practitioners and historians of medicine? What nineteenth-century contingencies gave rise to a particular medical-historical interpretation and how did it prevail? And what of these constructions has remained active until this day? Consequently, the book explores how our views of medieval sexuality may continue to be influenced by the Middle Ages imagined by the nineteenth century, while necessarily also being modified by our modern notions of sexuality, race, and nation.

Sex and Love

Part III reverses the order of previous chapters where I look at medical discourses in order to understand the medical heritage in medieval studies and popular knowledge. Here I begin with the *histoire des mœurs* to understand the parallel, but separate, developments of the medieval studies and medical discourse. Not only does the medicine of history draw information from the *histoire des mœurs*, but the medicine of history can be transformed into an *histoire des mœurs* by necessity: "The history of diseases that have died out, abated, or been transformed is always captivating. It is even more so when, because of the horror and fear that a disease instilled in populations at different times, it prompted prophylactic measures and protections that were inscribed in past laws and institutions. Thereafter, the history of an ailment is no longer a simple chapter of pathological history but a chapter of the *histoire des mœurs*, history of laws and customs."[49] The tradition of the *histoire des mœurs*, in which academic medievalists were raised and trained like their medical counterparts, provided two options: a civilizing French mission of courtly love and chivalry, which co-existed with sexuality, unrefined language, and obscenity of the fabliaux (Chapter 5). Chivalry and courtly love were considered as a hallmark civilizational indicator identified with the French nation, its original civilizing mission, its original civilizational export. This issue of national importance will play out in the marginalization of the fabliaux as non-French (but instead Oriental), and thus the exclusion of sexuality, favoring instead the chivalric epic and courtly love. Courtly love was preferable to the unbridled and contaminating influence of sexuality of the fabliaux that could not be contained within national, class, or racial boundaries. Chapter 5, however, brings up the question why courtly love, widely perceived to advocate heterosexual adultery, was not ostracized by professional medievalists or doctors as a form of deviant behavior. The answer given could be that adultery is the lesser of all evils of deviant behavior. But another historical-cultural reason, explored in Chapter 6, exists. The precise moment in the early 1880s when professional medievalists defined and embraced the concept of "courtly love" (rather than the unbridled sexuality of the fabliaux) corresponded to the height of the legislative and public battle for the reestablishment of divorce. The term "courtly love" is coined in 1883; the law on divorce is passed in July 1884, after eight years of efforts that began on June 6, 1876. A debate over the nature of marriage—whether it should be indissoluble or not—and an increasing pressure on the institution of marriage to become inviolable, not by denying the right to divorce, but by the very nature of the husband-wife relationship, created the opportunity for the doctors to transform indissoluble marriage into courtly marriage, that is, marriage based on love. Chapter 6 thus shows that courtly love, also known as "chivalric love" or "mystic love" (with which I began this book), had a wide cultural resonance throughout the nineteenth century thanks to the *histoire des mœurs,* and it lent itself to a series of transformations that made it productive to the social engineering of the French nation. Both doctors and academic medievalists inherited the term from the *histoire des mœurs* but appropriated it in two different and parallel ways. While philology appropriated it within a carefully contained and crafted literary phenomenon, doctors, in trying to discipline the nation, used it to redefine marriage, transform conjugal love and marital sexuality, and improve national demographics. Consequently, in the attempt to transform chivalric love into courtly marriage, and thus eliminate the need for the Republican divorce, this concept had a great influence on the marriage manuals written by doctors in the last decade of the nineteenth century and the first decade of the twentieth century, in the middle of the demographic crisis and pro-natalist debate in France.

A Pornographic Morality Play

The Epilogue narrates the history of the priapic exhibits (lead badges of profane and erotic content) at the Musée national du Moyen Âge-Thermes de Cluny in Paris. Although the first acquisitions date back to 1861–1862, the museum began exhibiting one of the items from its collection only in 1998. It would be easy to dismiss one of the objectives of this book, namely my desire to understand how philology and sexology parted ways, and why the Cluny Museum put the erotic badges on exhibit only in 1998, with an obvious reference to nineteenth-century morality and the invention of pornography as a regulatory category. I admit that one could well imagine nineteenth-century medical medievalism taking a moralizing turn. Indeed, nineteenth-century morality could be used to argue that doctors and historians of customs and morals believe they "recognize" in medieval literature and documents models of morality that justify their own morality, while at the same time imposing their grid of normative and moralizing sexuality on the Middle Ages. In contrast to this approach, I question the often-invoked explanatory power of nineteenth-century morality and normativity. For how do we account for the variety, breadth, and playfulness of nineteenth-century research on and imagination about medieval sexuality? Why did so many researchers focus so particularly on the Middle Ages? Instead of being shaped by nineteenth-century morality, would the Middle Ages have provided a space in which to escape—through fantasy of origins, a genealogy of sex, and heredity of vices—nineteenth-century codes and strictures? As Michel Foucault points out in *The Birth of the Clinic*, morality emerges only when there is an epistemological need (Foucault, *Birth of the Clinic*, p. 163 [p. 167]). Morality does not prevent one from knowing what there was previously no need to know; rather, it comes into existence in response to the need to know. The wish to observe and study what "one *cannot* see is shown in the distance from what one *must not* see" (Foucault, *Birth of the Clinic*, p. 164 [p. 168]). When a different need for knowledge presents itself, then there is a discovery of a moral obstacle (which can retrospectively be dated back to times immemorial). The moral interdiction to know is a part of the unfulfilled desire to know. In Foucault's view, it could be said that morality marks precisely the limit of what the scientist knows about what he wishes to know. Foucault further developed in *The History of Sexuality* the "repressive hypothesis": contrary to commonly held belief, "repression" produced the surfacing of sex in discourse and a proliferation of discourses on sexuality (Foucault, *History of Sexuality*, pp. 10–11 [pp. 18–19]). The will to know did not retreat because of a taboo, nor was it silenced; instead, it produced a *scientia sexualis*.

The repressive hypothesis implies that the Middle Ages constructed by the nineteenth century would have served the purpose of identification (recognition of strictly enforced models of normativity and morality). An outright opposition to the repressive hypothesis implies that the Middle Ages were marked by alterity and difference. But their opposition is not at stake for this book; rather, both identification and alterity are explored in the complex double transference between the Middle Ages and the nineteenth century: the Middle Ages were invented at the intersection of historical and medical discourses by nineteenth-century desires and fears and fantasies about sexuality. And medical medievalism simultaneously influenced the nineteenth century, its ideology and laws, its conceptions of the history and future of the race, while it served as a disciplinary power and regulatory technology of the race and its body.

The nineteenth-century medical history of the Middle Ages could be described as replete with guilty pleasures and voyeuristic fantasies. But whether guilty and voyeuristic or just erudite and scientific, this historicized and medicalized Middle Ages constructed the nineteenth century as much as the nineteenth century constructed medieval sexuality in a pleasurable interplay. *Pornographic Archaeology* unearths and recounts this dialectic of the forgotten and often phantasmagoric history of medieval sexuality—its "pornography"—that was fantasized into the very foundations of the French nation and, consequently, French medieval studies.

PART I

Sex and Blood

Chapter 1

"Pathologic Archaeology": An Introduction

> France is the classic land of military valor. . . . The French of the nineteenth century had not degenerated and, to carry out what is most admirable in audacity, most noble in sacrifice, most sublime in abnegation, they had only to follow the penchant of their generous nature, only to obey suggestions from heredity and instinct.[1]

Heredity and History

The principal nineteenth-century medical theory was hereditarianism, a theory of transmission of characteristics and dispositions in the process of organic reproduction. Hereditarianism was the scientific theorization of the permanence of form, meaning the reproduction of physical and mental traits transmitted over generations of descendants: "the tendency toward the regular transmission of traits . . . a legacy transmitted more or less loyally across an infinitely long chain of generations."[2] Laws of hereditary transmission were also derived from pathological deviations from normal heredity. Hereditary degeneration, which theorized this morbid, or pathological, heredity, thus became the companion theory of hereditarianism. Degeneration was understood as a process of irrevocable deterioration from generation to generation of the "race"—family lineage or lineage of the nation—resulting in degeneracy: sterility in the case of a family lineage and the decline of civilization in the case of a nation, both ending in extinction.

Although the concept of heredity, degeneration (process), and degeneracy (result) were scientifically formulated at the very end of the eighteenth century, hereditarianism and hereditary degeneration began to be theorized as domains governed by laws of their own in the 1850s in publications by Prosper Lucas (*Traité philosophique et physiologique de l'hérédité naturelle*, 1847), Benedict-Augustin Morel (*Traité des dégénérescences physiques, intellectuelles et morales*, 1857), and Jacques Moreau de Tours (*La psychologie morbide dans ses rapports avec la philosophie de l'histoire*, 1859).[3] The end of the series of revolutions that had shaken France since 1789, and the fading of idealism after the failure of the revolutions of 1848 and the proclamation of the Second Empire by Napoleon III (1851–1870), gave a real impetus to positivism and scientism, making room for materialism. The pressures of psychological problems created by industrialization and social dislocation, along with the economic depression of the 1840s, gave impetus to medicine as the science holding the key to social problems. Medical space coincided with social space as physicians took on the task of social prophylaxis through public hygiene policies and sanitary control.[4] Prevention of hereditary degeneration became the *raison d'être* for medical intervention on the body of the individual and in the body of the nation. By the 1880s and 1890s, hereditarianism and degeneration became highly influential and widespread medical theories in France and most of western Europe. At that time, French doctors began to reorient the debate away from biological, hereditary determinism and fatalism to environmental and sociological interpretation of criminality and degeneracy. Instead of degeneration, they emphasized the possibility of regeneration and rehabilitation, alongside reform of living conditions of population; eugenics, racially positive selection and improvement of the quality of the human race, shifted from negative to positive, that is, from attempts to prevent and eliminate morbid characteristics and pathologies to efforts at encouraging reproduction of healthy and desirable traits with higher birth rates.[5] However, throughout the nineteenth century, the precise cause of transformation from normal to morbid heredity remained unidentified and most problematic. Theorists of degeneration were unable to establish the etiology of morbid heredity and how precisely a race—family lineage or nation—degenerates to the point of extinction: Were the causes biological (heredity) or social (milieu)? Was there an (organic) morbid hereditary predisposition and, if so, what was the determinant cause (*la cause déterminante*) that triggered its physiological or psychological manifestation?[6] Consequently, what treatments and methods were possible and necessary for the regeneration of the race? Statements such as "the great question of heredity is far from having been elucidated"[7] were repeated in virtually every work written on heredity in the second half of the nineteenth century and the turn of the twentieth century. The insolubility of these questions and theoretical inconsistencies brought an end to hereditarianism in the early decades of the twentieth century. After August Weissman argued in 1885 that the germ cell nucleus (chromosome) is the repository of hereditary material, and Gregor Mendel's work on the gene as the carrier of heredity was discovered in 1900, hereditarianism was definitively replaced in the 1930s and 1940s by modern-day genetics (a term coined in 1905 and used for the first time in 1906).[8]

A complex interaction of history and medicine around the concept of heredity introduced the Middle Ages into medical discourse. The driving force behind the rehistoricization of clinical medicine, which had left behind its origins in natural philosophy in the first decades of the post-Revolutionary period, was the ancient link between heredity and history. Until the end of the eighteenth century and the early nineteenth century, heredity had referred to the legal concept of inheritance, the transmission or the succession of patrimony and title. The first definition that Pierre Larousse's *Grand dictionnaire universel du XIXe siècle* gives for "hérédité" is "transmission by means of succession," whose synonyms are "inheritance, succession."[9] The terms "inheritance" and "succession" refer to the transmission of patrimony and title over a period of time, according to a set of rules, which could but did not have to be patrilineal. In the 1820s and 1830s the noun "heredity" came into use, derived from the adjective "hereditary" and referring to organic reproduction and transmission of physical and mental characteristics and dispositions over generations of blood descendants (López-Beltrán, "The Medical Origins of Heredity," p. 106). In other words, the medical-biological concept of inherited physical and mental traits emulated the social phenomenon of succession; transmission of hereditary traits from generation to generation was considered very much like the transmission of inheritance of patrimony and title.

The shift from the social to the biological, and from the descriptive to the causal, could occur once medicine recognized the reproduction of form in lineage over a long period of time, as opposed to analyzing form as an isolated, individual occurrence. This distinction between production and reproduction aligned heredity with history. Heredity and its laws were seen as profoundly historical, both in that the conditions of possibility of their theoretical articulation lay in the length of time, the *longue durée* view, and in that the doctors relied on the historical evidence of genealogies for proof of scientific tenets. For medicine, the truth about the laws of heredity lay implicitly in a lineage's longevity and history; hence, the truth about the laws of hereditary transmission and transformation of pathologies could be found only if heredity were studied over an extended period of time in an uninterrupted succession of descendants: "Genealogies had to be constructed, distinctions drawn, and connections made among individual generative acts to produce the kind of structure that could meaningfully be addressed by the metaphor of heredity" (Müller-Wille and Rheinberger, p. 6). Preferably, heredity had to be analyzed all the way back to its origins: "It is good to trace back the farthest possible in a family's medical history and search for the original archetype of health, so to speak, among antecedents [*parents ascendants*]."[10] If for individuals that meant going as far back as possible into the family past and its antecedents, when it came to the genealogy of the French race—or nation (from Latin *natio*, meaning "birth, origin")—the point of origin was the Middle Ages. Although doctors looked to antiquity as an example of the civilization that declined and perished because of degeneracy, they saw the Middle Ages as the origin of the French race and thus as the seat of vices and diseases inherited by nineteenth-century France that threatened the French nation and European civilization with extinction. In short, the principal nineteenth-century medical theory could only be practiced with the assistance of history. Clinical observation had to be complemented by historical study. For medicine to succeed in its hygienist and prophylactic vision of regeneration and eugenics of the race and nation, history had to be entirely at its service.

Medicine was thus a historical discipline from the outset. The specific articulation of medieval history and hereditarianism was made during the consanguinity debate (1856–1866) immediately following the first major publications on hereditarianism around 1850. Consanguinity became a flashpoint of medical discussions on pathology and genealogy, medical intervention in marital policies and legislative matters, and notions of nation, race, and class. The debate over the positive or negative role of consanguinity in heredity inevitably began with the aristocratic and dynastic Middle Ages, as physicians practiced a "pathologic archaeology" (*l'archéologie pathologique*).[11] In the wake of the French Revolution, the *ancien régime* and aristocracy as social class became the nexus of desires or fears, nostalgia or hatred. Families that claimed their lineages stretched back to the Crusades became the prime source for studies of heredity and degeneracy. While numerous historical examples of dynastic heredity and consanguinity provided historical proof for the laws of heredity, they also achieved something else: they turned history itself into a medical category. This was to be the ultimate historical outcome of the practice of pathologic archaeology.

The Consanguinity Debate (1856–1866) and the Middle Ages

The bulk of the consanguinity debate was concentrated over one decade, from 1856 to 1866, but it reared its head periodically for as long as hereditarianism continued to be the dominant scientific theory. The consanguinity debate began exactly in the period of the 1850s when the scientific foundations of hereditarianism and hereditary degeneration were posited by Prosper Lucas and Benedict-Augustin Morel. It also corresponded in duration to the debate on the *droit du seigneur* (see Introduction), which raged from 1854 to 1884. Publications appeared regularly in journals such as the *Gazette hebdomadaire*, *Bulletin de l'Académie de médecine*, and *Bulletin de la Société d'anthropologie* well into the 1900s. A substantial number of doctoral theses in medicine and books intended for the education and hygiene of families were published: "Emotion overtook the public; all the newspapers addressed the subject; the Minister of Commerce and Public Works made it the topic of a letter circulated among the prefects; theses were defended in medical schools; numerous research papers published; a discussion took place during the Medical Congress of Lyon."[12] The public range of the debate was due to the fact that consanguinity not only concerned the scientific aspects of the role it played in heredity, but was also seen more broadly to play a vital social role in marriage and procreation, which raised complex questions pertaining to religion, politics, and medicine. The intensity of the debate was certainly linked to the increase in endogamous marriage practices since the mid-eighteenth century as alliances previously considered incestuous (with the brother's widow or a niece) or simply prohibited (first and second cousins) became acceptable in urban and rural milieus by the mid-nineteenth century.[13] The purpose of marriage being "to perpetuate society, not only by the number of individuals, but more importantly by beauty, vigor, and the perfect organization, in so much as possible, of the races,"[14] marriages *with or among* the sick were not advisable. The goal was to avoid unions that were hazardous both to society and the prosperity of the nation-state: "If the State needs citizens gifted with all of the strength, the perfection of organization, then society demands the same rights. Families constituted on a solid foundation benefit from the hope and certainty, physical as well as moral, of a longer existence."[15] It was indeed the presumed morbid causality between consanguinity and heredity that fueled the push for medical intervention in matters of public hygiene and social policy; the ultimate failure of anti-consanguinists to prohibit consanguine marriages that had become socially acceptable or to institute a mandatory prenuptial medical exam of hereditary antecedents showed the limits of the legislative reach of medicine in matters of public health.

The main question was the role that consanguinity played in morbid heredity. Was consanguinity "the cause of causes"[16] of morbid heredity? The anti-consanguinists "want to demonstrate by the facts that the consanguinity of parents, *ipso facto,* beyond all pathological heredity and all foreign influence on inherent organic traits, increases, in notable proportions, the chances of infirmities and sickness to which generations are exposed."[17] If in fact "consanguinity elevates heredity to its highest degree,"[18] it could be responsible for all kinds of mental and physical diseases, "the most serious, most numerous, most irremediable sicknesses":[19] "mental retardation, deaf-muteness, scrofula, rickets, albinism, teratologic malformations."[20] Consanguine marriages thus came to be viewed as an especially malignant cause of morbid heredity: "From now on, consanguinity can be classified as morbid etiology in regards to the human species."[21] Prosper Lucas was the first major medical authority to affirm, in his *Traité philosophique et physiologique de l'hérédité naturelle* (1847), that consanguine marriages are dangerous, because they are a source of physiological morbid heredity. Lucas believed that consanguinity increases morbid heredity; therefore, "the selection of *person,*" "the selection of agents of reproduction," is determinant in the prophylaxis of morbid heredity: "The methods of preventing the seminal transmission of illness derive necessarily from the laws and formulas of reproduction."[22] Lucas thus endorsed the mixing of different families with the same healthy predispositions: "There must be sought, in physical connections, within another family, only *similar* conditions,"[23] which later also came to mean curing morbid heredity by cross-breeding of the healthy with the sick, "to give as a spouse to an individual who has morbid hereditary antecedents only a husband or wife of irreproachable constitution, in a state of perfect health."[24]

The leader of the anti-consanguinist camp was Dr. Francis Devay (1813–1863), professor at the Lyon School of Medicine, author of multiple works, who, even before Lucas, warned of the dangers of consanguine unions in his 1846 *Hygiène des familles ou perfectionnement physique et moral de l'homme.* Devay was of his own admission a staunch Catholic: "What good does it serve to speak of my *well-known* religious orthodoxy? . . . I can tell you that I have always seen and that I still see in Christianity the only remedy to the physical, moral, social, and, I would add, . . . scientific misery of humanity."[25] For Devay, consanguine unions were "as contrary to the growth of the population as to its validity,"[26] and he therefore advocated and championed a "preventative and perfective medicine" (*médecine préventive et perfective*).[27] The best method of prophylaxis would be exogamous marriages: "A second principle emerges from the application of medical science to the cross-breeding of races: the antagonism of illnesses and original predispositions."[28] In his 1857 *Du danger des mariages consanguins au point de vue sanitaire*, Devay extended the dangers of consanguinity to social heredity and condemned not only sexual endogamy but also social endogamy of consanguinity: "Thus, our customs, our social habits generally bring male heads of family, often with honorable intentions, to seek for the establishment of their children what is generally called a *position identical* to their own. . . . This is what we call *factitious consanguinity* [*consanguinité factice*]."[29] Biological and social consanguinity have the same perilous result of physical and moral degeneration. While anti-consanguinists promoted "cross-breeding, mixing" (*le croisement, le mélange*) with "foreign blood," consanguinists defended in-breeding, "the pure race and individual type,"[30] and focused on biology rather than on social endogamy. They derived their most important arguments from the evidence of zootechnology, based on the breeders' method of "in-and-in," known to preserve the best qualities in an animal race. Consanguinists argued that the lessons of zootechnology taught that cross-breeding cannot remedy morbid heredity, while in-breeding strengthens normal heredity.[31]

The consanguinity debate illustrates the dependency of medicine on history. Medieval dynastic genealogies, notorious for marriages of consanguinity, provided the historical evidence that both anti-consanguinists and consanguinists manipulated to their advantage. The first author to speak of the dangers of consanguinity was Charles Girou de Buzareingues. In his *Philosophie physiologique, politique et morale* (1828), he attributed the fall of the *ancien régime* to social consanguinity, the "factitious consanguinity" identified by Devay: "Consanguinity in the marriages of kings becomes, after several generations, detrimental to peoples . . . : inheritance . . . is transmitted without alteration, . . . there is no neutralization."[32] Consanguinity was detrimental to the families of the *ancien régime* that intermarried and thus recycled the same mentality, which hereditarily formed the same character, morally degenerate by virtue of power and flattery: "The passions that are born of authority, resistances, and flattery, as well as the character that results from them, are passed from father to daughter, and from her to the son, and are ultimately the common inheritance of all the children called on the reign."[33] Only social exogamy, marriages outside of one's limited social caste, can form a sound base for healthy procreation, avoiding any risk of moral degeneracy—that is, morbid heredity of passions—from social consanguinity: "It is by the choice of his companion that one can be assured of transmitting, with life, positive inclinations to one's children."[34]

For Prosper Lucas in 1847, consanguine marriages endangered the health of the family lineage physiologically as much as psychologically. He likewise derived his conclusions from the example of nobility: "History testifies . . . : aristocracies, reduced to recruiting in their own midst, die out . . . often, after falling into degradation, folly, insanity, and retardation."[35] Devay in 1857 introduced a whole new section on "Historical Proofs" (*Preuves historiques*) to support his claim that consanguinity, and prolonged consanguinity over several generations, is the sole cause of morbid heredity. Medical legislative intervention, he claimed in 1860, is justified by history:

> And yet, historians who have dealt with the causes of the decline of the noble classes in France have ignored precisely that cause which had the greatest impact, and which, literally, contaminates the blood, atrophies the race, when a race can be born at all from monstrous combinations. And yet, nothing would have been easier, for no century provided a more faithful mirror of its physiognomy by its numerous memoirs and voluminous books of correspondence; in no other century have the obligations of conjugal relationships been so strangely misunderstood; in no other century has the word marriage been so synonymous with the word *establishment.*[36]

Devay considered the *ancien régime* a fallen civilization: "The spirit fades away, forces ebb, and the race is bastardized. . . . We know what happened at the end of the following [eighteenth] century, when the upper classes, numerically quite diminished, being neither intelligent enough to improve, nor strong enough to resist, dissolved into the Third State."[37] Armed with such historical evidence of "degeneration" (*dégénérescence*) of the *ancien régime*, the authors called for a "regeneration" (*régénérescence*) of the French race by cross-breeding: "History teaches us that it [consanguinity] was one of the causes of degeneration in certain royal and noble races (family of Charles-Quint, etc.), and it shows us at the same time the positive effects of cross-breeding, which lead to *regeneration* by introducing better blood into lineage."[38]

Eugène Dally and the consanguinist camp rejected such historically based claims by Devay and anti-consanguinists and introduced their own historical counter-arguments, based on Louis-François Benoiston de Châteauneuf's immensely important *Mémoire sur la durée des familles nobles en France* (1846). For a long time, Benoiston de Châteauneuf says, historians assumed that noble families of the *ancien régime* were degenerate because "the flaw of cross-breeding in noble families had to lead to weakening, and thus extinction. The horror of misalliances, they claimed, and the care with which the upper classes always intermarried, largely contributed to their degeneration."[39] Benoiston de Châteauneuf arrives at exactly the opposite conclusion on the role that consanguinity had played in the downfall of aristocratic families and the *ancien régime*. He observes the banal fact that most medieval families ceased to exist long ago, that their duration rarely exceeded 300 years, and that the average lasted 230 years (Châteauneuf, pp. 51, 54). An able statistician, he proves that the reason for the extinction of noble families was neither the primogeniture rule nor consanguine mariages, since noble families had plenty of offspring: "According to a study completed on 300 noble families, I found that 2,398 heads of household . . . had 10,490 children, or 4.3 each."[40] He shows statistically that the social and sexual endogamy of medieval nobles, "the lack of cross-breeding" (*le défaut de croisement*), had no effect on their physiological capacities of reproduction: "The matrimonial bed of nobles, rather than being mostly sterile, as it has been said, was on the contrary very fertile."[41] Since consanguinity is not the cause of sterility, it therefore cannot be blamed in the extinction of noble families. Rather, mortality (of infants and adults) and celibacy account for some of the decline: "First, the military state, then the ecclesiastical state; among these two professions embraced by the majority of nobles, one required them to sacrifice their lives, the other condemned them not to give life to others. I would add to these two causes yet a third: the great number of children that they lost or who, grown to adulthood, died without having been married."[42] The Middle Ages are known for the physiological vigor and strength in both men and women, and even if "the mores of the Middle Ages were likely not exemplary," medieval men and women were sound of body in their sexual activity "when it comes to a question of bodily vigor rather than honest actions."[43] Thus, above all, it was the gradual loosening of morals *after* the medieval period that brought about the decline of the nobles: "Morality and medicine, noble sciences . . . have long associated the decline of the upper classes with the decline of morals, much more than the lack of cross-breeding."[44] Moral, not physical, degeneration led to extinction. Benoiston de Châteauneuf thus rejected consanguinity as the cause of physical morbidity and extinction, blaming instead mortality, celibacy, and moral degeneration of the noble races in the post-medieval period, a time of "unbridled debauchery" (*libertinage effréné*). His study remained the most quoted and attacked source throughout the consanguinity debate, as doctors in search of incontrovertible historical evidence returned over and over again to medieval history and its dynastic genealogies.

History and Medicine: "Pathologic Archaeology" and "Historical Pathology"

Privileged in their relationship to state power and committed to the matters of social policy and the future of civilization, doctors' theories of hereditarianism and degeneration were a constitutive force in the construction of the French nation. Medicine laid claim to the discourse of truth, but neither consanguinists nor anti-consanguinists could legitimate their claims without looking into the nation's medieval past. Heredity and degeneration could not be understood without history and the *longue durée*. As seen in the example of the consanguinity debate, the immediate consequence of medicalization of the social space was the interaction of history and the medical sciences. This happened in two ways. First, history was used as evidence, as "historical proof" (*preuves historiques*), to substantiate medical theories. Second, and more significantly, history itself became a medical category, that is, a category of medical analysis. As soon as medicine started to organize and monitor the social space, history became its target: in 1839, Brierre de Boismont affirmed that "the remarkable events that are the salient traits of a century belong not only to history, but also to the domain of medicine."[45] As late as 1895, Émile Dusolier could exclaim, "By what right would the historian portray himself as supreme judge and would he be able to condemn without having listened to the doctor?"[46] Just as nineteenth-century medicine was a historical discipline, history was a medical category. These complex interactions between medicine and history took center stage in three related but different medical-historical disciplines: the history of medicine, medical history, and the medicine of history.[47]

The History of Medicine and Positivism

Charles Victor Daremberg (1817–1872), the best-known figure among "erudite doctors" (*médecins érudits*) along with Émile Littré, started teaching the history of medicine at the Collège de France in 1846.[48] Daremberg drew together history, medicine, and philology, though he did not draw crowds at the Collège de France.[49] As a discipline it its own right, the history of medicine had a particularly fraught history, a history of "ostracism."[50] Established in 1794 as one of eighteen chairs at the École de santé, the chair of the history of medicine was abolished in 1822–1823, to be reinstituted only in 1870 at the École de médecine.[51] History of medicine had to battle the materialist physiology of dominant figures such as Magendie, Cabanis, Bichat, and Broussais; as Daremberg explained, in the aftermath of the Revolution, there was an urgency to separate medicine from the dominance of natural philosophy, hence the sudden turn to clinical medicine in order to prepare physician-practitioners rather than philosophers. The clinical rupture between natural philosophy was as complete as the break with the *ancien régime*, but its unfortunate consequence was that the history of medicine, which studied natural philosophy, could not be undertaken or supported because this kind of study of medicine "was itself only history."[52] In 1870, Daremberg lamented that "on all sides, there is a place for the history of each science, while the history of medicine is officially represented, in France, only by the chair at the Collège de France."[53] Five years earlier, in 1865, in the midst of student demonstrations for a chair in history of medicine (Wonhlich-Despaigne, p. 16), José Miguel Guardia argued that, cut off from its roots in history, clinical medicine looks like "a doctor who would suppose to expound upon the pulse without knowing how blood circulates."[54] Clinical medicine, observation, empiricism, cannot alone comprehend the social fabric that needs to be disciplined, controlled, and, above all, regenerated. Without history, medicine is not a science but just an art, a practice devoid of its tradition: "To appreciate medical doctrines, it is necessary to know their authors, their precise origin, their successive modifications, their diverse transformations. . . . And, in fact, positive science, which is born of reality, remains incomplete if we ignore how it was founded, developed, and perfected by an uninterrupted series of efforts and attempts, how men proceeded in their intellectual evolution."[55] By this account, observation and history go hand in hand; clinical observation must be complemented by historical observation. Medicine without history does not possess the view of the *longue durée* that alone justifies social and even legislative intervention. Medicine must benefit from history to check its observations against time-honored examples and thus to prevent the spread of doctrinal errors. History makes medicine a social science and the insights of the past can help regulate and improve the body of the nation.

These were the views especially of positivist doctors and thinkers, such as Émile Littré (1801–1881), a disciple of Auguste Comte and chief proponent of his positivist philosophy.[56] Positivism legitimated and reinforced medicine as a social science. An overarching philosophy that claimed total knowledge, positivism theorized the classification of sciences that form total science: mathematics, astronomy, physics, chemistry, biology (used alternatively with physiology), and sociology, the latter of which was in the highest position, immediately above medicine, "biology or the science of living bodies, and sociology or the science of societies."[57] They were in a relationship of double transference: sociology would not exist without physiology, which was "an indispensable support in the constitution of sociology,"[58] but since sociology was the master science of society and history, doctors also needed sociology to guide them in their interpretation of history: "The word [sociology] would not be included here if the men who develop the sciences, and, in particular, doctors, were not in need of a philosophy to guide them."[59] Sociology and medicine were seen as mutually dependent: "to heal degeneration in avoiding and in opposing its propagation is a question of pure sociology," but only doctors could "identify the illness"[60] to be cured. Positivism supposed the perfectibility of the human race, its progress: "Science includes the life of societies, the laws of that life, including progress, among others."[61] But the future depended on the past: "By studying the political past of societies, you will learn their future, and not only the future conceived as probable outcome, but as right and duty."[62] The discipline of history was downgraded to a mere art of narration of events, rather than a science: "History, as it has been written until now, is not a science, but simply the knowledge of events that happened within societies."[63] In fact, positivist medical history aimed to rid the study of history from superstition: "Too large an importance has been granted to poison and magic in all royal and princely deaths."[64] It is then the medical corps alone that can furnish the true interpretation of history; without medicine and sociology, history is nothing but raw data of events, devoid of comprehension, and, worst of all, it is utterly unscientific: "As long as the ways in which events lead to one another, derive from one another, will not yet have been demonstrated, there will be in the field of history only material for erudition, but without any scientific theory."[65] History itself is a part of natural and organic phenomena that are precisely the original object of the study of medicine, it is animated by heredity, the very principle of life: "Heredity intervenes in the evolution of civilization, in *sociology*. Without it, history would lack one of its essential organic or biological conditions."[66] Medicine can be imposed as a historicist practice because it alone can produce scientific principles that make sense out of history.

Thus, in the view of proponents of positivism, if medicine certainly needs history, history needs medicine even more. What can a historian do "in the presence of an action or a person, which must be deciphered like an enigma, if he fails to possess the knowledge which allows the psychiatrist and the doctor of legal medicine to establish an accurate diagnosis or to enlighten the judges"?[67] If a historian wants to resolve great problems in the philosophy of history, "does he hold the right to disdain numerous facts, anecdotes, documents, in considering only general ideas?"[68] In other words, the historian should neglect the philosophy of history in favor of physiology. The positivist argument for the medicine of history was made. This passage from the history of medicine to the medicine of history also definitively confirmed medicine as a social science.

Pathologic Archaeology: Medical History and the Medicine of History

Alexandre Lacassagne (1843–1924), professor of legal medicine at the University of Lyon and one of the founders of criminal anthropology in France, lent strength to the medicine of history via his numerous students.[69] In his 1901 preface to Dr. Cabanès's *Les morts mystérieuses de l'Histoire, souverains et princes français de Charlemagne à Louis XVII*, Lacassagne wrote of "the importance of legal medicine for history. It is by using its biological knowledge that the doctor of legal medicine succeeds in resolving certain problems to which historians alone could not find the solution. . . . Our knowledge of psychopathology authorizes us to understand and to explain some strange natures that have remained like historical puzzles."[70] Without the help of medicine, he continued, "Historians are like the blind or infirm who, from their bed or chair, with the theatrophone, listen to a drama or an opera. . . . There are a few highlights, sometimes clear speech, often brouhaha, sounds, noises. But where are the sets, the costumes, the acting of the performers, their attitudes, and their gestures, the movement on the set?"[71] History needs to be auscultated, but without the stethoscope historians resemble patients, "blind or infirm," more than they do doctors. Doctors, on the other hand, "have conducted themselves like experts before the court; they have reported, by their honor and conscience."[72] Under the microscope, history suddenly reveals itself to be overflowing with "morbid types to bring to light."[73] Medieval history especially seems to be taken out of the textbook for legal medicine: "Have we not heard of male and female saints, great mystics, who were hysterical! Brave men of war who had the moral anesthesia and the audacious courage of epileptics! Undisputed monsters because they were well placed enough to incur popular indignation and who ought to be given the dignity of the mentally ill!"[74] In another work, Dr. Cabanès will give this rejoinder: "Legal medicine could undertake so many rehabilitations; there remain so many trials for it to revise!"[75] Thus, in the course of the nineteenth century, history became a medical category and history's mysterious events as many occasions for the practice of medical diagnostics and new interpretations of historical turns of events. Medicine even looked for pathologies in the history of art: "In the past several years, we have seen affirmed an admirable awakening to the taste for studies of *retrospective pathology* [*pathologie rétrospective*]. We have begun to examine from various angles, in museum and church paintings and in sculptures affixed to porticos or cathedral pillars, anomalies, deformities, nervous contortions. . . . So much so that the careful analysis of artworks, accomplished from a medical standpoint, has brought together a set of documents, not only curious, but also sometimes very instructive, touching upon the history of certain afflictions."[76] Alfred Maury, who wrote extensively on the physiology of hallucinations, judged in 1845 that the advent of physiology in history was revolutionary: "Until that time, it had seemed that between the constitution of societies and the temperament of individuals, there were only remote influences; no one would have dared to suggest, without fearing ridicule, that the health of a given historical figure had exerted considerable influence on the ideas, the doctrines of an age. . . . There is thus nothing unreasonable about asking physiology to find the solution to certain problems in the philosophy of history."[77] Doctors changed the philosophy of history into the physiology of history. In other words, if medicine needed to be historicized, in order to accede to the power of regulating the society, history needed to be medicalized in order to provide for the first time a full picture of the past: doctors can perform "*pathologic archaeology*" (*l'archéologie pathologique*), Lacassagne affirmed.

This "pathologic archaeology," or "retrospective pathology," could be performed in two ways: as medical history and as the medicine of history. Both had their origin in the advent of physiology in the study of history, but the implications for each were very different. A distinguished early practitioner of medical history was Émile Littré. His 1853 toxicological study of Alexander the Great's death argued that it was not poison that killed the ruler, but "intermittent fever," or malaria.[78] Morbid psychology was also thought to influence the course of historical events: "It is obvious that insanity has exerted a great influence on the destinies of peoples."[79] The works of Dr. Auguste Corlieu, *Étude médicale sur la mort de Charles IX* (1871), *La mort des rois de France, depuis François Ier jusqu'à la Révolution française, études médicales et historiques*

(1873), *La fistule de Louis XIV* (1874), *La mort de Louis XVII* (1876), and *Le Roi François Ier est-il mort de la syphilis?* (1880), were clearly identified as "medical history" (*l'histoire médicale*).[80] One of Lacassagne's most prolific disciples and an advocate of pathologic archaeology, Dr. Augustin Cabanès (1862–1928), set out to write medical histories of medieval fashions: long-toed shoes (*à la poulaine*), invented by Henry Plantagenet, served "to conceal a growth on the foot," and the ingeniousness of women hid their pregnancies: "In certain eras, they adopted an outfit that made them all appear pregnant; as a result, those who actually were had an excellent means to hide their condition."[81] Medical history was classified as a topical solution "to unravel enigmas" and "to explain certain controversial deaths in history."[82] Indeed, medical history did not have higher aspirations or pretensions: "That we be well understood: far be it from us to feign to reduce all of history to a series of problems in morbid psychology; . . . we aspire only to provide a contribution, the scientific contribution."[83] Medical history could not attain the breadth that the medicine of history had, since it did not pursue a hermeneutics of causality: "Medical History collects the pathological documents relative to an individual, his or her illness, doctors, funeral details, etc. It is historical Pathology in a fragmented, anecdotal, isolated state."[84] The fragmentary nature of medical history tended to produce only localized information that here and there resolved the famous enigmas of history and provided a piece of the puzzle, but it did not give a global, causal view of history and its events. Doctors believed nevertheless that this would be accomplished by the medicine of history, whose breadth and depth of insight promised to replace the philosophy of history: "The doctor alone . . . may undertake this work, in a word to accomplish: the philosophy of history, the medicine of history."[85] Only a doctor will be able to perform "the daunting task of commentating on history," while historians are sidelined: "Historians remind us of accomplished facts, nothing more; they seek not to explain them: they cannot do so."[86]

Just as medical history had its precursor in Littré, so the medicine of history had its origins in two important near-contemporaries: historian Jules Michelet (1798–1874) and psychiatrist Louis-Francisque Lélut (1804–1877). Both Michelet and Lélut may have been influenced by Philippe Pinel (1745–1826), the founder of French psychiatry, who was interested in the "melancholics of history" (*mélancoliques de l'histoire*). He admitted in 1797 wanting to write a book on Louis XI, "model of the most perfidious and most refined politics," whom he compared to Tiberius: "A somber despondency, a harsh and repulsive gravity, the bitter inequalities of a character full of capriciousness and outbursts, the desire for solitude, an oblique look, the timid discomfort of an disingenuous soul, betray, as early as youth, the melancholic disposition of Louis XI."[87] Be that as it may, doctors overwhelmingly credited Michelet with being the first to adopt a medical approach in the study of history: "He was the first to search for the origin of events in morbid constitutions of people, the causes of political intrigues in the health of their actors."[88] Michelet had an organicist view of history, "my historical question posed in terms of *the resurrection of life in its integrity*, not its appearances, but its profound interior organisms."[89] His vision of history had possibly been shaped by the fact that he grew up in a private sanatorium, managed by his father. Michelet's influence was considerable on the future generations of historians and doctors; it was summarized in the famous formula, "before and after the fistula": "Such are the bizarre phases of personal rule. The reign of Louis XIV divides into two parts: *before the fistula, after the fistula* [*avant et après la fistule*]. Before, Colbert and the conquests; after, Madame Scarron and the defeats, the banishment of five hundred thousand of the French. Francis I changes in the same manner: *before the abscess, after the abscess*. Before, the alliance with the Turks, etc.; after, the rise of the Guises and the massacre of the Vaudois, events which mark the end of his reign."[90] While Louis XIV suffered from an anal fistula that was operated on repeatedly, Francis I was struck down in 1525 with an abscess in his head, a consequence of old wounds and syphilis.[91] After these health problems, the behavior of both kings became erratic and inconsistent. For Michelet, the psychology of events depends entirely on physiology—the body makes history, literally. His history is humoral, physiognomist, and essentialist, "it describes a complexion."[92] History and spirit are bound by the body, and health is the main, albeit not the sole, factor determining the course of historical events: although it does not by itself direct the course of events, "the role of sexuality in History is far from negligible."[93] One of the leading historians of the Middle Ages at the turn of the century, Gabriel Monod, summarized the judgment of several generations of historians and historians of medicine in claiming that Michelet "gave the portrayal of mores and character the status that it warranted, and he showed how the driest documents may become instructive for those who know how to analyze them; he insisted on the influence of physiological and pathological causes hitherto neglected in history, and opened up a new path of investigation—quite dangerous, it is true, but fertile with curious discoveries."[94] If Michelet was nonetheless "dangerous," it was because "he had no method that he was able to teach or pass on, since he proceeded merely by intuition and divination":[95] "His method indeed worries those of a positivist bent, who care to set foot slowly on only solid terrain; they have asked themselves if they should believe in those inductions, those hypotheses, those physiological analyses, so new."[96] Following his intuition and flights of fancy, he painted entirely new portraits, "guessed at based on a chronicler's sentence, a portrait or an engraving," and made completely new and original histories, but such a so-called method did not provide "models to imitate."[97]

Michelet was a "physiologist," obsessed with sexuality and the influence of physical ailments on historical events; at the other end of the spectrum was Lélut, a psychologist, "doctor-philologist."[98] While in Michelet's world the body and its ailments influenced the course of history, in Lélut's "the psychiatrist will come efficiently to the aid of the historian. Henceforth, it is no longer a question of penetrating character through temperament, but of showing the consequences that the cerebral health of a leader can have on the destiny of the nation that he leads."[99] Lélut advanced the theory of "the psychology of History."[100] In *Du démon de Socrate* (1836), he gave for the first time a credible account of hallucination to replace the superstitious legend of demons.[101] He thus promoted "historical psycho-pathology" (*la psycho-pathologie historique*) and a theory about how hallucinations and insanity influence history, sparking a medical debate over the origin and meaning of hallucination—arising from an external or internal cause, as a symptom of genius or insanity—that lasted for most of the nineteenth century. It pitted Lélut's followers against those of Brierre de Boismont who refused to "subordinate doctrines, beliefs, convictions to the sickly state of the body, which may be true regarding character or humor, but which is entirely false in respect to the soul."[102]

Historical Pathology

The medicine of history was perfected by Auguste Brachet, a disciple of Émile Littré and a model figure for Doctor Cabanès. Brachet was a recognized philologist and linguist, author of *Grammaire historique de la langue française* (1868).[103] But he is also the author of *Pathologie mentale des rois de France*, in which he coined the term "historical pathology" and gave the term this definition:

> *Historical Pathology* is properly speaking the explanation, through biological science, of the data furnished by historical texts, data collected and verified according to the rules of scientific criticism, in the dual goal of serving, at times, medical science and, at other times, historical

> science. It should not be confused with the *History of medicine*, which recounts the doctrine or the life of a doctor . . . nor with *medical History*, meaning the pathological biography of an individual. *Historical Pathology* ignores the former and includes the latter, which serves it as scaffolding.[104]

Neither a medical history nor a history of medicine, historical pathology is a rigorous methodology for the medicine of history. It is also yet another attempt at establishing the laws of heredity that remain uncertain. Toward the end of the nineteenth century, the successful determination of the laws of heredity in history is seen as a solution to a double problem. While historical pathology will resolve once and for all many historical mysteries, "historical *loci desperati*,"[105] the historical study of pathological heredity as *longue durée* will identify once and for all the laws of heredity. The method of historical pathology is much more systematic and broader in reach than medical history: "By neglecting among these facts the ones that bear only anecdotal interest, historical Pathology, in the strict sense of the term, connects pathological pieces of *data* that it receives in fragments from medical History in order to establish their succession, and to draw from them the pathological law of the individual or, as they say in clinics, the patient's pathogenic, and, hence, mental, formula."[106] Brachet identifies the goals for this new historical science as "the reconstitution of the biological formula (somatic and mental) of each of these *representative men* within humanity. We will need to justify the usefulness of this type of research from the three-fold perspective of *political history* (the interpretation of the actions of historical figures), —of *pathology* (the history and evolution of illnesses), —of the *clinic* (the scientific constitution of heredity)."[107] The constitution and the interpretation of the "pathological file,"[108] "clinical archives of the past,"[109] is the *sine qua non* of the study of history. Because Brachet recognizes Michelet's error—his obsession with physical ailments, his focus on the individual rather than the *longue durée*, and his lack of method—Brachet privileges a "mental pathology" over "somatic pathology,"[110] that is psychopathology over physiopathology. In other words, *Pathologie mentale* is also a plea for the intervention of a psychiatrist, rather than physician, in the work of a historian: "Physiology is not *all* there is to psychology."[111]

Pathologie mentale was Brachet's direct engagement with historians to show the usefulness of historical pathology for history. Brachet attacked historians who dismissed psychopathology, such as Gabriel Monod (p. lv), while he praised Charles-Victor Langlois and Léopold Deslile (p. cxxii), the eminent French philologist, who collaborated with Gaston Paris, founder of the discipline of academic French medieval studies. According to Brachet, "historical Pathology must be purely clinical and not philosophical."[112] It should avoid "a psychological pathology that would be purely fanciful [*fantaisiste*],"[113] like that of Michelet, who speculated that Louis XI had all the birds from his gardens caught on Saturday, November 19, 1468, because they supposedly chattered the name "Péronne," the place of his recent humiliating meeting with Charles the Bold, duke of Burgundy, who imprisoned the king despite guarantees of safe conduct and dictated the terms of a new peace treaty between the kingdom of France and the dukedom of Burgundy in exchange for the king's freedom. While this hypothesis could work at best "for magpies and jays, garrulous birds, it is improper for cranes and owls, all the more so for stags and does, which could hardly have cried 'Péronne' upon the passage of the king."[114] If Louis's motive was vengeance, then it is incomprehensible—and Brachet underlines with delight that Michelet avoided this problem—why the king "instead of wringing the neck of his winged insulters, after having begun to have them plucked alive, encircles them with care, confers upon them, in exchange for their jibes, royal treatment in the parks of Amboise."[115] Rather, there are three other solutions to this conundrum: one, "the (highly improbable) hypothesis of chance"; two, "the moralist solution according to which Louis would have chosen out of pride this moment to enact a measure without precedent, precisely to make Parisians delighted by his failure in negotiations in Péronne understand that, although they believed him defeated, he was still their master"; or three, the solution "of the neuropathologist."[116] Indeed, the correct diagnosis should be "the irresistible impulse of zoophiliac kleptomania."[117] Since Louis XI was "a hereditary degenerate, no doubt the facts cited above, far from being an aristocratic idiosyncrasy, are a very clear manifestation of one of the classic marks of degenerates: zoophilia."[118] Michelet's "fanciful pathology" (*pathologie fantaisiste*) was a result of the lack of method and precision, but rigorous medical-historical work could shed light on Louis XI's reign.

Brachet also wished to prove the usefulness of historical pathology for medicine, especially in deciding the meaning of hallucinations in the study of genius or insanity. He criticized Moreau de Tours for his analysis of hallucinations in great, "representative men": yes, there exists indeed a "fact of morbid psychology," but that morbid psychology is not the result of "special organization" in "superior men,"[119] but of heredity. As other doctors had shown, the study of heredity in families of intellectuals is difficult to begin with because "many great men seem not to have had ancestors and, too often, they do not establish a family line,"[120] and because, Charles Féré is categorical, genius is hardly transmissible.[121] Moreover, study of those families is counter-productive according to Brachet (p. ix) because of the unquantifiable number of undocumented and unknown cases of transmission of heredity. Some of the best scientific studies of hereditary genius, in Brachet's judgment, by Francis Galton (*Hereditary Genius*), Jules Dejerine (*L'hérédité dans les maladies*), and Théodule Ribot (*L'hérédité psychologique*), prove nothing because they lack "all counter-evidence. It would have been necessary to note, alongside the cases of transmitted talents, all the known cases in which paternal talents were not transmitted."[122] In Brachet's view, since there has been no method to control the data, another influential thinker of heredity, Paul Jacoby, was able to write his absurd thesis that genius leads to degeneracy: "Intellectual development leads to neuropathies, and, as their consequence, degeneration and extinction of the race."[123] To isolate one case of Joan of Arc, as Moreau de Tours did, for instance, is simply to be doing "a medical history"; but even to study intellectual lineages is to falsify the findings because of incalculable unknown hereditary data of larger family kin. Instead, heredity has to be observed and analyzed, in order to have any value, over an extended period of time and it requires an assembly of "family archives."[124] The richest resources for such an archive are "royal dynasties" (*les dynasties royales*), followed by the archives of "aristocratic families" (*les familles aristocratiques*). Pathology will become visible only from observations of historical sequence: "The goal of my book is to reconstitute the fact and the symptomatology of the kings in texts that history and this certain symptomatology offer us, to reconstitute the pathological chain of the Capetian family."[125]

Brachet's monumental work is the culmination of decades of reflection on history, heredity, and degeneration that brought the Middle Ages into the medical discourse and national consciousness. Theorized by Littré and Brachet and tirelessly promoted by Lacassagne and his disciples, like Dr. Cabanès, pathologic archaeology provided historical proof both of the scientificity of method (studying heredity over multiple generations) and the exactitude of its conclusions; and, most important for physicians engaged in social and national prophylaxis, it drew political as well as social lessons from the *longue durée*: "If the history of medicine has a great usefulness, the medicine of history has an importance no less considerable. It is in certain cases a profound lesson for politics; it shows us, for the Valois lineage, the irrefutable application of the great laws of general pathology."[126] This dichotomy of history that taught medicine (*histoire de la médecine*) and of medicine that alone could understand history (*médecine de l'histoire*) is a double transference: on the one hand, medicine found itself justified as a legislator of social behavior based on the information gathered from genealogies of the Middle Ages and the *ancien régime* and, on the other hand, medicine treated history as a medical category and displaced it in its role of philosophy of history, "the philosophy of history, the medicine of history." Even though, ultimately, the vision of the past that the medicine of history put forward did not become the dominant historiographical version or disciplinary model, this double transference had a profound impact on the nineteenth-century constructions of the *ancien régime* and, in particular, of the Middle Ages that we have inherited, as we will see in Chapter 2.

Chapter 2

"Pathologic Genealogy": Biological Heredity and Medieval Kinship

"From the entirety of our study, the role of the doctor emerges then in all of its importance. . . . Knowing the hereditary antecedents of his clients, he will be able to foresee what their children will be; in his perspective, the adage: *Pater est quem morbi filiorum demonstrant* will offer many more guarantees than the famous saying of Roman law: *Pater est quem nuptiae demonstrant.*"[1] These words, written by Dr. Fernand Debret in his 1901 doctoral thesis, *La sélection naturelle dans l'espèce humaine*, synthesize the major conceptual and intellectual shift in the conceptualization of genealogy that occurred under the influence of medical discourse on hereditarianism. The identity between the children's and father's disease(s) confirms alone the father's biological paternity. Marriage in and of itself cannot prove the blood relation of the father to his offspring; it merely establishes his juridical recognition of and responsibility for them. If, before the naturalization and medicalization of heredity in the earlier part of the nineteenth century (see Chapter 1), filiation had depended on marriage law—a husband is necessarily the father of the child, since the marriage bond carries an implicit paternal recognition of his offspring—once medicine theorized heredity, filiation was defined by biology. Medical discourse on heredity introduced a difference between biological paternity (natural law, *cognatio*, *parenté naturelle*) and juridical paternity (civil law, *agnatio*, *parenté civile*):[2] the biological father is the one designated by the hereditary diseases of his offspring, while the father by law is the one designated by marriage. Juridical and medical paternity were now two separate paradigms, subjects of two distinct discourses of law and medicine.

In addition to the demarcation of the juridical from biological paternity and filiation, medical discourse also transformed kinship and consanguinity from a juridical to a biological notion: "After having belonged exclusively to juridical language, the word [consanguinity] was adopted by doctors who made it into the synonym of kinship."[3] Kinship now referred to blood relatives, "the kinship of individuals born of the same trunk," and consanguinity meant direct blood descent: "Coming in direct descent from a common genitor, consanguine relatives are attached to the same family by *blood* relation."[4] But prior to this definitive conceptual shift of the mid-nineteenth century, consanguinity had a juridical and social meaning tied to the institution of marriage: it identified and restricted eligible marriage partners. Marriage partners were defined by proximity, that is, degrees of kinship (*les degrés de parenté*). While ancient societies allowed marriage among even directly related blood kin, after the advent of Christianity an extended notion of proximity restricted marriage eligibility. A juridical definition of consanguinity was the heritage of the ecclesiastical Middle Ages and the Lateran Council of 1215. Antoine Furetière's *Dictionnaire universel* (1690) still used the medieval definition of consanguinity by degrees of proximity among kin who were not eligible to marry: "Marriage is forbidden by the Church up through and including the fourth degree of consanguinity, but it is not an obstacle to marriage according to the law of nature, except in direct lineage. Consanguinity ends at the sixth or seventh degree, except for the Crown, in which case it carries on to infinity."[5] A broad notion of proximity made all who were kin, whether by blood (biologically, *consanguinitas*) or by alliance (nonbiologically and/or spiritually, *affinitas*, in-laws and godparents), linked metaphorically by the concept of blood: "This word [consanguinity] is also used in a more general way to denote all kinds of kinship."[6] Medieval genealogies were thus not always reliable since their members may or may not have been blood descendants of their ancestors. Consanguinity was malleable, and its capacity to designate blood kin as well as in-laws and godparents exposes the powerful rhetoric of "blood kinship" as precisely a discursive effect, subject to (re)definitions by jurisprudence.[7]

In the seventeenth century, as seen, consanguinity was still a juridical concept that was subject to social convention and canonical prescription. But by the mid-eighteenth century, the juridical concept of consanguinity as proximity began to weaken. Then began an increase of endogamy, and alliances that had previously been considered incestuous—with the brother's widow, first cousin, or a niece—became acceptable in urban and rural milieus by the mid-nineteenth century.[8] The erosion of the juridical meaning of consanguinity meant that proximity was now increasingly defined in narrower terms of nuclear family, that is, those most directly linked by blood. This coincided with the emergence of the notion of "species"; it was understood that the reproductive power of individuals made the species. The focus on reproduction began to transform the metaphorical relationship between family and state of the *ancien régime* into a biological one of disciplinary bio-power over population.[9] Thus the epistemic space of heredity and consanguinity as organic reproduction, and not succession or proximity in juridical terms, could also emerge (see Chapter 1). In the mid-nineteenth century (1856–1866), the debate on consanguinity and its role in heredity sealed the notion of consanguinity as biological and finalized the linguistic and conceptual shift from kinship to blood kinship, from lineage to bloodline, from blood metaphor (proximity among kin) to biological materialism (blood kin). This difference between juridical and medical conceptualizations of consanguinity split the notion of kinship in a radical way, between old genealogies established on the basis of marriages that guaranteed filiation through the juridical (that is, potentially affiliative and fictional) bond and modern genealogies identified as a bloodline of hereditary characteristics (that is, biological and objective). Consequently, the meaning of genealogy was altered from lineage as proximity (juridical) to lineage as bloodline (biological).

At this critical juncture of the mid-nineteenth century, medieval and *ancien régime* genealogies provided the documentary evidence of the *longue durée* that medicine needed to theorize the problem of morbid heredity and consanguinity; conversely, medical discourses impacted our modern understanding of these medieval genealogies. "Blood" had been a genealogical concept in the French Middle Ages denoting both juridical filiation and biological bloodline, but nineteenth-century hereditarianism and the consanguinity debate narrowed its semantic field exclusively to a naturalized, medical concept of biological bloodline. Whereas genealogies in the Middle Ages had a juridical-social function, under the influence of nineteenth-century medical theories of hereditary transmission by blood, we have since understood medieval genealogies in narrower, quasi-medical terms, as biological *bloodlines*. This has had a long-lasting effect on the disciplines of history and literary history and on their analytical models of the medieval family, kinship, and the disposition and structure of medieval history, time, and narration. It is the aim of this chapter, then, to show how heredity as a medical, that is, scientific idea gained cultural ascendancy and ultimately was translated into medieval disciplinarity and a disciplinary logic that could produce a biological reading of medieval genealogies.

Blood and Genealogy

By the 1850s blood was undoubtedly a medical concept: on the one hand, the subject of hematology, on the other, the carrier of biological heredity.[10] Blood signified the fluid circulating in the body of an individual and the organic substance that transmitted the same physiological form from generation to generation, giving conformity to a species. In other words, blood was no longer a metaphor, a rhetorical device, whose discursive effect was to establish ties of kinship, regardless of biology, but an organic matter and medical term. A materialist, biological concept of blood replaced the rhetorical and juridical-social function that blood had held in the Middle Ages and the *ancien régime*.

The consanguinity debate essentialized the concept of blood as a fluid circulating in individual bodies and substance creating biological lineages: "The blood is the breeding-ground of *races*, it is the cement of the *organism*."[11] The circulation of blood holds together the "organism" of an individual because of "the considerable role played by nourishing fluid in organic renovation."[12] Blood is also the seed from which a lineage, the "race" (*la race*), grows and expands: "Blood is this grounding of *generations*."[13] Both in its individual physiological (organism) and collective genealogical (race) role, blood is a medicalized concept; in each case, in circulating in the organism or in the race, it performs "this double mission of blood": "It is from the blood that are derived the materials which repair, which heal wounds; it is also there that are found the materials which fortify."[14] Blood's double mission is either to "repair" and "heal" morbid heredity and stop degeneration, or to strengthen what is already healthy, to "fortify." But this double mission is undermined in the individual: "If the blood is infected by any particular virus, then the seminal matter will also be infected and, as a result, the germ will share in the dominant virus of the father."[15] Consequently, blood can regenerate as much as contaminate a lineage, the "race": "A head of family glorifies the superiority of his race when he speaks of the beauty of his blood; he sees it degenerate when the blood alters in it."[16] In consanguine marriages especially, "we accuse alliances between parents of the same lineage of bringing on, of creating, by the mere fact of the nonrenewal of the blood, a special cause of organic degradation, fatal to the propagation of the species."[17] Bad blood is the cause of extinction of lineages.

Medical discourses on consanguinity and heredity also naturalized representations of family lineage: genealogies. Genealogies were an important tool of historical analysis: "Beyond that they serve to distinguish historical figures of the same name and the same family, they reveal the ties of kinship, succession, the rights, claims."[18] But this instrument of historical research was now an object of medical intervention, where heredity and husbandry of the race played a major role. Articles on "Généalogie" and "Hérédité" in Pierre Larousse's *Grand dictionnaire universel du XIXe siècle* are a prime example of how medicine reduced genealogy to heredity and put it into use for eugenics. The article on heredity adopts the general idea espoused by anti-consanguinists that in-breeding causes degeneration of a family—"families that procreate among themselves degenerate and bastardize without delay"—and that an absence of cross-breeding (between the weak and the strong) only leads to further deterioration: "The union of two individuals with poor constitutions . . . gives birth to children who are weaker and more dim-witted."[19] It is logical then that neither intermarriage nor unions between degenerates can "be advised; on the contrary, it is necessary to renew constitution and temperament through well-reasoned cross-breeding."[20] The only way to arrive at "well-reasoned" regenerative cross-breeding is by the constitution and consultation of genealogies: Larousse gives here the example of animal breeding. Larousse's article calls for "the determination of *genealogies* for each race," that is, a "genealogical book" (*livre généalogique*), the equivalent of a "civil status [*état civil*] of individuals," and he laments that the *état civil* has been underused: "This institution . . . remains sterile in respect to maintaining and perfecting the races because man, who claims to be intelligent, is too proud, too infatuated with his individual freedom to apply to himself the principles of animal husbandry which should regulate all procreation."[21]

A number of books on the medicine of history also used medieval and *ancien régime* genealogies—most often dynastic—as raw material for the medical study of heredity. The first impetus to genealogical-medical studies was given in 1846, at the dawn of the consanguinity debate, by Louis-François Benoiston de Châteauneuf, who studied genealogies of 300 noble families over the course of five centuries in his *Mémoire sur la durée des familles nobles en France* (see Chapter 1). Alfred Bourgeois followed suit in 1859 with the study of sixty-eight consanguine unions in one family over 130 years.[22] Doctor Auguste Corlieu can be credited with signaling and exploring what would become the most fruitful area of the medicine of history: French royal dynasty. In 1872, in his *La médecine de l'histoire, étude médicale sur la dynastie des Valois*, Corlieu set the example by showing "for the house of Valois, the irrefutable application of the great laws of general pathology. We indeed see a head of lineage, Francis I, whose fecundity was never in doubt, unable to conserve the throne in his family for more than 74 years."[23] At about the same time, in 1873, Théodule Ribot published *L'hérédité psychologique*. For the medieval period, Ribot picked only Carolingians but his examples diversify for the *ancien régime* to include the Medici family, with Charles-Quint and the royal house of Spain, distinguished families, such as the Condé and the Guises, but also statesmen (Pitt, Walpole).[24] In 1881, Dr. Paul Jacoby published a widely read and contested book, *Études sur la sélection dans ses rapports avec l'hérédité chez l'homme*, in which he "made genealogy into a subsidiary of pathology."[25] His preference was for the genealogy of the house of Valois, from the fourteenth to the eighteenth century, "earlier eras not being well enough known in regards to genealogies."[26] In Jacoby's footsteps followed Jules Dejerine, who in *L'hérédité dans les maladies du système nerveux* (1886) published "70 genealogical tables," with a portion dedicated to dynastic genealogies, in particular the royal house of Spain from 1440 to 1700.[27] In 1888, Alexandre Cullerre reviewed the royal dynasties of Spain, England, and France over four centuries, starting from the premise that "the history of insanity is . . . the very history of civilization."[28] Émile Dusolier, a student of Alexandre Lacassagne, published in 1895 his thesis, *Psychologie des derniers Valois (François II, Charles IX, Henri III)*, in order to show "the influence of weighty heredity and a poor education on the mental state of these three sovereigns."[29] They had been judged with severity, he argued, when they should be "far more worthy of our pity, by which I mean the pity of doctors."[30] But the reason behind the harsh judgment of history is that "we have not looked for the secret of their degeneration in the far past of their antecedents and in their ill-fated birth."[31] On the tail end of the medical obsession with genealogical study is Dr. Victor Galippe's 1905 study *L'hérédité des stigmates de dégénérescence et les familles souveraines*. His argument on the Habsburg dynasty's "sign of degeneration" (*stigmate de la dégénérescence*)—prognathism—was constructed entirely on genealogies: "Reading the genealogies of sovereign families may have seemed dull, but they constitute, along with portraits, the very basis of our demonstration, allowing us to identify the exact filiations of the historical figure represented and to show, despite that we lack the portrait of his ancestors, the preservation across the ages of a hereditary familial type, or its resurgence."[32] In Galippe's case, genealogy justified physiognomy and physiognomy in turn was proof of biological bloodlines. Finally, works of another disciple of Alexandre Lacassagne, Dr. Cabanès, on the Habsburgs and Bourbons of Spain, served as a belated homage to nineteenth-century genealogical-hereditary thought.[33] All these voluminous studies used historical genealogies to show the harmful effects of consanguinity, in the effort to regulate the social practices of the French nation: "It follows that, in many cases, consanguine unions will be harmful, and all the more dangerous given that morbid predispositions, common to both spouses, will be more marked."[34] But they also comment on a connection between genealogy and politics, dynasty and (national) government, since they serve as a powerful anti-royalist argument: "Had it always been thus, France would perhaps not have had so to suffer under the administration of degenerate races, and pathological heredity is, in my opinion, one of the most powerful arguments against dynastic heredity."[35] Medical study of historical genealogies was the best political argument against the possibility of the return of the French royal family to power in Republican France. Doctors were thus putting new scientific methods in the service of providing material evidence for the claims about the decadence of the ruling classes long suggested by moralists and political opponents of hereditary regimes.

A book by Auguste Brachet, a philologist, linguist, and disciple of Émile Littré, was no exception to this genealogical-medical rule. Begun in 1880, originally published in 1896, *Pathologie mentale des rois de France* rejected the study of individuals, the great men or "representative men" that Moreau de Tours attempted in his *La psychologie morbide dans ses rapports avec la philosophie de l'histoire*.[36] Rather, Auguste Brachet argued, entire dynasties have to be painstakingly brought in as evidence. Pathologies need to be considered within the *longue durée* perspective of genealogical trees (*dynastie-chêne*) that no individual case, considered in isolation and compared to an annual plant, can provide: "How could an inpatient at Charenton, a simple annual plant, be compared with a lunatic like Charles VI, whose heredity can be traced back six hundred years, through a direct line of eighteen generations, not to transform them into as many insane rulers in the House of France, as historians might do in some Michelet-inspired way, but to see through what gradual morbid alterations they passed, and how the accumulation of morbid defects, either alone or through consanguine marriages, . . . ultimately led to one lunatic."[37] Thanks to dynastic heredity, one can study medical heredity. In other words, hereditary defects *must* be studied genealogically, as a "pathologic genealogy" (*généalogie pathologique*) (Brachet, p. viii). Brachet's book was entirely dedicated to the medieval period. He applied "genealogies of blood relations" (*généalogies de consanguinité*) to Louis XI's ascendancy and he was especially interested in Charles VI's madness. Historians had focused on "various accidents" to account for Charles VI's madness: "Some attributed it to the overindulgence in tournaments and festivities, others to his sickly childhood,

Michelet to an equestrian race that never took place."[38] But the answer, says Brachet, lies in biology, and more particularly in consanguinity: "Any degeneration of a race can only be the result of heredity, and one of the most important factors in heredity is precisely the parents' consanguinity."[39]

The genealogical method of studying heredity, and consanguinity in particular, reached its apex in Brachet's historical pathology. Brachet studied heredity and consanguinity in twenty-two generations over six hundred years. For him, this constituted overwhelming evidence in favor of the soundness of genealogy as an instrument of medical analysis. Nevertheless, Brachet resisted one subsidiary of medical-genealogical study, the claim that dynasties and aristocracies were a privileged site of degeneracy, resoundingly articulated by Jacoby: "All upper classes, all families that occupy exclusively elevated positions, share the fate of reigning families, although to a lesser extent, always in direct relationship to the height of their privileges and their social position."[40] For followers of Jacoby, hereditary transmission of power morphed into pathological heredity; "Cesaritis" (*la césarite*),[41] "the delirium of power,"[42] "the degenerative influence of power,"[43] led inevitably to degeneracy and extinction: "The greater it [the power] is, the greater the evidence of the moral, intellectual and physical decadence" and "absolute power . . . produces the deepest and most rapid degeneration."[44] But, Brachet asked, are we to locate in dynasties a particular type of degeneracy? Do they constitute a legitimate, isolated object of study and should they be studied in isolation? Is there anything inherent to power that predisposed dynasties to degeneration and, moreover, is this degeneracy proof of royalty? "How could we go so far as to affirm that degeneration is *proof* of royal origins?"[45] Brachet demonstrates that dynasties are as susceptible to degeneracy as any other lineage and that, indeed, the only difference lies in the quality and thoroughness of available documentation: "Royal dynasties offer the clinician the most precious sources of pathological documentation."[46] But there is no excuse for any doctor who, "studying *royal* families, and *not* having *others* at his disposition for scientific research, maintains that there are mental illnesses with *royal causes*," because there is nothing to which to assign the specificity of "the *royalty* of Louis XI's zoophilia": "*Every* degeneration of *a* race can be only the result of *one* heredity."[47] Degeneration is not exclusive to royal families; in Brachet's view, a member of any lineage, nobleman or noblewoman and commoner, can be the victim of hereditary degeneracy. Brachet thus relativizes the importance of genealogies to make them, not into the carrier of truth about dynasties and their (royal) degeneracy, but a tool for understanding medical laws of heredity: "Historical Pathology . . . connects pathological pieces of *data* that it receives in fragments from medical History in order to establish their succession, and to draw from them the pathological law of the individual or, as they say in clinics, the patient's pathogenic, and, hence, mental, formula."[48] In this process, historical pathology transforms historical genealogies into medical genealogies of pathology, but it also democratizes dynastic genealogies, making them equal to any other "individual pathology . . . of a community of individuals."[49] In that, *Pathologie mentale des rois de France* was an "attempt to constitute a partial science from human heredity," one that was judged as equal to "a revolution in history."[50]

The End of Genealogical Fictions

The medicalization of genealogies had its benefits for the discipline of history, as well; in the absence of paternity tests, nineteenth-century doctors and historians could use *morbus*, morbid heredity, to prove paternity: "*Pater est quem morbi filiorum demonstrant*." Evidence of pathological heredity could serve as evidence of a bloodline or biological filiation and provide a corrective to genealogical fictions.

Before biology, historical genealogies rested upon "the belief in paternity, meaning upon *an act of faith*," which, in turn, provided the foundation for "European families."[51] Paternity operated on a discursive act of recognizing oneself as the father of a child, an act grounded in the legal bond of marriage: "*Pater est quem nuptiae demonstrant*" (Brachet, p. cxcii). This discursiveness and malleability of juridical convention created the very fantasy of *ancien régime* genealogies that made the author of Larousse's article on genealogy dismiss them: "We must be on guard against preposterous claims by certain historians, who, out of adulation, trace back to the heroic times the origins of the houses or the princes in whose favor they are writing."[52] Patrilineal continuity was a fictional construct: "We see that noble families, and above all those that date their origins back to far reaching eras, have managed up through the present to continue not their lineage, but their name—far older for most of them than their descent—only through the aid of all sorts of fictions, and by grafting their names onto different families."[53] To perpetuate their lineage, all means were good for families, "infinitely multiplied substitutions," "marriages successively contracted two, three, and four times," "requests for annulment of vows," "transfer of names, by women, in different families, etc.," to the final desperate solution, "the legitimizing of a natural child."[54] But Brachet underscores the fundamental difference that "in political history, whether a king is the son of the man believed to be his father or not, if no one other than his mother knows, then politics follows its course."[55] If this uncertainty "suffices for the historian, it is insufficient for the clinician": "In pathology, the maxim *Pater est quem nuptiae demonstrant* has no currency, since the conclusions to be reached from the sequence and succession of facts rest entirely, before any study, on the fundamental premise that the sovereigns to be studied indeed descend from one another."[56] A major shift in the historical and cultural understanding of lineage had occurred: paternity was no longer "an act of faith" (a legal convention or a construct of discourse), but a biological given, formative of a *bloodline*: "The basis, the premise of this study is necessarily the reality of heredity."[57] Thus, if Brachet could not prove the biological paternity within Louis XI's royal dynasty, the entire study of his heredity was compromised: "Since the study of heredity is essentially based on the premise of *biological* certainty, every experiment on heredity is nullified as soon as one of the factors does not surely belong to the genealogical chain."[58]

Brachet circumvented the problem of uncertain biological paternity in an inventive and original way, "through knowledge of his heredity *absolutely neglected* until now by historians and doctors,"[59] Charles VI's maternal line. Louis XI's psychological troubles are in no way proof of his direct descent from Charles VI. If we take into account the notoriety of Isabeau's adultery, the statement that "any son of an insane father must be insane" means that his father could have been "Charles VI, Louis [d'Orléans, Isabeau's lover], or an unknown man."[60] But regardless of who was the genitor, Charles, Louis, or a "commoner" (*roturier*), "Isabeau's biological formula . . . allows us to affirm that, whoever the father was, this formula was in and of itself *sufficient* to explain the defects in Charles VII and Louis XI."[61] That is, the maternal line itself was saturated with enough morbid heredity to produce hereditary defects of the two kings. Every single author of genealogical tables, except for Brachet, neglected the mother's heredity as central to the bloodline, and thus conflated genealogy not only with heredity, but also with patrilineal heredity.

The passage from Benoiston de Châteauneuf's description of genealogical fictions to Brachet's medical use of genealogies marks the moment when blood was inscribed in "physiological genealogy": "*Historical Pathology* is properly speaking the explanation, through biological science, of the data furnished by historical texts."[62] Brachet's study shows us how history turns into biology, how the legal bond of filiation turns into a bloodline. Historical study from then on would no longer speculate on paternity and filiation as a question of faith, but had to accept the idea of genealogy as a bloodline.[63] Furthermore, since historical pathology could only use "the negative biography in the history of an individual, meaning it could only admit the existence of verified facts,"[64] it would establish with exactitude and scientific rigor the bloodline of dynasties. Finally, Brachet's study also shows us that doctors (just like historians before them) focused exclusively on the paternal side of families and neglected the maternal line.

The confusion of genealogy and heredity resulted in the naturalization of blood and patrilineal genealogy. Genealogy's primary signification was no longer based on faith and marriage law but on blood descent, whereby heredity put an end to medieval genealogical fictions (and replaced them with other fictions). While different genealogical strategies had protected the aristocracy from failures in the bloodline—if blood could not certify biological filiation, its metaphor could give a fictional continuity to the line—the nineteenth-century family no longer had this luxury of resorting to hereditary fictions because heredity, unlike the inheritance, could not be cheated with fictions. Genealogy, the very mechanism and mode of representation that organized, maintained, and represented the *ancien régime*, was adopted and transformed to suit the purposes of the medical corps in studying an individual's heredity, on the one hand, to determine hereditary predispositions and causes, and, on the other hand, to draw lessons from the body politic of dynasties and apply them to the body of the French nation. Paradoxically, then, genealogy did not die out as a relic of a bygone era; rather, it was resuscitated and rearticulated in the medical-biological discourse of eugenics surrounding the family and nation.

From Psychology to Biology: An Inheritance

As we have seen, medieval and early modern preference for choice and flexibility in defining kinship juridically was sidelined in favor of the material concept of blood; and the importance of maternal lines (and adultery) that could have pried open the dominant frame of the study of patrilineage was disregarded (see Chapter 6 for an explanation of why adultery was set aside). Modern historical study of medieval genealogies inherited these two types of presuppositions from the medicine of history and biology (a discipline which did not exist before the 1800s): the focus on patrilineage and bloodlines. Georges Duby's foundational work on the noble family and the institution of primogenital patrilineage can be read within the ongoing development of medical and anthropological discourses since the second half of the nineteenth century.

Duby's social histories from the 1950s and 1960s indelibly marked modern historians of the medieval period. Duby provided a model of the medieval aristocratic family so powerful that it gained immediate and widespread acceptance and dominated modern historians' work, as much as that of literary historians, until the twenty-first century. In his landmark 1953 study of the Mâconnais region, he described a reorganization of the medieval family from an amorphous horizontal kin group to vertical and linear patrilineage, and he grounded it in the social and political changes in the aftermath of the year 1000, and especially the institution of castle lordship. Lords imposed the vassality bond on the smaller lords and exercised tighter fiscal controls on peasantry. In order to preserve their overlordship, the nobles then began restricting the transmission of inheritance, patronymic, and marriage rights to the eldest son, creating the institution of primogenital patrilineage.[65] Duby's later studies, "The Nobility in Medieval France" (1961), "The Structure of Kinship and Nobility" (1967), and "Lineage, Nobility and Knighthood. The Mâconnais in the Twelfth Century: A Revision" (1972), increased the currency of historical study of medieval genealogical literature and strengthened the patrilineal model of family.[66] Medieval genealogy was understood as a transmission from fathers to first-born sons, from which sisters or younger brothers were excluded. This shift from an amorphous horizontal, highly unstructured kin group to a sudden appearance of a rigid primogenital patrilineage was once again reversed in the thirteenth century, Duby argued, and replaced with a relatively fluid structure more accepting of daughters and younger sons. If "social and literary historians have easily accepted the conflation of the medieval term [lineage: *linea, lineagium, lignage*] with its modern anthropological homonym to represent medieval descent conceived narrowly as agnatic or patri-lineage,"[67] it is so precisely because Duby's work articulated medieval genealogy in easily accessible medical-anthropological terms that had displaced the juridical-social definition of lineage into the background and that were a part of our modern discourse on family and nation. We can see the overdetermined influence of medicine play a hand in this formulation of medieval genealogies.

Based both on the work of anthropologists who were the first to call attention to the misrepresentations of the patrilineal model, primarily Jack Goody and Anita Guerreau-Jalabert, and on the rich twelfth- and thirteenth-century records from Champagne, Theodore Evergates makes the most up-to-date attempt to free the medieval aristocratic family from the stranglehold of modern patrilineal genealogy.[68] Evergates refutes the primogenital patrilineage as the model of aristocratic family—"The patrilineal model of the family fails utterly to explain the nature and customs of aristocratic families" (Evergates, p. 88)—concluding that "the more conventional understanding of lineage as kinship, including all consanguines, accords far better with medieval usage" (Evergates, p. 85). It is striking, Evergates remarks, "how infrequent the term 'lineage' appears in the twelfth-century sources from Champagne" (Evergates, p. 85). In one instance where lineage clearly means descent—Chrétien de Troyes's hero Cligés belongs to King Arthur's lineage—"lineage carries a general sense of kinship rather than patrilineal descent" (Evergates, pp. 85–86). David Sabean has been working on the long-term evolution of kinship, from the Middle Ages to the twentieth century, to reach the conclusion that the shift to patrilineal devolution of property occurred only at the passage to the early modern period and that it was "less general than earlier research had assumed." It was also "more specifically related to modes of linking political power to the possessions of certain goods such as castles, titles and offices."[69] Put differently, succession of rights and titles was more strictly regulated than other property that was inheritable by all children. Anthropologist Anita Guerreau-Jalabert's work on lineage as kinship (*parenté*), and not patrilineal descent, helped define medieval lineage as descent through both the maternal and paternal line (patrilineal or matrilineal), that is, as cognatic bilateral kindred. She also showed that the importance of agnatic lineage had been exaggerated in the study of medieval genealogies: "Latin and French terms . . . refer . . . to this or that manner of thinking about or fractioning the fluidly shaped whole that kinship constitutes."[70] For instance, the medieval Latin term *cognatio* does not differentiate between the paternal and the maternal lines, or between spiritual, "*cognatio spiritualis*, *cognatio spiritus*,"[71] and "biological" kinship.[72] And *parentela*, a term predominant in the Middle Ages, but nonexistent in classical Latin, refers to a mixture of "biological" and "alliance" relationships, organizing the kin group around an individual (Guerreau-Jalabert, "La désignation," pp. 81–82). *Propinquus* could mean "either a blood relative or a relative by marriage."[73] This functional view of lineage, constructed on the basis of need with relative disregard for blood relations, brings to light "the role of noble women, who could, through marriages with men with undistinguished ancestry, bestow exalted pedigrees upon their jointly produced children" (Bouchard, pp. 10–11). Women were not marginalized in medieval family histories or in medieval family organization; rather, they were marginalized by the later genealogical literature, the patrilineal reorientation of the medicine of history (with the exception of Auguste Brachet's study).

In the wake of Georges Duby's work, two important works, both published in 1983, took up the notion of medieval genealogy, which they established as one of the main methodological tools for understanding medieval culture and anthropology: literary historian R. Howard Bloch's *Etymologies and Genealogies* and historian Gabrielle Spiegel's landmark article "Genealogy: Form and Function in Medieval Historiography."[74] Bloch argued that genealogy organized etymology, that is, that all signification grounded itself in generation. Medieval grammar and literature represented the making of meaning as organically and naturally as fathers begot sons. Bloch described lineage that lent an organic cohesion to medieval linguistic structures, but without questioning from what structures genealogy itself made meaning. This further strengthened the underlying notion that genealogy equaled generation or procreation, that medieval genealogy necessarily meant biological, blood filiation. Bloch reversed course in a later, but less known, article, "Genealogy as a Medieval Mental Structure and Textual Form," where he attributed both genealogy and etymology to the "general problematics of the sign,"[75] thereby allowing for the possibility that not only structures of medieval lineage were molded by those of grammar but also that, if genealogy and etymology were signs, they were then concepts, or discourses—"mental structures" and not a natural given—that organized the medieval social space. But Spiegel's article is a flagrant example of the discursive slippage from juridically defined medieval discourse on paternity to the modern biological procreative descent: "Genealogy functioned to secularize time by grounding it in biology, transforming the connection between past and present into a real one, seminally imparted from generation to generation" (Spiegel, "Genealogy," p. 107). What interested Spiegel was to show how the "human, biological significance" of genealogy also functions as a metaphor for the relatedness of historical events, a sort of humanization of historical time, where "events stand in a filiative relation to one another that mirrors the reproductive course of human life" (Spiegel, "Genealogy," p. 108).

There is one problem, however: while medieval genealogies are certainly mirrors of social constructionism of family, they are not necessarily mirrors of biology, as Spiegel claims, "a biological model" of seminal connection, "seminally imparted from generation to generation" (Spiegel, "Genealogy," p. 107). As with primogenital patrilineage, a series of studies have pointed in recent years to an anachronistic identification of medieval lineage with our modern notions of genealogy, conceptually defined as blood lineage and visually identified with genealogical trees.[76] In the *Bloodless Genealogies of the French Middle Ages*, I analyzed the representations of genealogy in literature and art of medieval France and Francophone England to demonstrate that genealogy is a cultural and discursive construct, one that is essentially embedded in linguistic structures. Of those, medieval texts most frequently use the metaphor of blood in order to lend a semblance of biological continuity to lineage. Describing the development of the genealogical concept of blood, Jacques Le Goff declared that "blood *as definition of kinship* in the aristocracy appears . . . only very late. Starting only in the fourteenth century, direct descendants of kings are called 'princes by blood.' And only at the end of the fifteenth century does the notion of 'blood purity' appear in Spain, in opposition to the Jews."[77] This occurs at the same time that the notion of aristocratic "blue blood" (*sangre azul*) takes root.[78] Anita Guerreau-Jalabert found that the Old French word "*sanc* can include blood-related kin, but this usage seems to have developed particularly at the end of the Middle Ages."[79] Christiane Klapisch-Zuber offered evidence that "in the treatises on consanguinity and marriage prohibitions, theologians and canonists from the twelfth through the fourteenth century nonetheless make far fewer references to shared blood than one might expect."[80] Most recently, Caroline Walker Bynum confirmed that the meaning of blood as lineage or family continuity, one "sense in which 'blood' is used in modern parlance," is "infrequent in late medieval devotion." In a few instances when it seems to refer to family, "blood signifies primarily the individual, not group identity or continuity and certainly not lineage."[81] Instead of blood, a far more frequent term designating familial relationships is "flesh" (*char*), as in expressions such as "cognatio carnalis, carnis" and "ami charnel."[82] The prevalent metaphor of kinship in the literature of the twelfth and thirteenth centuries is the body: a corporeal metaphor that creates "the body of kinship" (*le corps de la parenté*) (Klapisch-Zuber). Hence, while it is certain that the Middle Ages had a notion of blood kinship, it is also true that filiation could be constituted in ways that did not presuppose a bloodline. The usage of the term "blood" did not signify only a biological tie, but rather "blood" was a juridical substance and therefore a metaphor of a possible biological tie.[83] If the concept of consanguinity referred both to bloodlines and lines of alliance, though it referred to them as consanguine—the case in point being the 1215 Lateran Council prohibition of marriage in the fourth degree applied to both blood and spiritual kinship—then these consanguinary concepts were not grounded in blood as a natural given, but rather in a concept of "blood" that was a construct of discourse and that upheld these consanguinary concepts. Genealogy was a narrative deploying (juridical) metaphors of blood, out of which sprang genealogical fictions.

Saying that blood was not a primary genealogical concept is not, however, to deny blood as fluid. Certainly, blood was omnipresent in the Middle Ages, associated in particular instances with heroism or impurity, birth (literal and symbolic), Christic devotion, giving cohesion to the body or spelling out a taboo.[84] The contradiction between the omnipresence and devotion of Christ's sacrificial blood and the taboo of blood(-shed) is one of the defining paradoxes of the Middle Ages: "good blood, a symbol of human life and divine sacrifice, and bad blood, an image of sin and death."[85] Blood, according to the historical-anthropological data, was first and foremost viewed as matter, before it also became a genealogical term at the end of the Middle Ages; there may even be a link between the prohibition of bloodshed and the scarce vernacular use of "blood" terminology to describe kinship ties.

It is important to note that Duby had underscored precisely this difference between medieval kinship and nineteenth-century bloodline. Historians' genealogies, he argued, are "biological," "those *reconstructed after the event* by historians patiently uncovering evidence of consanguinity and intermarriage. . . . This kind of genealogy, necessarily incomplete, and often uncertain, gives us a *true*, one might almost say a *biological*, picture of the family group throughout its existence." But medieval genealogies are "*psychological*," "differently constructed, but invaluable nonetheless," and offer "evidence of family psychology, and the reactions which at that period the ties of kinship produced."[86] Constance Bouchard challenged the patrilineal family as a manifestation of the mutations of the year 1000 and talked about the constructed family, "defined operationally" (Bouchard, p. 3), "an abstraction that fluctuated depending on time and circumstances, both for the individual over a lifetime and for successive generations" (Bouchard, p. 4). The constructionist nature of family, the psychology of lineage, explains medieval genealogical fictions, those fanciful constructions and representations of medieval lineage in genealogies. Biology is of no essence to such constructions: "'Family' does not and did not reside only in biological connections. The real question is which people noble men or women would have considered part of the same group as themselves—and how this shared group membership affected how they treated them, or how they sought to identify new members of the group with those who were already in it."[87] This means that being of "noble blood" was not about biological heredity but a metaphor of it (Stahuljak, *Bloodless Genealogies*), and the family structure "was continuously open to new individuals and new lineages but incorporated these into its preexisting structures" (Bouchard, p. 178). Though represented through a male-identified line, they were "embedded *through their marriages* into the web of alliance and blood-relationship" (Bouchard, p. 178), a web but not necessarily a bloodline. Strategies of incorporation and representation were not identical to actual bloodlines.

The modern discourse on genealogy as bloodline emerges in the transformation from the medieval "psychological" into the modern "biological." Precisely because of the nineteenth-century transformation of the notion of genealogy as heredity, medieval genealogy was no longer assimilated to the discursive *line*, often one of genealogical fictions, but to a biological *bloodline*. We have thus struggled with understanding medieval genealogy on its own terms, using the inaccurate shorthand of *bloodline* to express the medieval notion of lineage for which we have no adequate word today; the term "genealogical fictions" likewise conveys our modern perspective without embracing the medieval mindset. That is, only because we have been reading medieval genealogies from our modern perspective of a bloodline have they appeared to us as either discontinuous or simulations of continuity. Our own modern notion of genealogy as bloodline allowed us to locate the biological ruptures and discontinuities in medieval bloodlines and to identify strategies the Middle Ages deployed to secure continuity. But these strategies were fictions in as much as we take biology as our genealogical premise. Only if biology is the norm for a continuous genealogy can we call the continuity of medieval genealogies "fictions." Induced by biological materialism, we have assumed that medieval genealogy is based on the fluid of blood rather than the metaphor of blood. Genealogy represents the continuity of succession more than the continuity of a bloodline, pointing to the occasional divergences between blood kinship and succession of kingship or title. This is not to say that blood kinship is not represented or desired in medieval genealogies; on the contrary, it is represented but along with representations of other forms of continuity, for which blood serves as a metaphor but not as evidence of a material, biological bond. The continuity of medieval genealogy is as psychological as it is natural.

An Analytics of Genealogy: Medievalists and Foucault

Throughout this chapter I have been tracing the logic of the implementation of heredity in a narrow medical-biological frame and the formation of the consistency of this logic down to the medieval discipline in particular. I argue that it is important to retain the cognitive distance that is the very basis of our disciplinary knowledge and historicize disciplines in order to understand how certain ideas rise to dominance and remain embedded as uninterrogated assumptions. The fixity of biological kinship and the scientific assumptions derived from biological materialism have obscured other histories of medieval kinship that reemerge when the formation of the scientific logics is traced. Historicizing assumptions about the very categories deployed in the analysis of the medieval period illuminates one major, and sticky, issue: the fact that the naturalization of genealogy in the work of modern historians of the medieval period has also been the reason for rejecting Michel Foucault's genealogy. Gabrielle Spiegel outright refused any "utility of Foucault's postmodern theories for the analysis of medieval genealogical phenomena."[88] Certainly, genealogy as analysis and medieval genealogy are not the same: one is a paradigm, the other is a concept describing a phenomenon of family organization. Bringing to the fore other logics of medieval kinship therefore renders arguments like Spiegel's non-operative, and it shows that genealogy as analysis is a valuable tool of historical analysis for medievalists, historians and literary historians alike, especially when it comes to charting our modern reconfiguration of medieval genealogies in the wake of the nineteenth-century medicalization of family life.

Foucault's articulation of genealogy as analysis appeared in 1971 in his "Nietzsche, Genealogy, History."[89] It marked the paradigm shift in Foucault's thought from archaeology to genealogy; in an archaeology, "*episteme* is a specifically *discursive* apparatus," whereas in the genealogy, "the apparatus in its general form is both discursive and non-discursive."[90] His was a radical redefinition of genealogy as resolutely bloodless; genealogical analysis can be productive only if it is free from blood, from its symbolism. Nietzsche's *Herkunft* means "stock or *place of origin*; it is the ancient affiliation to a group, a bond of blood, tradition or social class" (Foucault, "Nietzsche, Genealogy, History," p. 145 [p. 1008]), but as analytics

> genealogy does not pretend to go back in time to restore an unbroken continuity that operates beyond the dispersion of forgotten things; its duty is not to demonstrate that the past actively exists in the present, that it continues secretly to animate the present, having imposed a predetermined form to all its vicissitudes. Genealogy does not resemble the evolution of a species and does not map the destiny of a people.

> On the contrary, to follow the complex course of descent is to maintain passing events in their proper dispersion; it is to identify the accidents, the minute deviations—or conversely, the complete reversals—the errors, the false appraisals, and the faulty calculations that gave birth to those things that continue to exist and have value for us. (Foucault, "Nietzsche, Genealogy, History," p. 146 [p. 1009])

Genealogy is rooted not in the immanence of stock or race, but in the contingency of accident, an essence made from the collage of alien forms. The symbolics of blood, which represents the line of descent, creates the illusion of an identity and continuity of instincts, passions, and truths. It thus misleads us into "recognizing" ourselves in history, even though tradition is not identical, nor is truth transhistorical and universal. Therefore, genealogical history has three uses: to revolt against reality, identity, and truth. This is not because genealogical history would be anti-historical, but because it is anti-metaphysical and anti-teleological. If genealogy "opposes itself to the search for 'origins'" (Foucault, "Nietzsche, Genealogy, History," p. 140 [p. 1005]), it does not mean that it opposes itself to a "genealogical analysis" (*analyse généalogique*), nor should a quest for origins be confused with this type of analysis.

Genealogical analysis shows that "at the historical beginning of things is not the inviolable identity of their origin; it is the dissension [*discorde*] of other things" (Foucault, "Nietzsche, Genealogy, History," p. 142 [p. 1006]). Genealogy de-essentializes and renders relative the truth-value of narratives of origin and descent; "it fragments what was thought unified; it shows the heterogeneity of what was imagined consistent with itself" (Foucault, "Nietzsche, Genealogy, History," p. 147 [p. 1010]). At its origin lie discord and conflict, a confrontation of forces not attached to a subject, a place, or an institution. Genealogical history records the overthrows, modulations, reversals of violence by another violence, the relationship of domination that gives rise to a universe of rules: "Humanity does not gradually progress from combat to combat . . . where the rules once and for all replace warfare"; rather, "humanity installs each of its violences in a system of rules and thus proceeds from domination to domination" (Foucault, "Nietzsche, Genealogy, History," p. 151 [p. 1013]). This play of domination (*le jeu de la domination*) only looks like successive identical configurations when in fact it is discontinuous and differential. Foucault thus turns genealogy into a method, into a paradigm that analyzes discourses that subtend history, effects of power that transform the play of domination, and, significantly, the play of forces in the set-up of a system of rules, that is, an apparatus (*le dispositif*). It is this analysis of effects of power that makes genealogical history an "effective" history (*wirkliche Historie*). This is the method for analyzing the system of rules of sexuality that also provides the specific genealogy of sexuality. It is a bloodless genealogical analysis, to which is added the categorical rejection of any claim to continuity. Foucault's genealogical analysis applied to reading medieval genealogy shows that medieval genealogies, the narrative representation of the symbolics of blood, are an effect of power. The endless play of permission and prohibition, of continuity and discontinuity, of alliance and misalliance, of origin and dispersion, of rupture and inauguration, papered over by genealogical fictions, is what a Foucauldian genealogist will find in medieval genealogies. Rather than continuity itself, s/he will find the will (*la volonté*) to it.

In a 1998 article, Paul Freedman and Gabrielle Spiegel took aim at Foucault because genealogical analysis would be the undoing of history as a discipline, since "the goal here is not so much an expansion, enrichment, or even complication of our understanding of medieval culture but rather its 'undoing.'"[91] Genealogical history would be contrary to the practice of social and institutional history. Freedman and Spiegel viewed negatively the flattening of history by cultural history because it reconfigured the study of alterity in medieval history in terms of proximity and distance. In "surface" terms, alterity now equaled marginalization: "For example, rather than seeing sexuality and its 'non-normal' practices such as homosexuality as repressed by the church, thus still lingering *within* the normal, postmodern scholars tend to see them as ejected or excised. Hence the characteristically postmodern feature of this work is its erasure of depth in favor of surface" (Freedman and Spiegel, p. 700). They prefer the study of alterity as repression—institutional and subjective—whereby "deep" history is possible, even if they distance themselves from another form of "deep" history, in the form of foundationalist narratives. They condemn the linguistic turn that opened up a new interpretative era, transforming "the understanding of documents as texts rather than sources" and that made it into "a science not of things and deeds but of discourses" (Freedman and Spiegel, p. 697). At stake is the former image of the Middle Ages "as progressive, pluralist, rational, and self-aware" (Freedman and Spiegel, p. 702). In a circular logic, Freedman and Spiegel marginalize the new cultural history of the Middle Ages by using Foucault against himself. They point out that he left undisturbed the master narrative of modernity, presenting the Middle Ages "as a free, untrammeled period," which therefore "seems to escape the fate of those knowledge-power systems so characteristic of the 'modern' world" (Freedman and Spiegel, p. 698). If the Middle Ages are outside the knowledge-power system, they conclude, the cultural turn in medieval history then goes against the grain of Foucault's own thought on the Middle Ages, and this justifies Freedman's and Spiegel's critical position.

Spiegel's 2001 article delivers a much narrower and more personal attack on Foucault to explain why "the historian dedicated to understanding the form and function of medieval genealogy would find Foucault's genealogical conspectus illegible" (Spiegel, "Foucault," p. 12). Spiegel starts from her analysis of form and function of medieval genealogies. Its form is one of "narrative order based on genealogical succession" (Spiegel, "Foucault," p. 6), admittedly a highly selective one. This is a narrative of "the transmission of lands, ignoring the remaining members of the biological family not included in the patrimonial inheritance" (Spiegel, "Foucault," p. 4). Medieval genealogy is made on a "biological model," and genealogies have "human, biological significance" (Spiegel, "Foucault," p. 6). It symbolizes "a family's claim, real or imagined, to political authority and prestige" (Spiegel, "Foucault," p. 4). Its function is to serve as "a metaphor of procreative time and social affiliation": "The procreative process by which human beings engender successive generations is the human shape of history generating events over time, events which stand in a filiative relation to one another" (Spiegel, "Foucault," p. 6). Medieval genealogy is symbolic and historical. It is a retrospective narrative of linear descent of identity; likewise, history is a succession of identical events, in linear progression. When analyzed as a narrative, medieval genealogy is "linear in structure, legitimizing in function, identitarian in politics, seminal in substance and metaphysical in meaning" (Spiegel, "Foucault," p. 8), taking the "self as a unit of history and/or analysis" (Spiegel, "Foucault," p. 12). Spiegel demonizes Foucauldian genealogy because it "does not legitimize, but rather de-legitimizes by de-naturing, by demonstrating the historically constructed character of persons and things, thus bracketing the bio-filiative core of the traditional model of genealogy" (Spiegel, "Foucault," p. 10). For biology is the condition of possibility of history, a guarantee of the continuity of historical time and "the very possibility of a historical 'science'" (Spiegel, "Foucault," p. 9). A biological model that is equivalent to human time ("human, biological significance") is a reassuring guarantee of identity and origin, for otherwise differential knowledge "deprives the self of the reassuring stability of life and nature" (Spiegel, "Foucault," p. 11).

But when Spiegel asserts that "the historian dedicated to understanding the form and function of medieval genealogy would find Foucault's genealogical conspectus illegible," she places on an equal footing medieval genealogy and Foucauldian genealogy, conflating phenomenon and paradigm, narrative and method, epistemology and technology, an effect of power (medieval genealogy) and an analytic (technique of power). Spiegel thus takes medieval genealogy to be a truth-narrative, rather than an effect of medieval discourse of power, a truth-claim. She does not see genealogy as a discursive practice, but as a representation of medieval filiation, where juridical base of filiation is taken to be biological. If Foucauldian bloodless genealogy as analysis is a "solvent of historical continuity and linearity," if it is "aleatory, contingent, potentially disruptive and delegitimizing" (Spiegel, "Foucault," p. 2), it is such precisely because it reveals medieval genealogy to be a narrative and a representation. It is a narrative that invents family's origin in "mythical ancestors" and asserts "its identity through the fictionalization of its past": "In effect, the family rejects its own (genuine) past in favor of an invented one" (Spiegel, "Foucault," p. 4), in order to produce "the myth of genealogical continuity" (Spiegel, "Foucault," p. 6). It is a representation, "representing, while regulating, the transmission of family name, title and patrimony from father to son" (Spiegel, "Foucault," p. 3). According to her, "the historical narrative is emplotted as a genealogy" (Spiegel, "Foucault," p. 5), yet it is precisely this organization into narrative that Foucauldian genealogy undoes as narrative, representation, and fiction regulating an otherwise disorderly, plural, nonbiological as well biological relation of forces within the medieval family unit. This further illuminates Freedman's and Spiegel's assertion that the Middle Ages are outside the Foucauldian knowledge-power system, a claim that misses the fact that Foucault's account of overlap of sanguinity and sexuality—"The bourgeoisie's 'blood' was its sex"[92]—provides a key for understanding how the nineteenth century constructed a biological Middle Ages that was inherited into the twentieth century.

Thus Foucauldian genealogy as analysis helps us examine medieval genealogy and helps us understand that what is "seminal" is really "semiotic," that the "bio-filiative core" that legitimizes does so as a fiction and only because it is de-legitimized from the beginning. Genealogies of noble families belonged to the juridical-discursive representation of power, which functioned on the axis of licit and illicit, permission and prohibition. Around the system of rules defining what is permitted and prohibited emerges the apparatus of alliance: "In a word, the deployment of alliance is attuned to a homeostasis of the social body, which it has the function of maintaining; whence its privileged link with the law; whence too the fact that the important phase for it is 'reproduction'"

(Foucault, *History of Sexuality*, p. 107 [p. 141]). There is then "a discipline of sexuality from the standpoint of reproduction" that functions within the apparatus of alliance, and this reproduction is legitimized and manipulated in its fictions by the juridical bond (Foucault, "Confession of Flesh," p. 226 [p. 327]). The function of medieval genealogy is to legitimize reproduction and to provide social stability. In aiming for homeostasis, medieval genealogy is a technology of domination of space and time, not immanence and truth of a naturalized vision of history. Genealogy seems to be organic to the Middle Ages, but it serves in actuality as a means to stability. Conceived as reproduction, it is a technology of the apparatus of alliance and a discourse of a will to biological reproduction. Precisely because medieval genealogy is (legitimizing) reproduction, it can be easily confused with biology—but reproduction is not what genealogy is or represents, it is what genealogy achieves as a disciplinary measure. With the help of Foucauldian genealogical analysis, medieval genealogy can be read both as a technique of juridical power and a fiction of social identity.

Foucauldian genealogy helps contemporaries see medieval genealogy for what it is, a discourse.[93] It also elucidates the indebtedness of modern history of the medieval period to nineteenth-century biological-medical reformulations of genealogy in the process of unmasking the fiction that only a bloodline can be a genealogy. As such, it affords us a historical practice of the Middle Ages, not so we accept fiction for truth, but so we understand genealogical fictions as a truth-claim, an effect of power, as strategies of stability. Genealogy is first a discourse of the juridical power specific to the Middle Ages, and later a discourse of the disciplinary bio-power specific to modernity. Foucault's analysis offers the genealogy of all genealogies.

PART II

Sex and Race

Chapter 3

Symbolic Archaeology: Sex in the Colonies

Fantasizing about Mohammed

The fall of the *ancien régime* in 1789 created a favorable climate for the revision of the fourteenth-century trial of the Order of the Templars. Charging them with heresy and sodomy—denial of Christ, spitting on the cross, obscene kissing, and idol worship—the French king Philip IV (Philip the Fair) had the Templars arrested in September 1307. Pope Clement abolished the Order in 1312 and gave its property to the Order of Hospitallers. During the lengthy proceedings, fifty-four knights were condemned to the stake as relapsed heretics in May 1310, and, in March 1314, Philip IV had the Grand Master, Jacques de Molay, and the governor of Normandy, Geoffroi de Charnai, burned.[1] In 1789, a French Premonstratensian, father Claude-Mansuet Jeune, published a revision of the trial in his *Histoire critique et apologétique de l'ordre des chevaliers du Temple de Jérusalem, dits Templiers*, followed in 1797 by an *Apologie des Templiers et des Francs-Maçons* by a certain J. M. Plane. New documents on the Templars were also published by German and Danish researchers, Cristoph Friedrich Nicolai (1782), Daniel Gotthilf Moldenhawer (1792), and Friedrich Münter (1794) during the last quarter of the eighteenth century.[2] The French military campaign in Egypt and Syria (1798–1801), in the footsteps of medieval Crusaders, only excited the public imagination that was already at a feverish pitch among scholars, especially Orientalists. Thus the Templars were among the first of medieval figures to enter the burgeoning imaginary of the early nineteenth century. In May 1805, *Les Templiers,* a play by François Raynouard, a writer and well-known Romance philologist, was performed and published to popular and critical acclaim: "[Their success] was immense because it revealed a national literature (and this revelation is wondrous after two centuries of classical literature)."[3] A month later the play was translated in Germany to general acclaim (Chambure, pp. 14–15). At the same time, Raynouard published *Précis historique sur les Templiers*, his defense of the Templars.[4] He expanded this defense in the 1813 reprinting of the play, claiming it a patriotic duty to clear the Templars of any charges: "Combined with the hope of resolving a historical problem was a more noble and more imperious motivation: I thought that I might add to my homeland's claims of glory, and to the just respect that national character inspires, if I brought together, with zeal and impartiality, proofs of the innocence of the Order and of the Knights."[5] Raynouard portrayed them not as renegades who had "gone native" in the Orient, but as "venerable victims"[6] of a political regime.

The ground was thus prepared for a curious archaeological quest: the Templars had been accused of adoring an idol named Baphomet that was "according to some, a statue, according to others, a head, and according to still others, a set of several heads or even a wood painting."[7] Baphomet represented not just an archaeological object but also a symbolic, esoteric philosophy, attributed to the Order of the Templars. Friedrich Nicolai speculated in 1782, in the *Versuch über die Beschuldigungen, welche dem Tempelherrenorden gemacht worden und über dessen Geheimniss, nebst einem Anhange über das Entstehen der Freymaurergesellschaft* (translated in 1783 in France as *Essai sur les accusations intentées aux Templiers et sur le secret de cet ordre, avec une Dissertation sur l'origine de la franc-maçonnerie* [Essay on the accusations against the Templars and the secret of this order, with a paper on the origin of Freemasonry]), that the name "Baphomet" mentioned in the deposition of two witnesses interrogated by the ecclesiastical authorities in Carcassonne "was not the name of the idol, but rather a hieroglyph imprinted on it. The head . . . had been a symbol, the image of the eternal Father in a state of rest, as the ancient Gnostics represented him."[8] Baphomet would thus have been the name for a Gnostic symbol. The Templars were supposedly introduced to Gnosticism in the Orient, since "the Orient, and above all Syria, was the refuge for all of the sectarians and heretics banished by the Western [Christian] Empire. What a breeding-ground for *Gnosticism*! Without a doubt, that is where the Templars saw their religious traditions corrupted, letting themselves give in to the lure of a supposedly *rectified Christianism*."[9] Historian Jules Michelet was convinced that "the order that had represented to the highest degree the symbolic genius of the Middle Ages died out due to a symbol not understood."[10] Baphomet, a mysterious name that stirred the imagination of archaeologists, historians, and Orientalists, stood at the intersection of an archaeology of objects and esoteric symbolism. It led to a symbolic archaeology, as scholars sought the existence of either an object or a symbol that would correspond and give meaning to the Templars' confessions. Not just a metaphor, these scholars resorted to actual archaeological research in Templar churches and inspection of their reliquaries, cups, coffers, etc. The search for the meaning of the symbol was continued by "mystical, heterodox, sects like Freemasons and Rosicrucians, who have glorified the knights of the Temple of a part of the crimes . . . in order to attach themselves to an ancient stem: they wanted to see depth in the inept symbolism that Temple accusers described."[11] Yet this symbolic archaeology merely contributed "to the forming of a ghost, which, after having been exploited against the orthodoxy of the Templars, caught the attention of more than one archaeologist."[12]

An Austrian Orientalist, Joseph von Hammer-Purgstall (1774–1856), was the principal culprit for this archaeological frenzy. Hammer-Purgstall participated in 1800 in the British campaign against the French in Egypt, as an interpreter for admiral William Sidney Smith, a Freemason. The young Orientalist published in 1806 the fruit of his Egyptian discoveries in the form of a translated book that, he claimed, had escaped the attention of French scientists: "The original of this translation was found at Cairo, where it had escaped the researches of the French *Savans* [*sic*], who, though successful in collecting many valuable Oriental books and manuscripts, failed in their endeavours to procure a satisfactory explanation of the Hieroglyphics."[13] Picking up on Nicolai's idea of Baphomet as name for a symbol, Hammer-Purgstall hypothesized in *Ancient Alphabets and Hieroglyphic Characters Explained* that "Bahumed or Bahumet [was] one of their secret and mysterious formulas, with which they addressed the idol of a *calf* in their secret assemblies" (Hammer, p. xiii). The discovery of the ancient Egyptian hieroglyph "*Bahúmíd*, translated into Arabic by the word *calf*" was proof for Hammer-Purgstall that "the Templars had some acquaintance with the hieroglyphics, probably acquired in Syria" (Hammer, pp. xii, xiii) (Figure 1).

Thus, according to Hammer-Purgstall, the Templars were idolaters of Baphomet, they denied God and spat on the cross, and they had been indoctrinated in the Orient, the source of all heresies, and more precisely in Syria. The French province of the Order of the Templars had been dissolved most violently under Philip IV, but other provinces—England, Aragon, for instance—were much more lenient in their repression of the order. Did this mean that the French Templars were particularly corrupt and degenerate?

Figure 1. Hammer-Purgstall's hieroglyph, from *Ancient Alphabets and Hieroglyphic Characters Explained* (1806). Rare Books Division, Department of Rare Books and Special Collections, Princeton University Library.

Stung in their national pride, the French philologists of the early nineteenth century resolutely stepped into the debate to absolve the Templars of any crime of idolatry, Gnostic heresy, and Islamization. In 1810, Antoine Isaac Silvestre de Sacy (1758–1838), France's great Orientalist and Arabist, shook the ground under Hammer-Purgstall's affirmations by pointing out the impossibility that knowledge of hieroglyphs was transferred from the ancients to the Arabs, accusing Hammer-Purgstall of failing to show "how knowledge of hieroglyphic writing, [which was] lost, or so it appears, long before the Arabs were the masters of Egypt, could have emerged from the shadows that covered it to pass into their hands."[14] It is more plausible, Sacy argued, that the relation between the word *Baffumettus* (as cited in the records of the Templar trial) and *Bahoumid* is purely accidental, without a grounding in etymology, "and that this name, in the minds of the Templars and their accusers, must have been the name of Mohammed."[15] François Raynouard, in the 1813 edition of his *Monumens historiques*, went even further and asserted that Baphomet was a simple scribal error or mispronunciation, and not an esoteric mystery: "In the deposition of two witnesses heard in Carcassonne, who speak of *Figura Baffometi*, it is obvious that it is by a mistake of spelling or pronunciation that this word is thus written instead of Mohammed, either because in the Southern provinces the name of Mohammed was thus pronounced, or because the scribe wrote *Baffometi* in error, as he also wrote *asorare* instead of *adorare*; and what must leave no doubt in this respect is that the witness claimed that he was made to pronounce *Yalla*, a Saracen word, which, he says, means God."[16] Raynouard made an informed conjecture on Occitan pronunciation of the Templar witness, Gaucerant. A Templar Crusade poem served him as evidence: "Enans fara Bafomaria / Del mostier de santa Maria. / . . . / *Bafomet* obra de son poder" (Rather, from the Church of Saint Mary he will make a temple to Baphomet [Baphomaria] . . . *Baphomet* will exercise his power [there]).[17] Thanks to Raynouard's philological expertise in Occitan, the Templars could be cleared of the charge of idolatry, though the suspicion that they perhaps were Muslim still hovered over them. Hammer-Purgstall's elaborate esoteric argument, based on a previously lost manuscript of ancient and secret hieroglyphic alphabets, collapsed under the evidence of error of scribal transcription or mispronunciation. Rather than proof of heresy, since Muslims do not worship Mohammed but Allah, such an error of substitution could point to an outright falsification by the office of the Inquisitor in the effort to turn Templars into followers of Islam: "Since the symbol that the Templars are supposed to reject and even defile is the crucifix, *Christ*, the symbol that they prefer and oppose to the former is dubbed with the name *Mohammed*, without considering that Muslims themselves had no images."[18] Despite these rebuttals, Hammer-Purgstall published several volumes of his journal *Fundgruben des Orients* (Treasures of the Orient) from 1809 to 1818. In volume 6, from 1818, he published his conclusions on the Templar idolatry of Baphomet in a paper entitled "Mysterium Baphometis revelatum," or "Mystère du Baphomet révélé, ou les Frères de la milice du Temple convaincus, par leurs propres monumens [*sic*], de partager l'apostasie, l'idolâtrie, l'impureté des Gnostiques et même des Ophianites" (The mystery of Baphomet revealed, or the Brothers of the militia of the Temple convinced, by their own monuments, to share the apostasy, idolatry, impurity of the Gnostics and even of the Ophianites).[19] Rather than rely on the Arabic language in his etymological analysis of the expression used in the trial "in figuram Baphometi," Hammer-Purgstall identified two Greek words that make *Bafomet*: *Bafo* and *Meti*. The first means baptism, the second spirit. Hammer-Purgstall could therefore make the leap to the Gnostic doctrine of "baptism of the spirit, Gnostic baptism, which was not carried out by the water of Redemption, but was a spiritual purification ceremony by fire: *Bafomet* thus signifies the illumination of the spirit"[20] or "baptism of the mind."[21] In support of his thesis, Hammer-Purgstall claimed that Templar churches still contain symbols and sculptures with figures of Baphomet and Gnostic symbols, and he studied objects, "Gnostic vases, chalices; . . . he asserts that the novel of the *Holy Grail*, or the holy Cup, is a symbolic novel which both hides and proves apostasy, the Gnostic doctrine of the Templars."[22] As before, François Raynouard responded immediately in 1819 in the *Journal des savants* with the same argument that *Bafometi* is a corruption of *Mahometi*. With painstaking attention to detail, Raynouard dismantled Hammer-Purgstall's scholarly hoax, showing that Gnostic and esoteric meanings were a random

Figure 2. Hammer-Purgstall's "androgynous Baphometic statue," from the *Mémoire des deux coffrets gnostiques du Moyen Âge* (1832). Bibliothèque nationale de France.

interpretation of "in figuram Baphometi" and contrary to archaeological and historical evidence. He criticized Hammer-Purgstall's defective scholarly integrity as "simple and . . . vain conjectures" and he exposed this as an effort on the part of a conservative Catholic to discredit Freemasons by linking them to the sullied reputation of the Templars.[23] But despite these renewed refutations from the scientific community, Hammer-Purgstall published in 1832 a summary of his thesis and conjectures from the "Mysterium Baphometis revelatum" in the *Mémoire sur les deux coffrets gnostiques du Moyen Âge, du cabinet de Mgr le duc de Blacas* (Memoire on the Two Medieval Gnostic Cases from the Collection of Monseigneur the Duke of Blacas).[24] They served as the basis of Prosper Mignard's history of Gnosticism in the 1850s that kept Templar esoterism alive.

Mignard gave a systematic presentation of all known Gnostic sects and divergences in their philosophy, sketching out possible links to the presumed anti-Christian doctrine of the Templars: "The solution to this serious question of *anti-Christian metaphysics* . . . is hidden in the thousand folds of the numerous sects of *Gnosticism*."[25] Most important, however, Mignard picked up from Hammer-Purgstall the idea that the Baphometic idol (*idole baphométique*), which the Templars had arguably worshipped, was androgynous (Figure 2). In his description of the Duke of Blacas's Gnostic cases, Hammer-Purgstall had conflated the previously identified Gnostic "idol" (*idole, figure*) with "the androgynous Baphometic statue."[26] Androgyny could invoke alternatively homosexual practice, "androgynous love" (the Templars were accused of sodomy),[27] and hermaphrodism, a description of dual sexual organs. Although the Gnostic idol that Hammer-Purgstall identified as Baphomet was not shown with both sexes, unlike "the androgynous Baphometic statue," the prejudice against sodomy was motivation enough for Hammer-Purgstall, Mignard, and authors following them to assume and perpetuate the idea that the two idols were the same (Figure 3).

Sodomy, "impure practices," and Gnostic heresy go hand in hand: "These serious disorders had been imported from the Orient and introduced into the Order by several of the most perverted leaders; and, even in respecting morals, it is impossible to refuse to believe it!"[28] This association of androgyny and sodomitic behavior with esoterism prepared the assimilation of Baphomet, "hermaphrodite devil" (*diable hermaphrodite*), into the occult in the work of Éliphas Lévi in 1856.[29]

The works of two great historians of France writing in the 1830s exemplify the fantasy of Baphomet/Mohammed's sexual alterity. In his *Histoire du Moyen Âge*, first published between 1833 and 1844, Jules Michelet did not

Figure 3. Hammer-Purgstall's Gnostic idol, from the *Mémoire des deux coffrets gnostiques du Moyen Âge* (1832). Bibliothèque nationale de France.

subscribe to Hammer-Purgstall's Baphomet, although he did believe that the Templars "experimented with some Oriental superstitions, Saracen magic."[30] Michelet also believed that "as they did without women, they also did without priests," and little by little "their true god, it seems, became the Order itself" and "the Order . . . fell into a fierce religion of itself, in a satanical self-absorption [*égoïsme*]."[31] As a result, the Templars were thought to have lost their moral bearings and renounced Christianity, symbolized in the ritual spitting on the cross that was an imitation of the denial of Saint Peter (Michelet, 3:160). They mistook this for an *imitatio Christi*: "Thus the Order that had represented to the highest degree the symbolic genius of the Middle Ages died out due to a symbol not understood" (Michelet, 3:161). Heresy and sodomy, decidedly, bleed into one another: "the infamy of mores" (*l'infamie des mœurs*), albeit not widespread, accompanied heresy, the Gnostic doctrines (Michelet, 3:160). In his *Histoire de France*, first published between 1834 and 1836, the other great contemporary historian of France and Michelet's rival, Henri Martin, went even further than Michelet in his symbolic interpretation, giving credence to the archaeological legend of Baphomet, "the androgynous and bearded god or genie."[32] Martin assumed Baphomet to be a remnant of Gnosticism, "an amalgamation of ideas and rituals half borrowed from recent Muslim heresies, half from old Christian heresies that had never completely disappeared from the Orient, an amalgamation where a sense of reaction against Christian *spirituality* dominated to the benefit of a religion based on power and matter."[33] For Martin, Baphomet was both real and symbolic: "If the Templars denied Christ, it was to exact revenge against him who denied them, because he abandoned the cause of his servants."[34] Baphomet also "tolerates revolting pleasures among his worshippers": "The mores of the Templars were no less abhorrent than their impiety: they cast aside natural as well as divine law; they were initiated by a repulsive and disgusting ceremony; they vowed to abstain from commerce with women, but they were allowed amongst themselves to sin against nature."[35] The fantasy of sexual alterity symbolized in Baphomet overdetermined historical interpretations.

As we saw, in the beginning, Baphomet was an erudite archaeological quest inspired by an error of transcription or mispronunciation of the name "Mahomet." Then, from the 1830s to the 1850s, Baphomet came to be thought of as the name of an androgynous figure associated with sodomy. This marked the transition from a philological quarrel to a symbol of a "deviant" and secret sexuality. The period from the 1830s to 1850s was also one of an increased contact with Islam and Muslims, especially in the wake of the French colonization of Algeria in 1830. Thus the stakes of the Templars' continued presence in the national imaginary—the collective imagination of the French people—were not just their rehabilitation, but also the very status of the Orient and the Middle Ages. The collective imaginary of the French nation formed itself in racial contrast to the Orient in the wake of the Napoleonic expedition: "The English who owned the place [India] and the Germans who studied it indefatigably had made India the *fons et origo* of everything; and then there were the French who had decided after Napoleon and Champollion that everything originated in Egypt and the new Orient."[36] Thirty years after the Egyptian expedition, in 1829, Victor Hugo wrote in the preface to his book of poetry *Les Orientales* that "the Orient, either as an image or a thought, has become for reason as much as for the imagination a sort of general obsession to which the author of this work obeyed perhaps without his knowledge." The future author of the novel set in the fifteenth century, *Notre-Dame de Paris* (1831), added that the Middle Ages were, alongside Orient, "this other sea of poetry."[37] The centrality of both the Orient *and* the Middle Ages in the understanding of world culture, Hugo continued, may as a consequence result in a rewriting of cultural genealogies: "Until now, the modern era has been too closely viewed in association with the century of Louis XIV and, antiquity, with Rome and Greece: Would our gaze not aim higher and further by studying modern times within the Middle Ages and antiquity in the Orient?"[38] Medievalism and Orientalism were parallel movements for the nineteenth century, giving two contrasting, and sometimes competing, visions of the past: medievalism of the birth of modernity, and Orientalism of the cradle of antiquity. However, to insist that the parallelism of medievalism and Orientalism defined only the nineteenth century is to neglect that writers—philologists, historians, and doctors—considered that the "Middle Ages" and the "Orient" were already formative of each other in the medieval period. The recent imperial French connection to the Orient had a medieval history of intimacy with the Oriental foreign that could be dated back to the very origin of the French nation. Inaugurated in the Middle Ages, after the break with antiquity, the French nation has been in contact with the Orient ever since the Crusades, which provided the historical lesson on what the consequences of intimacy with the foreigner may be. Baphomet's transformation from a philological quarrel to a symbol of secret sexuality reflected therefore the nineteenth-century effects of colonization: medieval androgynous Baphomet, a symbol of heresy and figure of sodomy, embodied the renewed fears of sexual contamination by proximity, as well as apprehensions about religious and cultural acclimation or acculturation in the new French colonies.

The fantasy of sexual alterity linked specifically to Baphomet will slowly die down after 1860 and Baphomet will become a province of modern esoterism, resurfacing occasionally in works such as Jules Loiseleur's *La doctrine secrète des Templiers*, or again in the first quarter of the twentieth century, for instance in the work of Thomas de Cauzons on *La magie et la sorcellerie en France* and Victor-Émile Michelet's genealogy of chivalry.[39] But while Baphomet slowly receded into the background as an object of scholarly pursuits, what remained was a general memory of the corruption of the Templars' morals by cohabitation with Muslims and a generalized fear of assimilation that would lead to sexual and religious deviance. Many authors called the Templars "secret Mohammedans"[40] and affirmed that "numerous relationships . . . existed between this secret sect [the Ismaelians] and the Order of the Templars."[41] Cohabitation with Muslims led to acculturation: "All the Crusaders that returned from Syria spoke only of the Templars' betrayals, of their attachments to the infidels. They were notoriously in relation with the Assassins of Syria; the people noticed with terror the analogy between their dress and that of the sectarians following the Old Man of the Mountain. They had welcomed the Sultan into their homes, permitted the Mohammedan cult."[42] With acculturation in manners and dress came, of course, the danger of adoption of "Oriental mores" (*mœurs orientales*): "The contact with Syrian pleasures corrupted their mores; pride, avidity, unrest, selfishness, replaced the austere devotion of the initial period."[43] In Syria the Templars had learned the ways of "listless and delicious life"[44] and so theirs was the fate of "a famous, but degenerate Order."[45] Pierre Larousse's article on "Le Templier," whose *Grand dictionnaire universel du XIXe siècle* (1866–1877) was a popularizing Republican medium of scientific theories in the last third of the nineteenth century, can serve as the illustration of the ways in which the views on the sexual and religious deviance of the Templars evolved in the course of the nineteenth century. In the tradition of Voltaire's attack against unfair political trials, most authors could clear the Templars of the trumped up political charge of heresy, even though the Templars had been officially charged in their trial only with heresy and not with the charge of sodomy. But the accusation of their sexual dissoluteness remained associated with the overall perception of cohabitation and colonization of Oriental, Islamic lands. Larousse rejected the official charge of the Templars' heresy, but he found it difficult to deny the unofficial charge of sodomy, hesitantly admitting that it was practiced among the Templars: "A thousand wretched rumors circulated concerning their secret acts, their initiations, their spiritual life, the alteration of their faith, mixed with Oriental superstition and Sarassin magic, concerning the degrading vices that they had brought back from the Orient, concerning the idols that they worshipped."[46] Larousse excused them from the sexual charge only partially: "In regards to mores, it is credible that the infamy of which they were accused was not widespread. As for the accusation of heresy, these knights, above all men of action, were doubtless little concerned with dogma, and the worship of an idol called Baphomet hardly seems realistic."[47] Larousse could then rehabilitate the Templars from the formal accusation of heresy, even reject Baphomet as historically not credible, all the while displacing the Templar's supposed crimes onto the purported, albeit not generalized, practice of sodomy. Thus what remained heuristic for writers of the second half of the nineteenth century was the notion of dangerous proximity and corrupting acculturation to Islam that inevitably results in sexual "deviance."

Early nineteenth-century fantasies about the medieval Orient were transformed, once there was a new imperial project in the Orient, into fears of contagion where history served as a lesson. Sexual contact with foreigners in the colonies produced a certain interpretation of the past. The fantasy of the Templars' sexual perversion because of contact with the followers of Mohammed remained active in the nineteenth-century imaginary of the Middle Ages in large part under the influence of imperial medicine, which watched carefully over the threats of racial miscegenation and pathological contamination of the French race in the new colonies.[48] Doctors who were interested in the genealogy of sexual vices looked to the Crusades as the moment when sodomy was imported from the Orient and re-actualized on the territory of metropolitan France. When they assigned the origin of sodomitical practices to the inhabitants of Oriental regions, doctors cleared the Templars of any original sin but accused them of the weakening of their morals: "We know that, since ancient times, these crimes against nature existed in the Orient within all of the idolatrous nations."[49] The memory of the Templars and the Crusades thus continued to feed the imagination as a memory of the ways in which sodomitic habits were acquired in warm climates, through contact with the colonized foreigner.[50]

Sex and Race

As seen in Chapter 2, by the mid-nineteenth century the symbolic meaning of blood deployed in the semiotics of genealogy evolved into its primary medical-biological meaning of physiological fluid, carrier of heredity. Medical discourse fused the symbolics of blood with the technology of sex, whereby blood(lines) not only organized a family and a nation, but also determined their physiological heredity. Notions of the symbolic purity of *blood as race* and the physiological purity of *blood as fluid*, free from morbid heredity, overlapped. This shift also meant that the notions of race were reinvested with significance. The French term "race" had two different meanings: its older meaning of "family" and "lineage," and its modern racial meaning. The meaning of race as "lineage" is exemplified in the common reference to the "three races" (*trois races*) of the French kings—the Merovingians, the Carolingians, and the Capetians—and it was generally employed to signify aristocratic lineage of the *ancien régime*. By metonymical transposition, and under the influence of hereditarianism, "race" extended its semantic field to "nation": "Transmission of the characters of races, of nation."[51] The second, modern, meaning of "race" resulted from the polygenist theory of the origin of species. In the 1850s, anthropologists fought over whether there existed an original unity of the human species (monogenism), whereby all humanity would have descended from Adam and Eve, or multiple origins for different species (polygenism), whereby different races were seen as separate but contiguous. Polygenism was built on the notion of racial essence, an essential, and hereditary difference that distinguished one race from all others: "We give the name *race* to an entire collection of individuals of the same species that present a set of differences of the same order as that of variety, clear differences which, once produced, reproduce themselves in a certain number of circumstances that are not entirely identical."[52] The racial difference and colonial contact furthermore contributed to the understanding of "race" as white and belonging to the European civilization, of which France was perceived universally to have always been one of the leaders: "In every era, we find our country at the spearhead of great movements which have led humanity to chase after fantasies, albeit fecund ones, whether it be a question of liberating the tomb of Christ or preaching to Europe the Declaration of the Rights of Man."[53] The notion of "race" was thus produced from national (French vs. other nations), civilizational (European vs. Oriental), and racial (white vs. black) differences.

The overlap of the symbolic and physiological idea of purity of blood meant that henceforth the genealogy of the nation depended wholly on its sexuality and the management of its sex practices. The French race was therefore to avoid any racial mixing of blood, *métissage*, that would weaken the nation or its civilization. In his *Essai sur l'inégalité des races humaines* (1853–1855), Arthur de Gobineau developed a theory of racial determinism by transposing the zootechnological term of *métissage* to human racial mixity (*croisement ethnique*) in order to argue that miscegenation inevitably leads to the degeneration and extinction of races: "Some writers have maintained that a physically and morally superior race, when united with an inferior race, declines without raising it [the inferior race], such that each of these alliances would be a loss for civilization."[54] For Gobineau, introducing the notion of race into the theory of degeneration—that is, the dilution of blood by foreign racial elements—explained the fall of civilizations:

> I therefore believe that the word *degenerate*, applied to a people, must signify and does signify that this people no longer has the intrinsic value that it once possessed because it no longer has in its veins the same blood, which was gradually modified in value by successive adulteration; that, in other words, despite the same name, the people has not conserved the same race as its founders; and, finally, that the man of decadence, who is called the *degenerate* man, is a different product, from an ethnic perspective, than the hero of great eras. . . . He will definitively die, and his civilization with him, the day when the primordial ethnic element will be so subdivided and drowned in the imports of foreign races, that the virtual influence of this element will no longer exert sufficient action.[55]

Belief in fatal effects of racial miscegenation was further fueled by doctor Jean-Alexandre-Napoléon Périer, who, informed by his colonial experience in Algeria in 1840s, wrote against "ethnic cross-breeding" (*croisement ethnique*), as did the founder of anthropology, Paul Broca.[56] Broca's theories were based on racial gradations that determined "the possibilities of fertile unions between different races. . . . Between races deemed to be further apart in a racial hierarchy held as axiomatic, Broca saw a corresponding decline in the potential for successful hybridization."[57]

These arguments had surfaced in the consanguinity debate, which revolved around the question of how and whether different family lineages (races) should cross-breed or in-breed in order to either avoid or cure morbid heredity (see Chapter 1). Anti-consanguinists were opposed to in-breeding and in favor of cross-breeding; *métissage* among family lineages was desirable: "We will indeed see in studying heredity by this mode of reproduction that a weak degree of *métissage* ordinarily raises the vigor and almost always the fertility of couples."[58] Eugenics encouraged the idea that progress and the improvement of races were possible by positive selection, refusing the idea of racial determinism. However, even for anti-consanguinists there were limits to *métissage*: "Extraordinary cross-breedings . . . *those from people to people, between races*, among families too opposed in respect to customs, institutions, individual genius, even physical characteristics, are dangerous, and cannot lead to this fortunate stability, this balancing of the human faculties and energies, which constitute *civilization*."[59] Thus what was desirable for a family lineage was less appealing when it came to improvement across racial and national lines; that kind of cross-breeding threatened civilization. Across the divide, consanguinists flatly refused the notion of cross-breeding based on zootechnological evidence of animal husbandry: "At any rate, it is pure speculation to suppose that foreign blood might regenerate families in a state of organic decline, and it would be surpassing the laws of medical science to affirm in its name that cross-breeding is the supreme remedy. One race substitutes for another, one type for a new type; but cross-breeding, mixing, are generally illusory. The pure race and individual type must subsist unaltered, or they must disappear."[60] Consanguinists brought in as supporting evidence the colonial experience: "The study of cross-breeding in human races, far from favorable to this theory, shows that those of mixed blood [*les métis*] are all the less fertile the deeper the differences are between their parents."[61] Sterility, with its end result of extinction, was thus the greatest proof of inefficacy of "ethnic cross-breeding."

Fear of miscegenation, degeneration, and extinction of the French race was thus dominating both the national and colonial discourses at the time of the deployment of the apparatus of sexuality. Put differently, genealogy was thought of both as metropolitan and national, and as colonial and racial. Sex and heredity had to be managed internally and externally. As Ann Laura Stoler has shown, race *and* sexuality were formative of the nineteenth-century French nation: "Both racial and sexual classifications appear as ordering mechanisms that *shared* their emergence with the bourgeois order of the early nineteenth century. Racial thinking was not subsequent to the bourgeois order but constitutive of it."[62] The racial and sexual politics in the colonies was thus not simply unidirectional, issued from the metropole to the colony, but it shaped metropolitan and national discourses of sexuality. The French national body had to be defended as much from the internal as from the external other, an immoral European working class, destitute whites in the colonies, an ambiguous population of mixed-blood origin: "The regulatory mechanisms of the colonial state were directed not only at the colonized, but as forcefully at 'internal enemies' with the heterogeneous population that comprised the category of Europeans themselves" (Stoler, *Race*, p. 96). The society had to be protected permanently against itself, from within and from without. French, and by extension European, civilization, and white superiority, were defended by invoking the logic of race. Medical and economic discourses were grounded in colonial distinctions of difference and they tied "fears of sexual contamination, physical danger, climatic incompatibility, and moral breakdown to the security of a European national identity with a racist and class-specific core" (Stoler, *Carnal Knowledge*, p. 46). Race, the colonial space, and sexual politics were integral to the biologizing discourse and historical imagination and fantasy; blood, sex, and race intersected and interfered with the national and imperial projects. Thus the preservation of the purity of blood was tied to the preservation of the race and its superiority.

Building on Stoler's work, my point here is different. I wish to show not just how the contemporary history of empire affects the modern history of European sexuality, but also how the past history of empire, one that looks back to the Middle Ages, affects the modern history of European sexuality. That is, as we saw with the Templars, I contend that the medieval past played a significant role because of colonial politics. Contemporary colonialism that was formative of the idea of welfare in France also imposed a new view of the past, and especially of the medieval past. The mutual dependency of metropole and colony went much further back in the past than the contemporary period that has been the focus of postcolonial scholars. In this case, it went back to the Middle Ages: the origin of the importation of "Oriental mores" (*mœurs orientales*) by the Templars but also of the epidemic of syphilis. The disease (syphilis) and the vice (sodomy) that defined the nineteenth century were both seen within their long historical genealogy, in which the Middle Ages played a crucial role. The medical knowledge of the past was thus instrumental in the establishment and maintenance of separate metropolitan and colonial identities. The perceived danger, whether real or imaginary, exerted an influence upon the fantasies of the foreign, not just in the present but also in the past. And the medical writing of this past participated in the building of the nation as much as it did in the construction of the Middle Ages.

From Fantasy to Fear: Medicine and Sex in the Colonies

The triangulation of the metropole, the colonies, and the past surfaced in two quintessential ills of the nineteenth century—sodomy and syphilis—that threatened the security and the future of the French population, as well as French superiority over the colonial foreign. Sodomy and syphilis were not considered to have the same origin: while sodomy was widely considered a homosexual practice, syphilis was generally associated with heterosexual sex and prostitution.[63] But the medical imaginary nevertheless associated the emergence of both scourges with the past medieval colonies. From the lessons of history, the vice of sodomy was observed to be particularly contagious and even less preferable than syphilis.

The Templars were in direct interface with the racial-sexual reconfiguration of the post-1830 colonial moment. The Templars symbolized the crusading movement: "The ideal of the Crusades seemed to be achieved in the Order of the Temple."[64] But in the pursuit of its conquest the West was uprooted during the Crusades: "We witnessed . . . the West tear itself away from its foundations in order to charge at the East."[65] Such an upheaval could only have resulted in the loss of cultural and sexual bearings. The uprooting from the cultural milieu of origin led to a "loosening of mores" (*relâchement des mœurs*), a statement repeated by many authors. By virtue of being separated from their cultural climate and infused in another sphere, the customs of colonizers changed: "Sovereigns who wish to maintain their subjects in their habits and original beliefs must be well careful not to permit a great number of them to stay for long in foreign countries. These travels are always fatal to old habits."[66] The further the Templars were from their origins and from the cultural and religious surveillance of the metropole, the closer they were to falling prey to temptation. The hot climate incited heated passions: "It was natural that the loosening occurred among warrior monks, noble youth, who carried out adventures far from Christianity, often far from the eyes of their leaders, between the perils of a fight to the death and the temptations of a burning climate, a country of slaves, the luxurious Syria."[67] The depth of acculturation to Islam and subsequent moral turpitude were directly proportional to the distance from Christianity and Europe. Intercultural contact was the obvious culprit in the loosening of religious and sexual morals. Moreover, this imaginary of earlier cohabitation was enmeshed with the contemporary experience of colonial contact, perceived as an increasing social and political problem of "a growing population of mixed-blood children born out of these 'mixed' unions, of men who had 'gone native' or simply veered off cultural course, of European children too taken with local foods, too versed in local knowledge" (Stoler, *Carnal Knowledge*, p. 2). Dangers of cultural, sexual, and racial contamination sparked the scientific and popular interest in the Templars, just as

Templar history during the Crusades informed nineteenth-century authors about the extent of the dangers of cohabitation.

The great myth that was rehearsed by popularizers and historians of customs and morals such as Paul Lacroix, from the mid-nineteenth century onward, is that sodomy was imported or revived during the Crusades, that is, in contact with the Orient: "the vice against nature, which the travels of the French in Palestine had acclimatized to France."[68] The fact that in order to make this argument historians had to disregard numerous clerical Latin and vernacular sources that referred to sodomy without any obvious connection to the Orient or the Crusades further underscores the latent "Oriental," colonial threat that had infiltrated the French national imagination by the 1850s. Regarding the causes of the moral corruption of the Templars, "we find them in the long stay of the Templars in the Orient, where the vice against nature is almost endemic, and where the fear of leprosy, the burning illness [*mal des ardents*], and diverse cutaneous or organic infections, is always tied to the commerce of women."[69] As shown, Templars were thought to be infected by "Oriental mores." In the work of historians and popularizers, such statements assumed that sodomy was an Oriental sexual custom. The imagination and erudition of reputable scientists and doctors, such as Doctors Albert Moll and Léon-Henri Thoinot, who were writing in the last quarter of the nineteenth century, were nourished by such work. The doctors followed up regularly with assertions that "the Orient, at that period, was largely delivered unto antinatural practices, and the book by Moll gives several crucial indications about the mores of the Turks and Persians in the Middle Ages."[70] In popular manuals, other doctors in turn spread similar opinions: "In the Orient, around this period, pederasty flourished. In Constantinople, brothels existed where one could procure men as easily as women."[71] Doctors obstinately attached the importation of sodomy to the Crusades and the Templars: "The condemnation of the Templars by Philip the Fair and Boniface VIII was but a terrible act of justice toward the vice against nature that the knights had brought back from Asia Minor," wrote one of the foremost authorities on sodomy, Doctor Julien Chevalier, in 1893.[72] The idea introduced in Lacroix's *Histoire de la prostitution*, that "the existence of society itself could have appeared threatened, when Philip the Fair . . . took it upon himself to stop the progress of sodomy. . . . Such was the principal cause of the trial of the Templars,"[73] had a long life in the doctors' understanding of political history and public health policy: "But it was above all after the crusaders' return from the Holy Land that the phenomenon spread. Philip the Fair rose up against these practices, and it is under his reign that the famous trial took place against the Templars, who were accused, among many other wrongdoings, of pederasty."[74] The doctors, in the wake of the historians of customs and morals, repeat that "the Crusades were the most beautiful moment of chivalry," but it is through contact with the foreigner "that this prodigious gathering of men of all ranks and countries heated in its bosom the corruptive germ of lust."[75]

In a few cases, the Templars were also accused of being carriers of yet another germ: syphilis. Some historians of customs and morals and certain doctors assumed that leprosy was an earlier form of syphilis. For them leprosy existed "since the ancient times" and it had been imported by the Romans from Asia and Africa, from the conquered peoples by "the atrocious Roman debauchery."[76] Syphilis, albeit in the form of "chancroid" or "non-infecting chancre" (*chancre simple*), "successively combined with other diseases, and especially with leprosy, which gave it an entirely new physiognomy."[77] This disease, "which was then looked upon as a new illness," was new only in appearance and would in fact have been "simply a final transformation of leprosy."[78] And for those for whom leprosy mutated into the dangerous and complex venereal disease of syphilis, the Crusades and the Templars played a significant role: "The most audacious debauchery gangrened the undisciplined hoards . . . the majority bringing back with them the bitter fruits of Oriental life: we can affirm that the venereal disease was then but one form of leprosy."[79] This hypothesis was considered already in the 1870s "an obvious error,"[80] and it was rejected by scientists, except a few doctors-popularizers—such as doctors Villemont, Bouglé, Fauconney, Désormaux—who wrote marital manuals at the fin de siècle and continued to believe in this opinion that traced syphilis to antiquity.

This hypothesis on the importation of syphilis from the medieval Orient, then, was a minority opinion on the origins of syphilis, but what it had in common with all other hypotheses circulating in the nineteenth century is, first, that syphilis was thought to have emerged in the late Middle Ages and, second, that its origins were not European. Among others, a widespread medical opinion attributed syphilis to the discovery of the Americas. Because of the striking coincidence of the outbreak of syphilis in 1494 with the discovery of the New World, "it was long held that syphilis originated in America and that it was unknown in antiquity."[81] The first outbreak of syphilis was recorded during the 1494–1496 invasion by Charles VIII of the kingdom of Naples. Alongside French and Italian troops, we find at the epicenter of the epidemic the Aragonese troops sent to defend the kingdom of Naples. This is significant since the disease had been recorded already in 1493 in Spain, albeit sporadically, coinciding with a return of Christopher Columbus from the Antilles.[82] Émile Littré and Charles Robin disputed early on the American origins of syphilis: "When we consider the date of the discovery of the New World and the date of the explosion of syphilitic accidents, it is impossible not to recognize that . . . communication by coitus would have been insufficient to propagate this illness which, shortly after its first appearance, was displayed in a formidable way in Italy, France, Spain, Germany, and England."[83] As the article in Larousse's *Grand dictionnaire universel du XIXe siècle* pointed out, the argument in favor of the importation of syphilis from the New World also ignored that the Aragonese troops from Barcelona were still in Sicily when the epidemic already broke out between French and Italian troops (Larousse, "Syphilis," p. 1337). Thus the outbreak of syphilis had to be unrelated to the discovery of Americas and to the contact with the colonized of the New World. In other words, imagining an American origin for syphilis was a kind of ideological prejudice: "American origin is a fable invented for a shameful political interest, a fable contradicted by the authority of dates and powerless to explain the almost simultaneous explosion of the epidemic in ten countries at once."[84]

The majority medical opinion thus agreed that syphilis was neither Greco-Roman nor imported by Christopher Columbus. If leprosy did not mutate into syphilis during the Crusades, then syphilis did not originate on European soil: "Our masters in Greece, Rome and the Middle Ages did not describe syphilis."[85] And if the outbreak of syphilis was simultaneous in numerous European countries and too close to the date of Columbus's return, syphilis did not originate on American soil. In an important article for Amédée Dechambre's authoritative *Dictionnaire encyclopédique des sciences médicales*, Joseph Rollet summarized the other medical opinions. The disease was not European, but it was ancient, imported from other parts of the world, "from ancient and in some sense primitive breeding-grounds of the disease."[86] Whether its source was Asian, Indian, Chinese (Rollet, p. 257), African (Rollet, p. 259), or American (Rollet, pp. 261, 281–83), its origin was resolutely to be found outside Europe: "This part of the world did not see it born in its midst."[87] Syphilis would have been imported to Europe from any of these various centers by colonial contact during the Middle Ages: "It came to us then by importation," as it spread by sexual contact "by contagion, and it is in that way that it landed in Europe."[88] A coincidence of an outbreak in the year 1494 in a number of European countries would then be due to the travels to and from the colonies and the contact with the Antilles and the Orientals. In short, Joseph Rollet's genealogy of the disease was colored by the generic contemporary colonial experience. The scientific debate over the origins of syphilis illustrates the need felt in the defense of metropolitan France from the contemporary diseases and vices of the New World, from colonial contamination, and shows a striking degree of concordance between historical research and contemporary medical observation: "It is moreover a doctrine consistent with what we know today about the development of contagious diseases, which are no longer believed to be spontaneous, and with the importation of exotic infections and the ease of their dissemination beyond the original breeding-ground."[89]

Rollet does acknowledge that these colonial hypotheses merely explain "the final dissemination of syphilis in the entire world" instead of giving "the reason for its forming and concentration in limited points on the globe."[90] The doctors find it difficult to confine particular venereal diseases to one location and one nation. Sexual contagion is not about geography but about morality: "[Venereal diseases] have no nationality and . . . their origin is as ancient as the world, or at least the excesses of mores."[91] Sexual contagion is the result of moral contamination. Thus, just as Templars fell into sodomitic vice because of their heresy, syphilis spread from its foreign sources to Europe because of moral degeneration: "It was Prostitution which . . . served as auxiliary energy for this scourge. . . . That is how venereal leprosy erupted at the same time, with the same fury, in France, Italy, Spain, Germany, and England, at the time when Christopher Columbus had hardly returned from the first voyage that he made to Hispaniola."[92] These historical examples echoed the degenerate lifestyle of colonists and the importation of disease and vice to the metropole: "Through sexual contact with native women, European men 'contracted' disease as well as debased sentiments, immoral proclivities, and extreme susceptibility to uncivilized states" (Stoler, *Carnal Knowledge*, pp. 67–68).

For doctors and historians, diseases and vices were external, imported threats to European civilization, and its borders needed to be carefully policed. There was a need for sexual morality in the face of imported moral turpitude and venereal disease. It was felt that both syphilis and sodomy could be controlled by an appropriate sexual morality, and even by accommodating different heterosexual moralities to avoid homosexuality; hence, adultery (see Chapter 6), prostitution, and concubinage were preferred, at home and in the colonies: "Heterosexual unions based on concubinage and prostitution across the colonial divide were defended as a 'necessary evil' to counter those deemed more dangerous still—carnal relations between men and men" (Stoler, *Carnal Knowledge*, p. 2). Heterosexual sex could never be as immoral, as dangerous, and as degenerate as homosexual sex, and the Middle Ages served as a caution and reminder of that idea.

The Sexual Contact Zone: Degeneration and the Decline of Civilizations

By the last quarter of the nineteenth century, the theory of hereditary degeneration was at the peak of its influence and the medical research on inversion, the new term for sodomy, was gaining momentum. It was not long before inversion was experienced as an attack on European civilization, and inversion and degeneration became mutually explanatory and reinforcing notions.

After Ambroise Tardieu's groundbreaking *Les attentats aux mœurs* (1857), sodomy (or pederasty) was considered for a long time to be a question of morality. For Tardieu pederasty was a vice, meaning that it was not innate but a question of loose morals and lack of will: "perversity, that is to say vice in the strict sense of the term, where no force imposes sophisticated debauchery on the subject because he is conscious of his immorality and can avoid making himself guilty, if he so desires."[93] This medical-moral conceptualization of pederasty changed drastically in 1891, when Doctor Albert Moll proposed a new analysis in *Die conträre Sexualempfindung* (The perversions of the sex instinct),[94] followed by Doctor Julien Chevalier's respected and very influential work, *Une maladie de la personnalité: L'inversion sexuelle* (1893). Inversion was a new term that introduced the idea of pathology into pederasty; inversion meant going against the "natural" heterosexual instinct. Next to perversity (a vice), there was now perversion (a disease): "The deranged, degenerates, inverts are sick people."[95] This "perversion of instinct" (*perversion des instincts* in Moll) or "perversion of genital orientation" (*perversion du sens génital* in Thoinot) was perceived as due to the physical weakening of heredity, while vice, "perversity" (*perversité*), was the result of environmental and moral influences.[96] Yet doctors agreed that inversion was indisputably a sign of degeneration: if acquired, it was a sign of moral degeneration; if congenital, of physical. It thus became essential to understand the role inversion played in the degeneration of a race.

In 1857, Benedict-Augustin Morel gave the impetus to the theory of degeneration with his identification of "degenerations [*dégénérescences*] of the human species."[97] He defined degeneration as "an unhealthy deviation from the normal type of humanity," "an unhealthy deviation of a primitive type."[98]

Morel understood degeneration as an irreversible deterioration of a primitive type, which normal heredity was supposed to transmit but failed to do. Degeneration was an autonomous morbid state of an individual that could not be reversed. Morel's view was pessimistic since it implied an inevitable hereditary decline from the primitive man (that is, biblical Adam). His definition marked the medical sciences for several generations. If the organs and nervous system in individuals can deteriorate, and with them lineages, species can degenerate too. Since degeneration is irrevocable for individuals and family lineages, then with individuals the nation as well as civilization decline without the possibility of regeneration: "It goes without saying that the same causes which determine the extinction of individuals must necessarily determine the extinction of the groups to whom they belong."[99] A degenerate race was simply not viable; degenerate races were thought to go sterile and then extinct: "Degenerates therefore do not constitute viable races."[100] Degeneration described the pathological state of the collective body of the race and nation, and its decline and extinction. It thus became a theory of the rise and fall of civilizations. And the study of the past became the condition of the possibility for the survival of the present.

Before looking at the particularities of its geographical-historical evolution, doctors often began with a commonplace misconception that inversion was timeless. Doctor Fauconney, one of the most prolific writers on inversion past and present at the turn of the century, wrote in his capacity of "former sanitary doctor in the Colonies" (*ancien médecin sanitaire aux Colonies*) that "the history of vice against nature in its various manifestations becomes lost in the shadows of time."[101] Doctor Laupts, a disciple of Alexandre Lacassagne and an authority on "perversion and perversity," maintained that "it is necessary to recognize that deviation has existed at all times, in every country," "and we may add that it has attacked all races and occurred in all countries."[102] Doctors identified the Greek and the Roman civilizations as the apogee of pederasty: "Starting in ancient Greece, poets have taken it upon themselves to enlighten us about the highest depravity of their fellow citizens. It is especially in Rome that one became a master in the art of refinement of lechery."[103] In fact, the Romans were corrupted by the Greeks (Thoinot, p. 288). The Middle Ages perpetuated this habit, making "the uninterrupted chain, linking ancient Rome to the Renaissance, where all of the Greek and Roman traditions re-blossomed and flourished."[104] But the doctors were careful to specify that what looked like a continuity of pederasty was actually not a direct inheritance from the Greeks and the Romans. Rather, "it is to the establishment of the Norman races in the

Gallo-Frank provinces that we must attribute its appearance in France; it was later revived by the Crusades, and the French stay in Palestine," "the influence of the year 1000, Norman invasions, Flagellants, Crusades, and the Order of the Templars."[105] In tracing the genealogy of vice, doctors considered that in the Middle Ages the vice was always imported into the indigenous French space, either from a foreign territory (the Crusades, the Templars) or by foreign agency (the Normans, the Flagellants, where both groups came from the north). Even though doctors in general espoused the notion of French nation and European civilization as one and the same in contrast to the colonial foreigner, nationalist rivalry within bounds of Europe at times forced them to modify and attenuate the racial argument with general considerations of foreignness, where priority was not given to racial difference but to outsider status. After all, the other nations accused the French of having been the carriers of syphilis, "the French illness" (*le mal français*).[106] Despite such nationalist bickering, doctors invariably maintained that the Middle Ages, especially the Norman invasions and the Crusades, made this vice prominent once again and that the vice was not a direct continuation of antiquity.[107] There was thus a clean break between antiquity, marked by the fall of the Roman Empire, and the medieval revival of pederasty imported from elsewhere onto the territory of France.

Antiquity was an extremely important example for theorists of hereditary degeneration, though strikingly different from the Middle Ages. From the early nineteenth century, doctors, and especially psychiatrists, like Philippe Pinel, the founder of modern psychiatry, Étienne Esquirol, and Louis-Francisque Lélut, were obsessed with antiquity.[108] The rise and the subsequent decline, fall, and disappearance of the ancient Greco-Roman civilization set the paradigm for the present French nation and European civilization: "Thus peoples advance. Their origin is obscure; they grow, show the greatness of their force, then, fatally, they arrive at this period where they no longer belong to history."[109] This paradigm of rise and fall was explained by the theory of hereditary degeneration, in which morals held a principal role. If "the Caesars' Rome thus became one of the most striking breeding grounds for sexual inversion that history has preserved as an example,"[110] this was because of lack of moral direction: "From this review, it emerges that the principal causes of the depravation of morals in pagan antiquity were the opinions in philosophy and ethics and the absolute authority of the master over the slave."[111] Paul Jacoby dedicated the first half of his study on degeneration to the imperial dynasties of the Romans to find "intellectual as well as moral phrenopathies, neuropathies, singularities and aberrations, vices of conformity, sterility, degeneration, premature death and finally extinction of the race."[112] The history of antiquity taught a lesson that, exhausted and drained of their creative potential, the races and nations collapse into oblivion because of moral degeneration: "In her decadence, Rome had so abused the senses that it took long years of spiritual teaching to reestablish the cult of propriety."[113]

On the basis of this historical evidence, Paul Jacoby without hesitation condemned all the nations of Europe to extinction: "Such was the fate of the great civilized peoples of antiquity, such will be that of the civilized nations of Europe, and many symptoms already indicate that we have entered into this period of exhaustion and decadence."[114] Others were just as alarmed by the historical example of antiquity: "Europe may not yet be very old, but it is at the very least in its maturity and the day will come when, like all that submits to the evolution of life, it will be necessary to *make way for youth*."[115] It was considered a fact that selection acted on a race in a double way. First it elevated it and then it destroyed it: "Degeneration is only, in sum, the consequence of our evolution":[116] "By the process of selection and the fatal law of extinction of the privileged races, people first become civilized, rise up to the summit of greatness, then rapidly decline and disappear exhausted, overtired, extinguished, [they] fall back into barbarism and are replaced by younger peoples."[117] Were France and the European civilization of the nineteenth century going to imitate the example of the Ancients and of the noble races and decline, or were they going to reverse this decline? Must they perish, "the privileged peoples . . . the group of nations that march at the head of civilization . . . : *Old Europe*";[118] "Our Modern States must they perish like the great primitive nations?"[119] The answer lay in historical research on the causes of antiquity's decadence, its decline of morals, institutions and character. If the causes of decline could be determined, then Europe and its superior economic, cultural, and political position may be safeguarded.

After Charles Darwin, degeneration became a principle of natural selection: the weak are cast off and die, while the strong survive and can be husbanded through eugenics. Another important thinker of hereditary degeneration of the late nineteenth century, Valentin Magnan, influenced the late nineteenth-century medical thought on degeneration and eugenics. For Magnan, degeneration is a disease, a pathological process: "Degeneration is thus a *pathological* state and not a *regressive* state, it is a *reversive anomaly*, as certain authors understand it."[120] The difference between Morel's atavistic morbid type and Magnan's pathological state is that pathology, unlike Morel's regression, can be reversed. Morbid heredity in and of itself no longer endows degenerates with the ability to regenerate, "a morbid type . . . no longer possesses in himself the means to regenerate the self"; but if degenerates are cross-bred, healthier specimens can restore degenerate races to health: "Perfectibility is a quality of every being who evolves normally."[121] For Magnan, degeneration was not a pessimistic view of the world: whatever its cause, heredity or milieu, each cause could be treated prophylactically: the "preventive and perfective medicine"[122] could improve the health of the nation and increase its prospects in the future.

For eugenics to work, the great project of reeducation of the population on the moral grounds of the defense of civilization had to take place. The answer to the question as to whether "our modern Nations must . . . perish like the great primitive nations" depended on "their degree of morality."[123] History taught that to preserve the race one should preserve morals: "Incessant progress is a condition of duration, and a nation can live indefinitely, upon the sole condition of being worthy of it."[124] This dignity of the nation could be achieved by a good administration of moral inclinations, "a new development of more powerful attitudes, more delicate tastes and inclinations that are better regulated in the acts that they provoke."[125]

Since pederasty was primarily a vice before it was a disease, morality was to be a prophylaxis for those perverted by sodomitical vice, and perhaps even for inverts who were encouraged to overcome their in-born tendencies. Morality was the cure to this vice that was widely perceived to be the sign of modern times. For what are the factors in "the morbid exaltation of sexual instinct which is one of the marks of our time," and in "the corruption of mores"?[126] The answer is a lack of spiritualism and "instability in our political institutions."[127] Yet despite an alarmist medical discourse on inversion, by all appearances European civilization was not in an immediate danger, as already Tardieu claimed: "Without wishing to affirm, as I have often heard it done, that vice may be more and more widespread."[128] While it exists "still too often in modern society, it is certain that it was far more common in antiquity."[129] Doctors believed that "although, in every era, sexual perversion existed, it was more common before, and, in summary, public morality is better today,"[130] "that in these days it is relatively rare and that, never, perhaps, was it regarded with more disdain."[131] If the Greeks and the Romans celebrated this supreme vice, which was for them "a quite ordinary passion," the vice is less frequent today because "they take care to shroud themselves in shadow and mystery, they blush even to be suspected."[132] Good morals, disdain, and shame are the methods of containment of the vice, and the doctors felt that for the time being they controlled successfully the spread of this vice by means of morality and shame. But if the metropolitan borders are more or less protected from within, they are under perpetual assault from without: "Europeans . . . often refuse to give faith to the reports of the most truthful travelers. . . . We know also how much these horrific passions are in favor among the Arabs and their neighbors. Kæmpfer says that in Japan there exist, to facilitate these monstrous couplings, brothels that are tolerated by the government; circumstances are about the same in China."[133] Thus, sustained contact with the Orient where the vice is widespread and easily accessible may corrupt the good European mores. And for those who were incredulous of travelers' reports, the myth of the importation of sodomy and/or syphilis into France by the agency of the Templars' and the phantasmagoric fear of colonial contagion could serve as a prophylactic and disciplinary measure.

It is in this context that the example of the Templars and the Crusades resonated in the way that antiquity—where pederasty was integral to culture and society—did not. The Crusades were the apogee of chivalry, its high point after which the medieval nobility began to degenerate: "The Crusades and the emancipation of the communes that marked the culmination of feudal government are the two events which most contributed to destroying it" because chivalry in the Crusades lost its "virtuous independence, its simplicity and its morals."[134] Chivalry became too ritualized, too codified and therefore too dissolute to maintain the purity of its moral core (see Chapter 6). Moral degeneration had led to the extinction of noble classes and the French Revolution: "Surveying history, I have been able to see that each royal race, after having presented great political men, wise administrators, brilliant captains, or profound thinkers, the most often, ended with lowly princes, without virtues but not without vices."[135] The extinction of noble families was the result of the softening of their moral core, and the dissolution of the medieval warrior civilization triggered their decline: "Then, with the course of centuries, ideas, institutions, habits were modified, and, along with them, morals. They were all warrior-like, all imprinted with the bitterness, the harshness of the camps. They shed them bit by bit; they became softer, more polite, and ended up being languid, effeminate, licentious to the point of scandal, prodigious to the point of ruin."[136] A life of privilege and the subsequent loss of a moral compass were the causes of degeneration in nobles: "Morality and medicine have long since situated the decay of the upper classes within the decay of customs."[137]

The French nation, whose aristocratic caste declined and perished because of moral degeneration imported from abroad during the Crusades, had survived into the nineteenth century thanks to the fact that degeneration was circumscribed to just the noble class, whereas the French race—its people—was otherwise vigorous and strong. But imperial expansion brought dangers already experienced at the dawn of the nation back to the door of the metropole, and it threatened wide swaths of the French modern society with civilizational decline. On the one hand, because of imperial politics, doctors feared racial degeneration of the nation and exerted pressure on sexual morality. On the other hand, doctors also feared moral degeneration, a precursor of decline and extinction. The Templars and the Crusades could provide no better example. Moral depravity could be inflamed in the colonies and revived in France by importation from there. Both the internal, metropolitan, and the external, colonial, threats of physical and moral contamination, political and economic decline, needed to be addressed in order to safeguard European superiority. Economic viability and cultural reproduction depended on heterosexuality, sexual morality, and racial superiority, and for this the national borders had to be policed externally and internally by inciting fantasies and fears of foreignness, the colonial other, and aristocratic degeneration that were contrasted to the willed vigor and health of the metropolitan modern population.

Chapter 4

Gilles and Joan, Criminal and Genius: Medical Fictions and the Regeneration of the French Race

> The Kingdom of Satan was skillfully divided.
> Against the Witch, his daughter, his wife,
> was armed his son, the Doctor.[1]

The task of imperial medicine was to police the metropole's borders on the outside and to prevent racial *métissage* and importation of vices and diseases from the colony. But another fear of racial degeneration, the decline of family, French nation, and European civilization, permeated medical discourses as doctors and writers focused on degeneration brought on by internal exhaustion and moral corruption. What lessons about the degeneration of the French race could be excavated from the history of the nation, from within its metropolitan borders? In their search for answers, doctors subjected well-known historical figures to scrutiny and detailed medical diagnosis and treatment. Two major medieval figures in particular were the object of an overwhelming fascination throughout the nineteenth century: Joan of Arc and Gilles de Rais.

Joan of Arc and Gilles de Rais are among the best known historical "couples" of the French Middle Ages. Nineteenth-century historians represented them as companions in the liberation of the city of Orleans in May 1429, which enabled the crowning of the Dauphin as Charles VII of France, in July 1429 in Reims.[2] For centuries, the lifting of the siege at Orleans has been considered as the turning point in the liberation of occupied fifteenth-century France, which had been colonized by its hereditary enemy, England, during the final third, ca. 1415–1436, of the Franco-English Hundred Years' War (1337–1453). Yet Joan and Gilles each left to posterity a radically different, but equally fascinating, heritage: Joan, the savior of France, was sanctified in 1920 by the very Catholic Church that condemned her in 1431 to the stake as a heretic, apostate, idolater, and relapse. Her trial was nullified already in 1456, soon after the end of the Hundred Years' War. Gilles, her counterpart and evil double, a would-be alchemist and worshipper of the devil, a mass murderer of young, mostly male, children, was condemned in 1440 as a heretic and sodomite, sentenced to death by hanging and burning, and never rehabilitated. In short, since their condemnation and execution, they have represented the dialectical relationship of ultimate good and ultimate evil. These cursory descriptions of Joan and Gilles depict them as figures larger than life. Indeed, their stories have been not only the fabric of history but also the stuff of fiction and fantasy.

This chapter thus focuses first on the amalgamation of fiction and science, and especially on the way fiction in the nineteenth century facilitated the adaptation of Gilles's story from historical to medical discourse, transforming him from monster to patient and criminal. The medical elucidation of Joan's pathology, moreover, legitimated and solidified the contrast between Joan and Gilles, where Joan was a degenerate genius and Gilles a degenerate criminal. Just as they represented two sides of the same medical coin, these two cases, I contend, were inextricably bound together, and to the point that the shining symbol of national regeneration that Joan of Arc came to be could not stand on its own merits, but only in contrast to the degenerate Gilles de Rais. The more Gilles was vilified, the more Joan was inspirational; without Gilles's degenerate criminality, Joan's degenerate genius could not be fully grasped, for Gilles defined Joan and vice versa. Thus the dialectical relationship between Gilles and Joan, which historical, fictional, and medical narratives helped to build, was an illustration of the cyclical tension at the heart of the theory of degeneration: the cycles of moral degeneration and regeneration of the French race that had been ongoing since the Middle Ages. The fundamental tension between decline and renewal of the French race was symbolized in Gilles—decadence of medieval nobility—and Joan—the subsequent regeneration in the people. Their medieval examples of degeneration of nobility and regeneration by the people served as a historical lesson for the late nineteenth-century French nation-state that was battling its own demons of successive revolutions and regime changes—from monarchy to Republic to empire and all over again—and which were bringing with them fears of moral degeneration of the race, and national and civilizational decline.

Degeneration Theory: Genius and Criminal

Nineteenth-century historians studied civilizational rise and fall through the circumscribed example of the French nobility of the Middle Ages. The rise and decline of the medieval nobility furnished an excellent case study, one that limited civilizational decline to only one social class. It therefore also enabled the historical view that, despite this localized decadence among nobles, the French nation was perpetuated by and renewed in its people. The Crusades had been the apogee of chivalry, its high point after which the medieval nobility began to degenerate. The permanent state of war of the fourteenth and fifteenth centuries during the Hundred Years' War created a trauma that translated into a thirst for blood among the warlords defending France from the English enemy and occupier: "These men of blood who gradually returned to their manors after the war, continued it still more atrociously, against poor people without defense."[3] The total dissolution of morals and the "degree of perversity reached by some nobles, some great lords, who abandoned themselves to all of debauchery's aberrations"[4] were exemplified by Gilles de Rais (Retz), marshal of France. The example of this "former companion to Joan of Arc" showed "in what barbaric depths the upper classes themselves groveled,"[5] into what abyss the apex of depravity could descend: "The most monstrously deprived imagination had likely never dreamt of what the debates of his trial revealed,"[6] "the most criminal depravity of morals," "revolting wantonness," "abominable pleasures," "horrible debauchery,"[7] "atrocious crimes," "unchecked liberties," "sublime horrors," "the most vile instincts," "the crime against nature,"[8] "taste for blood, suffering and death,"[9] "atrocious and senseless sensual pleasures," "the delirium of ferocity."[10] Gilles de Rais was representative of the moral degeneracy of the noble caste, since "others perhaps were also guilty," "these *flayers* [*écorcheurs*] of the campaigns, these terrible little kings of the castles."[11] In fact, Gilles became an almost banal example of the bestiality into which the feudal age returned: "The case of Gilles de Rais only merits study because it is not exceptional."[12] As the institution of chivalry declined morally, the nobles returned to the barbaric behavior that characterized the time before its rise:

> Barbarism reappeared, without what good it had earlier, simplicity and faith. Feudalism returned, but not its devotion, its loyalties, its chivalry. These feudal ghosts appeared like the condemned who brought back unknown crimes from elsewhere.[13]

> Feudalism, which had been great, no longer had force or dignity, and crimes were committed with impunity in the light of day, or hidden in the castles' shadows.[14]

At the same time that the historians describe a steep curve of decadence in the fifteenth century, limited to the nobles, they also identify the regeneration and prosperity of the French nation in its people, embodied in "this daughter of the people,"[15] Joan of Arc: "Patriotism is to nations what the love of life is to individuals, for a country is the life of a nation. . . . Whenever such a feeling rises to enthusiasm in a nation, the women experience it to an equal, and even to a greater degree than the men."[16] "Angel, woman, people, maiden, warrior, martyr," Joan was led by her "patriotic faith" (Lamartine, p. 118 [p. 195]), not yielding to "the sword of the stranger" (Lamartine, p. 50 [p. 122]). She was "aroused by love for the homeland,"[17] "her enthusiasm" (*son enthousiasme*), of which many authors speak, was "like the effect of an exaltation of patriotic and religious sentiments."[18] Her religious inspiration and visionary gift saved the French from the English conquest of the entire territory under the French crown and brought the nation from the brink of disaster to become one of the most prosperous states and a paragon of European civilization.

Informed by historians' writings, physicians and psychiatrists probed the cases of Joan and Gilles. Their conclusion was that both were degenerate, the only difference being the nature of their character: while Joan's hallucinations were found to be the degeneration of genius, Gilles fell into the category of criminal degeneracy. Medical preoccupations with genius, criminality, and degeneration were born of their nineteenth-century sociopolitical context of cyclical revolutions. Moreover, the pressures of psychological problems created by industrialization, the economic depression of the 1840s, and the subsequent social dislocation and unrest, along with the general instability of Europe and the failure to realize the revolutionary ideals in the aftermath of the revolutions of 1848 and during the Second Republic (1848–1851), all gave a mandate to medicine as the science holding the key to major social problems of moral and racial degeneration and civilizational decline.

Benedict-Augustin Morel, the first to theorize degeneration in 1857, defined "degeneracy [*dégénérescences*] of the human species" as "morbid deviations from the normal type of humanity."[19] At the same time that Morel was working on the articulation of the theory of degeneration, Jacques-Joseph Moreau de Tours was developing the idea that certain mental disorders could be a sign of an exceptional intellectual capacity. He started from the premise that "various morbid states of the nervous system" were "the root cause of exceptional mental faculties": "*Genius*, meaning the highest expression, *the ultimate* in intellectual activity, but a mere *neurosis*? Why not?" he argued.[20] Moreau de Tours continued the work of Louis-Francisque Lélut, who in his 1836 book, *Du démon de Socrate*,[21] wanted to solve the question of Socrates' claim that "there was a genie or a demon whose voice he heard, and who directed him in his actions," a voice that would also lift him into a form of ecstasy.[22] Lélut's medical explanation of Socrates' demon was nothing other than hallucination, a form of madness; for Lélut, "all of those who have false perceptions, whether they be Socrates, Joan of Arc, Pascal, etc., are visionaries, hallucinators, insane, crazy."[23] But hallucination—this "delirium of the intellect [*de l'intelligence*]"[24]—did not make Socrates a lunatic to be committed to an asylum; for this to happen, a person experiencing hallucinations would have to believe that "these ideas that unexpectedly emerge from our being and that besiege our spirit despite ourselves" exist in reality. Only in those cases, hallucination "is complicated by a troubling of reason."[25] For Moreau de Tours, genius was not madness, though it had the same origin as the delirium; "delirium and genius have common root" since both are produced by "overexcitement of the nervous centers."[26] The proof that they are two sides of the same pathology lay in "the elevated ranks of society, in families that count among their members exceptional men," where manifestations of "neuropathological disorders of all types, insanity, hysteria, etc." are precisely the most common.[27] This kinship of genius and madness made for an easy explanation of flights of fancy among men of "exceptional" intelligence, who often experienced states of hallucination: "Higher energy in the functions of this organ, but also higher chances of inconsistency and deviation in those functions."[28] There was therefore nothing supernatural or divine about hallucinations experienced by Joan of Arc, Loyola, Luther, or Savonarola; their hallucinations were merely material manifestations of their cerebral activity, "the phenomena of morbid psychology."[29] Genius was simply a variation of a neuropathological disorder, "genius, unusual talent are members of the extended neuropathic family."[30] In linking genius and madness, Moreau de Tours wanted to formulate a scientific law for exceptional historical figures; this in no way implied that all hallucinations were a sign of genius, though many hallucinations were a sign of madness. Thanks to Moreau de Tours's work on their common origin, genius and madness came to be thought as signs of degeneration. In his *Traité des maladies mentales* (1860), Morel analyzed madness as such: "When this illness is the result of pathological phenomena by heredity, it constitutes a veritable degenerative state."[31] In other words, if madness is a sign of hereditary degeneration, and genius and madness have the same cerebral origin, then genius is also a sign of degeneration. In the continuity of Morel's thought, in the last quarter of the nineteenth century, Valentin Magnan diagnosed the genius as a "superior degenerate" (*dégénéré supérieur*), belonging to the group of those "imbalanced of the intellect."[32] This was the case of Joan of Arc.

In the 1870s, preoccupations with criminality, criminal responsibility, and penal law brought to the surface the links between madness and crime: "Among intellectual illnesses we understand criminality, which it is impossible to distinguish from insanity: criminals are degenerates who, at whatever age we observe them, differ from others neither by their biological nor by their anatomical characteristics."[33] If a criminal is necessarily degenerate, a degenerate has all the potential to become a criminal: "By the very fact of his mental state, *the degenerate often becomes a delinquent*."[34] But there was disagreement whether a degenerate was or not an innate criminal. In 1876, Cesare Lombroso proposed his theory of biological determinism of born criminals. Criminal pathology was explained through hereditary atavism in his *L'uomo deliquente* (Criminal man).[35] Alexandre Lacassagne and the French legal-medical school fought against Lombroso's atavistic interpretation of criminality and criminal degeneracy with an environmental and sociological interpretation, arguing for the possibility of regeneration (public hygiene) and improvement of the quality of the human race (eugenics). The sociological interpretation of the French medical body, and its belief in the improvement of the human race, defeated in the 1880s the Italian scientific materialism and atavistic fatalism of Lombroso. Therefore, according to Magnan, a category of criminals known as "moral madmen" (*fous moraux*), or the morally insane, "are born not as criminals, but with the maximum aptitude to become so."[36] These "moral madmen" suffer from "alterations of the sense of morality."[37] This was the case of Gilles de Rais.

Gilles and the Medical Corps

Before examining the medical diagnosis and treatment of Gilles, let us take a look at the historical facts of the case from the trial record. Gilles, baron of Rais, Marshal of France, was summoned to the canonic tribunal of Nantes on September 13, 1440, for "the murder of children and sodomy, the invocations of demons, the offending of divine Majesty and heresy."[38] He was arrested on

September 15 and indicted on October 13. The accusation, numbering forty-nine articles, was explicit on two major counts:

> Procuring children, as many boys as girls, to nab them and bring them to him, so that Gilles de Rais, the accused, could perform the sin of sodomy on them, cut their throats and kill them, or have their throats cut and have them killed. (Bataille, p. 175 [p. 215])
>
> Item, that the said Gilles de Rais, the accused, set his purposes, his hopes, and his belief in the invocation of evil spirits, divination, the murder of the said innocent children, the sin of sodomy, and unnatural lust; thus it transpired, and this is a true rendering. (Bataille, p. 176 [pp. 216–17])

On October 21, faced with the prospect of torture and excommunication, Gilles made his first confession out of court and, on October 22, his first public confession in-court, requiring in addition that it be translated into French and "published in the vernacular language for any and all of the people present, the better part of whom did not know Latin" (Bataille, p. 196 [p. 242]): Gilles "voluntarily, freely and grievously said and confessed that he had committed and maliciously perpetrated on numerous children, the crimes, sins, and offenses of homicide and sodomy; he confessed also that he had committed the invocations of demons, oblations, and immolations, and made promises and obligations to demons, and done other things that he had confessed recently" (Bataille, pp. 192–93 [p. 238]). In both instances the trial record reports Gilles's confession in indirect discourse. He was declared guilty on two counts on October 25: "guilty of perfidious apostasy as well as of the dreadful invocation of demons" and "guilty of committing and maliciously perpetrating the crime and unnatural vice on children of both sexes, according to the practice of sodomites" (Bataille, p. 208 [p. 259]). Excommunicated and immediately absolved by the ecclesiastical court, Gilles appeared before the civil court on the same day, where his confession from the ecclesiastical court was read, "which he acknowledged and confessed to be true" (Bataille, p. 283 [p. 334]). He was condemned to death by hanging and burning and executed without delay on October 26, 1440.

Before the nineteenth century, Gilles's case, although reported in histories of Brittany and France, had raised very few commentaries. Jean Bodin associated the murders with demonology in his *De la démonomanie des sorciers*

(1580). This interpretation by demonology and alchemy was refuted by the Benedictine of Saint-Maure, Dom Lobineau, who advocated in the *Histoire de Bretagne* (1707) the loss of reason as the principal cause and saw demonology as a fraudulent motive in the trial. Voltaire, in his *Essai sur les mœurs* (1756–1763), likewise assumed that the accusation of magic was a pretext for the Church authorities, and he argued that religious fanaticism and ecclesiastical prejudice condemned both famous medieval victims, Gilles de Rais and Joan of Arc. The Benedictine *Art de vérifier les dates* (1783–1787) adopted Voltaire's line of defense of the victims who were only "supposedly seers and magicians."[39] However, between the end of the eighteenth century and 1855, attention turned away from the medieval trial procedure to a fascination with the murders. The mass production of excessive detail and layers of fiction created so much misinformation that Armand Guéraud, one of Gilles de Rais's first biographers, bitterly complained: "We could carry on with quotes from the works of Daru, Mellinet, Leboyer, Massé Isidore, Verger, Loudun, Chevas, etc.; but we would learn nothing, except that modern writers have uncritically copied one another."[40] In the 1857 revised edition of his 1855 text, Guéraud added Sourdeval and Pître-Chevalier to the list of plagiarists and spoke of "the numerous errors committed on the topic of Gilles de Rais," and of the tradition identifying Bluebeard as Gilles as "completely lacking foundation."[41] This proliferation of historical and quasi-historical narratives of Gilles's crimes had a double cause: on the one hand, the neo-Gothic revival and, on the other, the revival of regional traditions.[42] Both contributed to a permanent confusion between history and fiction: "How false is history as the fantasy of novelists has made it."[43] Eugène Bossard, Gilles's most important biographer before Georges Bataille in 1959, admitted in 1885 that Gilles's story seemed like fiction, but it was a fiction that was absolutely true: "Fictions! They will say, pure fictions very purposefully made to frighten the imagination—Well, then! It must be affirmed: there is nothing but truth here, based on the least contestable proof. . . . The man became a demon."[44] J.-K. Huysmans's publication of his novel *Là-bas* in 1891 was in part a reaction to this continuing conflation of history and fiction, since "at the present time . . . history serves no purpose but to quench the literary thirst of provincial enthusiasts [*des hobereaux*]": "In the wake of Monsieur Taine, they [the historians] are busy gluing notes, pasting them down one after another, keeping only those, of course, that support the *fantasy* of their narratives [*la fantaisie de leurs contes*]. These people shield themselves from any kind of imagination, any kind of enthusiasm, by claiming to invent nothing—which is true—but they're faking it nonetheless through their selection of documents, their choice of stories."[45] Unlike these historians, Huysmans was not "afraid to dishonor [him]self by touching writers or painters whose lives had been marked by scandal" (p. 33 [p. 46]), for "history was the most imposing of fictions [*le plus solonnel des mensonges*]" (p. 32 [p. 44]). Following Huysmans, other writers found that the facts of Gilles's history were ready-made fiction: "This book is History cut up into a novel."[46] In 1926, Émile Gabory could only write with a degree of frustration that "the legend manufacturers still persist."[47]

At the root of the conflation of history and fiction lay the immensity of Gilles's crimes, considered "incalculable"—authors assume anywhere from 46, 136, 200, or 800 children perished[48]—and the savagery with which they were committed. Historians scrambled to find the words to describe them, and often their descriptions verged on the hyperbolic and fictional accounts. Gilles was unique: "Without example and without imitation, these crimes are such . . . that . . . Gilles de Rais occupies a unique place, almost without analogy, neither with those that preceded him, nor with those that followed."[49] His crimes were "horrors impossible to say,"[50] "things impossible to tell,"[51] because they "defy all expression."[52] The acts committed were so extreme that

> I will spare myself from repeating the atrocious or obscene details of this trial.[53]
>
> We have therefore left aside the majority of the horrific acts.[54]
>
> For the complete truth of the story, we must warn the reader that we necessarily gloss in silence over certain details of such immorality that only the judges could hear them. We would not dare repeat them, even by citing the somber documents of the proceedings.[55]

The more the authors insisted on the unspeakable nature of Gilles's sexual crimes, the greater the fiction and the mystification that surrounded it, the greater the rhetorical excess of writers. For instance, Stendhal's view was that the major problem for this historical event was "resembling a novel both atrocious and cold."[56] In response, he turned Gilles de Rais's story into a narrative of "the life of that first of Don Juans."[57] The horror of crimes had to be silenced and embellished.

Along the same lines as Stendhal's "making of a heterosexual," one of the most impressive strategies was to present Gilles de Rais as the historical original of the terrible Bluebeard of Charles Perrault's Mother Goose stories (Figure 4): "This character, who made himself sadly famous by his horrible cruelties, partly inspired Perrault's tale *Bluebeard.* In the Bluebeard article, we recounted the life of this ferocious maniac, [who] has remained a legendary figure in Brittany."[58] Thus, at the same time that the writers demonize Gilles in their descriptions, they also make the effort to make him into Bluebeard, even though "they resemble each other in only one respect, cruelty."[59] Gilles's portrait was more acceptable as a fable of the heterosexual serial murderer of women, silencing the fact of his homicidal pedophilia and eliding his homosexuality. One of the first developments of the myth of Gilles as Bluebeard appeared in the regional travel accounts of the Romantics, Édouard Richer (*Voyage pittoresque dans le départment de la Loire inférieure*, 1820), Prosper Mérimée (1836), and Stendhal (1838).[60] Eugène Bossard contributed greatly to this idea with his ethnographic research for his 1885 biography of Gilles de Rais:

> In the countryside, indeed, that he terrorized, all of the authorities, history, poetry, that is, written traditions—and above all oral traditions which, in themselves, must be definitely trusted on these topics—affirm to us, by unanimous, universal, and constant testimony, that the true Bluebeard is the hero of this book, Gilles de Rais, throat-slitter of women and children, judged for his crimes and burned in Nantes. That is what survived after the appearance of Perrault's tale and what it left intact. That is what dates back well before the seventeenth century.[61]

Bossard justified the disparity between history and Perrault's story by attributing the fictional element to the oral traditions of the people: "The close tie that exists between Bluebeard and Gilles de Rais comes from the people and is found within them," "but the novel invaded history."[62] The historians' voluntary confusion of Gilles with Bluebeard speaks to the importance of questions of sexuality and moral degeneration in defining the nation and its inhabitants.

The writers' other strategy was to make Gilles inhuman. The horror of the crimes he committed could only be the stuff of fiction, for they surpassed human capacity, "human perversity," and entered the realm of "bestial perversity."[63] "Bit by bit the man had come to destroy in himself all that he had of man";[64] "One can hardly pretend to adapt him to the classification of men, since he is perhaps not a man."[65] Gilles exceeded and transgressed all bounds of human behavior, falling into an "infinite evil," "an unnamable lewdness," and an "infinite depravity."[66] In Gilles's case, history had to be fiction, since

Figure 4. Gilles de Rais (1404–1440), companion of Joan of Arc, marshal of France in 1429 (1835), by Éloi Firmin Féron (1802–1876). Châteaux de Versailles et de Trianon, Versailles, France. Réunion des musées nationaux / Art Resource, NY. "This unique beard, which resembled no other, was black, although the hair was blond, but it took on, under certain influences of light and position, almost bluish tints, which had given Sir of Rays the nickname *blue-beard*, a nickname still popular in Brittany, where his story was transformed into a fairytale and a fantastical legend. . . . At times . . . his complexion took on pallid and cadaverous shades . . . and his beard, which seemed to stand on end, was ordinarily rather more blue than black" (Cette barbe singulière, qui ne ressemblait à aucune autre, était noire, quoique les cheveux fussent blonds, mais elle prenait, sous certaines influences de lumière et de position, des reflets presque bleuâtres, qui avaient fait donner au sire de Rays le surnom de *barbe-bleue*, surnom encore populaire en Bretagne, où son histoire s'est metamorphosée en conte merveilleux et fantastique. . . . Par moments . . . son teint prenait des teintes blafardes et cadavéreuses . . . et sa barbe, qui semblait se hérisser, était ordinairement plutôt bleue que noire) (Lacroix, *Curiosités de l'histoire de France,* p. 53). Émile Gabory qualified this as "a sensational portrait—really that of a madman," "this caricature . . . invented of whole cloth" (p. 180 [p. 145]).

only fiction could reach such dizzying heights of atrocity: "He sought beyond reality, beyond nature and possibility, a world of atrocious and senseless pleasures."[67]

Gilles's crimes were beyond words and because they were unspeakable, they led to an improbable identification with Bluebeard (or Don Juan) and rhetorical and fictional excesses. Gilles's sexuality was an exceptionally fertile ground for fantasy, as writers produced and reinforced a "shameless mystification" and "false erudition."[68] Their writings led Émile Gabory to ask in 1925: "To conclude, one may wonder if the glorious halo of companion of Jeanne d'Arc, the luxurious spectacles of Orléans, the melodious choirs, his sadism, his cruelty, the mystery of the evocations, the candles and the incense—one may wonder if all that has not contributed to build an ultra-romantic legend [*une légende ultra-romantique*] about him, and if, stripped of these artifices, Gilles de Raiz would not appear a vulgar assassin, although an artist and a *grand seigneur*."[69] Gabory wanted instead to produce a narrative stripped of all Romantic embellishments and hyperbole. It was precisely the kind of objective and scientific analysis that medicine was believed capable of providing. Medical diagnosis would furthermore carry the weight of incontrovertible authority. The example of Fernand Fleuret is the case in point. Under the pseudonym Dr. Ludovico Hernandez, "surgeon of the Lusitanian Marine," Fleuret published in 1921 a defense of Gilles de Rais.[70] Inscribing himself in the lineage of numerous doctors who wrote about Gilles de Rais between 1880 and 1920, he hoped to give unassailable scientific authority to his attempt to rehabilitate Gilles de Rais. In 1933 he admitted that the pseudonym Dr. Ludovico Hernandez "was . . . a whim like any other, an innocent whim." Above all, taking a pseudonym was an act that "spared him the insults,"[71] for a doctor's name imposed an objective, authoritative, and irrevocable judgment.

Doctors had access to Gilles's case thanks to the work of the historians whose "competent unanimity designates Gilles de Rais as the greatest criminal in history."[72] Since these sources were given over to fiction and fantasy, it is not a surprise that medicine participated in the reification of Gilles's ultra-romantic legend; rather than demystify it, medicine gave it an indelible stamp of scientific authority. What was true for writers and historians was equally true for fin-de-siècle doctors: with Gilles de Rais, doctors indulged in a veritable "pathologic archaeology" enriched with lively medieval fantasies. The hereditary defect (*tare héréditaire*) of Rais's inversion or pederasty, as homosexuality was still called at that time, sparked the imagination of a long list of doctors, as they subscribed to Michelet's well-established historical portrait of Gilles de Rais as "the beast of extermination" (*la bête d'extermination*),[73] and the literary portrait of "the beast of lust," the "monster," the "vampire."[74]

The most influential medical authority regarding the case of Gilles de Rais is without a doubt Richard von Krafft-Ebing. In his work *Psychopathia sexualis,* published in 1886, Krafft-Ebing recognized the link between lust (sensual pleasure and formative satisfaction) and cruelty and established two axes, sadism and masochism. For sadism, alongside the examples of "the degenerate Caesars (Nero, Tiberius)," Krafft-Ebing cited

> the history of that monster, Marshall [*sic*] Gilles de Rais (Jacob, *Curiosités de l'histoire de France*, Paris, 1858), who was executed in 1440, on account of mutilation and murder practiced for eight years on more than eight hundred children. According to the monster's confessions, it was from reading Suetonius and the descriptions of the orgies of Tiberius, Caracalla, etc., that he acquired the idea of locking children in his castles, torturing them, and killing them. This inhuman wretch confessed that in the commission of these acts he enjoyed inexpressible pleasure. . . . The bodies of the unfortunate children were burned, and only a number of heads of particularly beautiful children were preserved—as memorials.[75]

Gilles committed "lust murders" (Krafft-Ebing, p. 85) in imitation of Caesars, of whom he read in a book by the Roman historian Suetonius. The reading triggered an innate sadism that had lain dormant, it provided the determinant cause. Because the trigger came so late in Gilles's life, his sadism looked acquired, but Krafft-Ebing believed virtually all cases of sadism were congenital:

> A differentiation of original and acquired cases of sadism is scarcely possible. Many individuals, tainted from birth, for a long time do everything to conquer the perverse instinct. . . . Later, when the opposing motives of an ethical and aesthetic kind have been gradually overcome, and when oft-repeated experience proves that the natural act provides incomplete satisfaction, the abnormal instinct suddenly bursts forth. Due to this late manifestation of an originally perverse disposition, the appearances are those of an acquired perversion. (Krafft-Ebing, pp. 83–84)

In 1891, Albert Moll took up Krafft-Ebing's account of Gilles's reading of Suetonius in *Die conträre Sexualempfindung* (The perversions of the sex instinct).[76] In 1893, in his respected and very influential work *Une maladie de la personnalité: L'inversion sexuelle*, Julien Chevalier asserted that Gilles wrote to Charles VII of France a letter of confession, detailing the history of his criminal passions that were triggered by the reading of Suetonius. Chevalier likewise suggested that Gilles's inversion was of a much more ancient origin. He speculated that homicidal inversion was the motive for Gilles's withdrawal from the court, reporting from the same purported letter from Gilles to Charles VII that "I must nonetheless confess that I was induced to retire to my lands of Rais by a certain furious passion and covetousness that I felt toward your own crown prince, so much so that I almost slew him one day as I have since slain numerous young children by secret temptation from the devil."[77] On the basis of a fictional letter, Chevalier thus argued in favor of Krafft-Ebing's original sadism that would prove that Gilles was a born invert (*inverti-né*). In 1898, another important work appeared, *Attentats aux mœurs et perversions du sens génital* (Offense against public decency and perversions of genital orientation) by Dr. Léon Thoinot. Thoinot highlighted "this peculiar taste for licentious images."[78] He placed Gilles de Rais in the framework of sadism or "bloodthirsty perversion," and we again find him in the company of "Tiberius, Nero and other Caesars [who] were known sadists, and also perfect examples of other genital perversions (inversion, etc.)."[79] However, Thoinot was not satisfied with simply referring the reader to Krafft-Ebing's account of Gilles's sadism. He reported Gilles's exact words:

> "I know not . . . but by myself and my own volition, without the advice of others, I got hold of these imaginings to act thus only by sheer pleasure and delight in lust; in fact, I found in them *incomparable pleasure* [*jouissance*], likely at the instigation of the devil. Eight years ago this diabolical idea came to me. . . . And yet, being by happenstance in the library of the aforementioned castle, I found a Latin book about the life and mores of the Roman Caesars by a learned historian whose name is Suetonius; the book in question was decorated with very well-painted images, in which the behavior of these pagan emperors was visible, and I read in this beautiful story how Tiberius, Caracalla, and other Caesars reveled with children and *took singular pleasure in torturing them.* Upon which, I wished to imitate these Caesars, and that same evening I began to do so by following the images of the lesson and of the book."[80]

Such accounts of Gilles's sudden metamorphosis into a sadist and murderer upon reading Suetonius, sometimes accompanied by Gilles's confession, can also be found in works of scientific popularization written by doctors, for instance in Dr. Hayès's 1891 book *La pédérastie, historique, conséquences funestes de ce vice honteux* and Dr. Cabanès's 1899 *La flagellation dans l'histoire et la littérature.*[81] However, it is only in 1903, in Dr. Fauconney's *Physiologie du vice, son histoire à travers les âges,* that we find the exact title, previously not mentioned, of Suetonius's book that inspired the killings: "Reading the *Vie des douze empereurs romains,* by Suetonius, aroused this powerful lord to imitate their monstrous debauchery."[82]

From Krafft-Ebing to Fauconney, doctors drew their information about the onset of Gilles's homicidal frenzy and copied his confession from the same source: an 1861 article on "La folie héréditaire" by Benedict-Augustin Morel.[83] But where did Morel find this information and this citation, since it was not contained in the trial record? The trial record is written in the third-person indirect discourse, as opposed to Morel's rendition of Gilles's confession in direct discourse. Moreover, the trial record contains only the first part of the confession, without any mention of Suetonius: "Item, interrogated by the said Lord President as to who had persuaded him to the aforesaid crimes and taught him how to commit them, he responded that he did and perpetrated them according to his imagination and idea, without anyone's counsel and following his own feelings, solely for his pleasure and carnal delight, and not with any other intention or to any other end" (Bataille, p. 193 [pp. 238–39]). When the presiding judge insisted, in French this time, on hearing the motive for the murders, that "I desire and would like to know the absolute truth from you for the reasons I've already told you often," Gilles de Rais's additional response was likewise recorded in French, in the first person, possibly reflecting the intensity of their exchange: "Truly, there was no other cause, no other end nor intention, if not what I've told you."[84] Gilles's statement from the historical trial record contains no mention of Suetonius and even contradicts the possibility of an external influence and trigger: "There was no other cause."

Morel had actually used the translation of the trial record by Paul Lacroix, also known as Bibliophile Jacob (see Chapter 5), the same source to which Krafft-Ebing referred. Paul Lacroix decided to publish in 1858 his translation of Gilles de Rais's trial because "the trial of Marshal of Rais has never been published; it is summarized in collections of famous causes; it is merely mentioned in histories of France: M. Michelet is the only writer who dared to give an account of these crimes that condemned to the pyre a relative of the Duke of Brittany, and a Marshal of France."[85] Lacroix claimed furthermore that he did everything "to represent dutifully the physiognomy of a trial."[86] According to Lacroix, only the level of depravity of the "private life of Roman emperors" paralleled that of the Marshal "and it was probably from that unclean source that Gilles de Rays had borrowed his hideous crimes."[87] Following this logic, Lacroix introduced Suetonius in his translation, in reality an adaptation of the trial record. After numerous descriptions of acts of sodomy with children and messy bloody murder rituals, Lacroix, who was a talented Romantic writer, paved the way for Gilles's moving first-person confession, in which he would have publicly explained the motives behind such atrocities:

> "I know not . . . but by myself and my volition, without the advice of others, I got hold of these imaginings to act thus only by sheer pleasure and delight in lust; in fact, I found in them incomparable pleasure [*jouissance*], likely at the instigation of the devil. Around eight years ago this diabolical idea came to me; it was the same year in which my deceased grandfather the Lord of Suze left this earthly life. I then left the King of France's camp to travel to Chantocé, in order to lay hands on the wealth and inheritance of the deceased; and yet, being by happenstance in the library of the aforementioned castle, I found a Latin book of the life and mores of the Roman Caesars, by a learned historian named Suetonius; the book in question was decorated with very well-painted images, in which the behavior of these pagan emperors was visible, and I read in this beautiful story how Tiberius, Caracalla and other Caesars reveled with children and took singular pleasure in torturing them. Upon which, I wished to imitate these Caesars, and that same evening I began to do so by following the images of the lesson and of the book."[88]

Lacroix embellished the trial record by rewriting Gilles's confession in the first person and inserting the reference to Suetonius, a book that was not in the inventory of Gilles's personal library.[89] But he also exaggerated the reference to Suetonius by making it an illuminated manuscript, a detail that was historically inaccurate, as "none of Suetonius's manuscripts were illustrated with indecent miniatures" and no contemporary historical document ever mentions one.[90] In fact, the first edition of Suetonius appeared only in 1470, thirty years after Gilles's death (Meurger, p. 132). Salmon Reinach, one of the biggest critics of Lacroix's harmful invention, was right in qualifying it as a "shameless mystification."

Moreover, Lacroix's fiction of Gilles's reading of Suetonius fueled the development of an intricate (but incorrect) portrait of Gilles, summarized by J.-K. Huysmans: an "erudite Latinist, a witty conversationalist, and a generous and reliable friend. He possessed an extraordinary library for the time. . . . We have descriptions of several of his manuscripts: Suetonius, Valerius-Maximus, and an Ovid on parchment, covered in red leather and with a silver-gilt clasp and key."[91] Literate, for some even an accomplished artist (manuscript illuminator), Gilles was gifted with "rare," "exceptional," "remarkable," and "high" intelligence: "Historians have spoken of his great intelligence, his erudition, his refinement, and his infallible sense of aesthetics."[92] Among doctors, the portrait of an intelligent, erudite, artistic mind fit well with the idea that only a man of superior intelligence, albeit degenerate, could commit such intricate and sophisticated crimes. Gilles was "the supreme artist in evil," "a genius . . . conscious of that fact."[93]

Lacroix thus served as Morel's only historical source and indirectly established what gradually became a tradition perpetuated by Krafft-Ebing, Moll, Thoinot, and many others.[94] Fiction therefore facilitated Gilles's passage from historical to medical discourse. Moreover, I contend that the success of Lacroix's literary hoax escaped the scrutiny of pathologic archaeology and slipped into medical discourse because only this hoax gave "some scientific explanation of his conduct" (Hamblin-Smith, p. 11) to "the inexplicable crimes" of a Marshal of France that had made him "the greatest criminal in history."[95] Medicine seized on Lacroix's historical fiction that finally allowed Gilles's classification: on the one hand, it found in Suetonius the determinant cause that triggered Gilles's sadism, which had not been visible during the time he had been Joan's companion in arms, and, on the other hand, it grounded in his literacy, refined artistic tastes, and high intelligence, the diagnosis of superior degeneracy.

Gilles de Rais was "a murderer for pleasure [*par volupté*], a bloodthirsty 'sadist.'"[96] According to Krafft-Ebing, sadism was congenital, not acquired. After the stimulation of the hereditary terrain by "a sedentary lifestyle, a carnivorous diet, strong drinks, and spices [which] have repercussions on one's sex life," "ceaselessly aggravated by succulent dishes, alcoholic liquors, wakeful nights and emotions,"[97] all it took was an external trigger:

> The result awaits only a provocation, an impression coming from the exterior, in order to occur. For Moreau de Tours, the organism, the *physical and moral terrain*, is everything; as for the causes that determine

> the blooming of the vice, they are only secondary. . . . In a word, heredity plays a great role in the question, since the objective causes, the determinant provoking causes, which have a certain role that must be recognized, do not act indiscriminately on all individuals; they need a special terrain, a receptive terrain.[98]

Doctors found the incidental cause in "the fatal and harmful reading": "It is the simple mechanics of an obsession beating down on a ripe terrain poorly defended by a retarded or perverted will, gradually taking root."[99] Dr. Cabanès repeated what many others had said before: "The harmful reading in question awoke the sleeping beast in heat, and led the faltering will towards the abyss into which it let itself slip. Isn't it usually a trivial event that triggers latent neurosis?"[100]

The question remained as to whether Gilles de Rais had the excuse of being criminally insane or whether he had criminal and moral responsibility for his acts. Was his "a case for a specialist in mental aberration, or was Gilles simply a bloodthirsty sadist? In other words, to what extent may he be held responsible for his actions?" (Gabory, p. 153 [p. 123]). The assertion of Gilles's madness originated in the fifteenth-century text of "Mémoire des héritiers," when his family tried after his death, on the grounds of Gilles's unaccountability, to recover the possessions that he sold during his lifetime. The answer to the dilemma—morally insane and not criminally responsible, or morally *and* criminally responsible—lay in his intelligence. Gilles was a superior degenerate (*un dégénéré supérieur, un dégénérescent*).[101] His artistry in sadism placed him in the category of superior degenerates:

> His acute intelligence, his other attributes immediately eliminate the vast under-class of the daft and dim-witted, and at the same time his sentimental inclination towards power, riches, the Afterlife, and the mysterious lead us to [recognize] imbalance, the superior degenerate clearly defined by Magnan.[102]

> Yet, Gilles, due to his high intelligence, his aesthetic awareness, his culture and his erudition, differs as much from the average and inferior degenerates as he resembles them because of the nature of his murders.[103]

What this authoritative medical diagnosis showed was that Gilles, because he was a superior degenerate, was not insane. Some doctors did allow for the possibility of "lucid madness" (*folie lucide*). According to Ulysse Trélat's definition from 1861, "These sick people are mad, but do not seem crazy because they express themselves with lucidity. It is their actions and not their words that are mad. . . . They are lucid even within their delirious notions. Their madness is lucid."[104] However, in either case, doctors agreed that Gilles was fully morally and criminally responsible.

The medical adjudication of Gilles's case was linked to the larger contemporary social context: the long and ongoing debate at the end of the nineteenth century between psychiatrists and jurists on criminal responsibility of the insane.[105] The example of Gilles's criminal homosexuality was also connected to the contemporary perception of homosexuality as a menace to the health of metropolitan France. Homosexuality was perceived to threaten it from within because of moral degeneration: "How is *pederasty* possible in an advanced civilization like ours . . . with all of its disgusting and ignominious horrors? Unbridled debauchery, jaded sensuality . . . ; but it is difficult, in many cases, not to admit a true alienation of the faculties for moral judgment."[106] Doctors agreed: moral degeneration leads to homosexuality (see Chapter 3). Moreover, Gilles's imitation of the historical example of Roman Caesars offered in turn a historical example for imitation by contemporaries. Black masses, the Sabbath, "a Renaissance of mysticism"[107]—these were very much in fashion in the late nineteenth century. Huysmans's novel *Là-bas* or the writings by doctors concerned with such phenomena, such as Dr. Regnault's *La sorcellerie*, are witness to this social phenomenon, which pushed the conjunction of religious mysticism and eroticism to the limit. Doctor Gabriel Dromard's *Les fausses 'messes noires'* on the Fersen affair is particularly informative of the convergence of imitation, moral degeneration, and homosexuality in fin-de-siècle France.

In November 1903, Baron Jacques d'Adelswärd-Fersen, twenty-three years old, was charged with and tried for public lewdness (*outrages publics à la pudeur*) and the corruption of minors (*excitation des mineurs à la débauche*). He was eventually sentenced for the latter charge. Schoolboys, ages twelve to fifteen, were brought to his apartment in order to stage *tableaux vivants*, performances of classical literature. Soon, however, the press publicized these reunions of aesthetes as "Black Masses," turning the aesthetic, literary preoccupations into Satanic and, thus by definition, sexually aberrant rituals. For instance, Jules Bois, author of *Le Satanisme et la magie* (1895), probed in his article in *Gil Blas* the question of the link between inversion and Satanism: "'unnatural acts' and the 'inverting . . . religion of the Devil.'"[108] Unambiguously, the prosecutor Lescouvé linked Fersen to the Middle Ages: "If our laws no longer permit us to punish the sin, as they did in the Middle Ages, we can still punish the wrong-doers, so that this will not happen again" (qtd. in Erber, p. 202). By February 1904, this fascination with Satanism and its link to Gilles de Rais were spelled out in the play *Les messes noires*, "where Roland Brévannes wants to show the hellish office in its period of supremacy, in the Middle Ages . . . and its last contemporary avatar, in Axel Wartz's [Adelswärd's] salons."[109] The first scene opens with Gilles de Rais in his alchemist laboratory. Doctor Dromard warned against too detailed a coverage of the Fersen scandal because, like Gilles de Rais, whose readings of Suetonius supposedly taught him to kill, readers of the Fersen scandal might be susceptible to imitating such acts because "mental imbalance is frequently accompanied by a very singular tendency to imitate": "Literature and history participate in large part [in this tendency] because they provide the famous lessons that unrestrained imaginations know how to exploit marvelously. . . . In moving from the fifteenth century to the twentieth, and from the lands of Machecoul to Parisian boudoirs, we encounter this same influence from literature and history."[110] Doctor Bernelle likewise drew the explicit comparison between Suetonius's book and contemporary graphic images: "It is a sad psychic process that our modern life emulates and reproduces without reserve: depraved youth, future cynics, hypnotizing themselves with a murder scene, violently colored, on display at a newspaper stand, and receiving from foolishly illustrated everyday news [*fait divers*] the harmful shock which sooner or later will put—irresistibly—the blade in their hand."[111]

Nineteenth-century black masses, a revival of medieval witches' Sabbath, were further suspect because they also reactivated the association of two medieval phenomena, religious mysticism and erotic madness: "In the fourteenth, fifteenth, and sixteenth centuries, an era of exploding religious mysticism and erotic madness [*folie génésique*], we also see passions against nature appear."[112] If homosexuality implies artistic refinement and superior intelligence, it also goes hand in hand with mysticism and exaltation: "Just like religious emotion, aesthetic emotion is not so far as one might expect from carnal emotion. Let us clarify: this aesthetic emotion is realistically nothing but a derivative of sexual instinct, from the general viewpoint of evolution."[113] Gilles's exceptional erudition and artistry, his aestheticism, all due to his superior intelligence, had propelled him into "a prolonged search for mysticism and asceticism," "the cause of all this erotic madness [*folie génésique*]."[114] Gilles's mystical delirium was the explanation for his "erotic delirium [*délire érotique*]."

This same religious fervor was also responsible for past phenomena of hallucination: "It is not astonishing that religious and erotic monomania [*les monomanies religieuse et érotique*] should be signalized by hallucinations the most strange and frequent."[115] In the Middle Ages, these hallucinations had been interpreted as witchcraft and magic: "Demonomania, sorcery, belief in incubi and succubae, in the evil spells and vile acts of the devil pervade the age, along with widespread priapism. The law that supposes asceticism and religious mysticism cannot exist without sexual disorders of all sorts found a shining confirmation."[116] But rather than sorcery, such beliefs were signs of madness for nineteenth-century medicine: "In the fifteenth century, madness bore above all the imprint of superstitious ideas, theological doctrines then in good repute."[117] Religious fervor reached a level of social delirium, "collective hallucination" (*hallucination collective*),[118] "generalized madness" (*folie générale*), either as theomania or as demonolatry.[119] Historian Jules Michelet even exclaimed: "Satan may be but an illness!"[120]

While Gilles was the ideal case study for religious mysticism and erotomania, Joan's case served for the study of links between hallucinations and religious mysticism, madness and sorcery. Nevertheless, for doctors there was little difference between Joan of Arc's hallucinations and Gilles de Rais's erotomania. Both were signs of degeneration, the only difference being the underlying cause: while Gilles fell into the category of criminal degeneracy by giving in to sexual instinct, Joan's hallucinations (not accompanied by an erotic delirium) were found to be the degeneration of the intellect in the genius.

Joan and the Medical Corps

As seen, Gilles de Rais's medical history is in large part based on a hoax. This does not change the fact or the significance of Gilles's crimes, nor does it invalidate his condemnation. Nevertheless, the history of this hoax represents the fictional bent that contaminated medical discourses and gave rise to rhetorical excesses carrying the scientific stamp of approval. Gilles's story had indeed come full cycle, from literary hoax to scientifically sanctioned literature: the writings of nineteenth-century historians and writers nourished the late nineteenth-century medical diagnosis, and the medical treatment in turn gave their writings an air of scientifically objective truth, which then fed twentieth-century accounts of Gilles de Rais's case, as it continues to do to this day.[121] Certainly, one of the most influential has been Georges Bataille's 1959 essay

"La tragédie de Gilles de Rais," which, based on the medical authorities who had derived their diagnosis from Paul Lacroix's Suetonius, passed judgment on homosexuality as spectacle, excess, and expenditure.[122] What I suggest, moreover, is that this amalgam of fiction and science supported the thesis of the degeneracy of aristocracy and provided the contrast for Joan's immaculate legend that symbolized the internal racial strength and regeneration in and of the French people. Joan's case illustrated the other aspect of degeneracy—genius—but her exceptional talent could not be elucidated without its pathological pair, vice and criminality.

As in Gilles's case, in the medical case of Joan of Arc, doctors based their diagnosis on historical writings: "The character of the noble girl whose hallucinations we will study has been well outlined by the most eminent historians of our times."[123] For historians, Joan's voices that led her to liberate Orleans from the English siege in 1429 had been the object of intense speculation for centuries because they opposed religion and science. Romantic writer Alphonse de Lamartine succinctly summarized the problem: "These inspirations, which popular credulity converts into miracles, are they indeed supernatural? Are they really divine calls summoning young girls from out of the crowd by name, to give them the mission of saving their country? Or are they simply natural phenomena, the voices of internal whisperings—the concentrated recoil and reaction of a whole nation . . . compressing its universal wail into the shriek of a woman" (Lamartine, p. 51 [p. 124])? Unlike the historians, who had the freedom to adopt a religious explanation of Joan's divine mission, under the regime of scientific secular truth doctors could not accept the idea of the supernatural and divine, "the absurd, the irrational, the contrary of what everyone can acknowledge in the domain of medical and pathological observation."[124] Instead of "voices," psychiatrists of all convictions adopted the scientific term of "hallucination."

It was the great psychiatrist Étienne Esquirol who posited in 1817 the link between mysticism and hallucination: "The austere writings of every people, the history of magic and sorcery in every age, together with the annals of mental medicine, furnish numerous facts in relation to the subject of hallucinations" (Esquirol, p. 93 [p. 160]). It was also Esquirol who established that hallucination "is one of the elements of insanity, and may be met with in all the forms of this malady" (Esquirol, p. 93 [p. 160]). But Esquirol did not just argue that hallucinations were always the sign of madness. He also believed that "the history of many celebrated men confirms this view [on the independence of hallucinations], and proves that a man may be subject to hallucinations without delirium" (Esquirol, p. 106 [p. 191]; also p. 110 [p. 199]). The link between mysticism and hallucination, genius and madness, was a radical revisionism of history, creating an upheaval in the way history was written because "the health of a given historical figure had exerted considerable influence on the ideas, the doctrines of an age."[125] Psychology not only illuminated but also rewrote history. Precisely because of "the often overwhelming role that hallucination played in political and religious events," determining the exact nature of hallucination was of extreme importance for the history of the nation:

> It has escaped no observer . . . that the hallucinations that we observe in the ill are absolutely of the same character, presenting themselves in absolutely the same way as those observed in historical figures. As a result, we find ourselves facing a dilemma: either all of these figures were crazy, just as crazy as those that we lock up in our insane asylums, or, on the contrary, they were not crazy and, in that case, we must admit that there exist two types of hallucinations, one of which is compatible with the most perfect reason, the other characterizing mental alienation.[126]

The stakes of the medical explanation of the phenomenon of Joan's voices were therefore extremely high, for the historical interpretation of the salvation of the French nation from the grip of the English occupier depended on the scientific interpretation of the phenomenon that was generally credited with the liberation of France. Moreover, in associating Joan's voices with hallucination, the question that had to be decided was whether Joan's hallucinations were of madness . . . or of genius. The controversy over Joan's hallucinations lasted throughout the 1830s and 1840s and principally opposed, on the one side, psychiatrists Claude-François Michéa and Alexandre Brierre de Boismont and, on the other, psychiatrist Louis-Florentin Calmeil and Alfred Maury, historian and archivist.

Joan's case was instrumental in deciding the exact nature of hallucination. For some psychiatrists, like Boismont, who was a professed Catholic, Joan was in fact the best case study for the phenomenon of hallucination: "Of all of the figures that can make this so controversial and yet so real phenomenon of physiological hallucination comprehensible, there is none other comparable to the illustrious Joan of Arc."[127] Already in 1837, Michéa, another psychiatrist of Catholic convictions, cited the example of Joan of Arc to illustrate his thesis that there are "spontaneous, incidental hallucinations without mental alienation and independent of all other illness capable of causing delirium," which manifest themselves "following a deep meditation, a concentration of thought, extreme attention of the spirit on an exclusive idea."[128] In 1844, Michéa asserted with the authority of his medical expertise acquired in the meantime that "hallucination can manifest itself with perfect integrity of reason. . . . Hallucination is a pathognomonic, but not a *constant and necessary* symptom of mental alienation."[129] Almost at the same time, in 1845, Calmeil, head of the Charenton asylum, published his study of madness from the fifteenth to the nineteenth century. After an extended analysis of the monomaniacs that in the Middle Ages had been mistaken for heretics, Calmeil dedicated a whole chapter to Joan of Arc, a living example that "one can, on the condition of possessing great character, heroism and genius, accomplish great things even while giving in to the exaltation, to the inspiration of true delirium."[130] For Calmeil, there was no doubt as to the pathology of Joan of Arc: "The success of her arms will never prove that she showed solid reasoning by putting faith in her own visions. A person who hallucinates [*un halluciné*] can perhaps have the greatest ambitions in mind; but what obliges us to consider that person ill is that he perceives what does not exist and he firmly believes that his own ideas are whispered in his ear by beings that he believes to be other than himself. Such is the pathological case of Joan of Arc."[131] Joan of Arc's success was an accident: "It so happens that by mistaking errors of the imagination and judgment for celestial favors, the kingdom was saved."[132] This radical disagreement between Michéa and Calmeil, divided along the lines of a genius who cannot be insane and a lunatic whose projects fortuitously but accidentally coincide with the liberation of the homeland, continued in the same time period between Brierre de Boismont and Maury. Maury accuses Boismont of a lack of scientific rigor and objectivity because of his Catholic convictions, "this vague, undecided, nuanced character in between the Catholic and the philosophical which has so thoroughly evacuated from his work the unity of doctrine and the strength of deduction which would otherwise have given it value," while Boismont protested against "the current-day rationalists."[133] For agnostic doctors, hallucinations were always pathological; for Catholic doctors, they were just a physiological manifestation, perfectly compatible with the full possession of one's reason.

Brierre de Boismont argued in 1845, as had Michéa, that "hallucinations can affect a great number of persons without them being suspected of madness," and he used the cases of "famous men" to back up his theories: Peter the Hermit, Loyola, Luther, Joan of Arc, Mohammed, Socrates.[134] For each of his examples, Brierre de Boismont asserted that hallucinations were "those of his/her century," "in the social body, and not in individuals." They were collective, "generalized madness," and were of the order of "the beliefs of the times"; today, they are replaced by "alienations specific to each person."[135] But because these hallucinations were products of their time, products of belief systems, these "erroneous beliefs" were also "hallucinations of reason."[136] The error lay not in the person, but in the age during which the person lived. In consequence, for Boismont, the question was not whether Joan hallucinated, but whether the simple act of hallucinating made her insane. In 1861, Brierre de Boismont reaffirmed that "hallucinations compatible with reason, and those that we observe in the insane, do not differ in their essence; but, simply because they have the same exterior manifestations, do we have the right to conclude that they form a single pathological species?"[137] For Boismont, the answer was negative; after all, Joan never lost the ability to conduct her everyday activities, and she never contradicted herself. Joan's hallucinations were thus merely physiological, while those of the insane are pathological. Alfred Maury refuted Boismont in 1846 with this argument: "If general error explains hallucination in these great men, it fails to justify it or to make it reasonable for this reason."[138] Maury's view was that, even if "hallucinators" (*les hallucinés*) were otherwise perfectly sane, their hallucination brought a grain of madness to the mix: "The hallucinations that we have observed in great minds have been reconciled with a strong and sound reason in other regards, with perfectly reasonable acts in all instances where hallucinations were not their underlying principle: this is what psychological medicine has observed and constantly observes."[139] In a reply published in the same volume, Boismont gave this refutation of Maury: "Nothing is more painful, indeed, than attributing insanity to sublime doctrines, heroic actions, glorious enterprises."[140]

This initial debate of the 1830s and 1840s did not resolve the question of whether Joan's hallucinations were compatible with reason. What it did represent, though, were two competing views of religion. Those who believed in the compatibility of religion and psychiatry, Michéa and Boismont de Brierre, considered the voices hallucinations but disputed them as signs of madness. For them, the voices were isolated moments of religious exaltation that did not impair reason or sanity. Those who wanted to secularize psychiatry, Calmeil and Maury, believed that all hallucination was a sign of madness, differing only in the degree. It was Moreau de Tours' influential *Psychologie morbide* (1859) that provided an elegant and convincing solution to the dispute by acknowledging great moral and intellectual acts in their dimension of intelligence:

> Yes, Joan accomplished great things, and she demonstrated, by their accomplishment, [her] genius, intrepid will; it is that she believed in her visions: she believed herself to be the instrument of God and her belief rested upon the unshakeable foundation of numerous hallucinations, on a morbid mental fact that harbors within it a force of impulsion, extreme, irresistible drive, a moral and intellectual stimulus, capable of moving mountains, and which, in reality, placed Joan, in relationship to those who surrounded her, in a superior moral sphere.[141]

In Moreau's book, madness and genius rub shoulders with each other, they "are co-generated."[142] Hallucination is a sure sign of morbid psychology but one that reveals "the true origin of high intellectual faculties," without implying "in any manner, the general or absolute disorder of the intellect."[143] Hallucination is best suited to appear "at the heart of the most sublime faculties of the soul."[144] In Joan's case, superior intelligence, her genius, differentiated her hallucinations from "ordinary intelligences, of those common hallucinating patients that populate our asylums."[145] After Moreau de Tours and Ulysse Trélat's "lucid madness" (*folie lucide*), there will be only one step to Valentin Magnan, and the notion that a genius is a superior degenerate.

For the historians and doctors, the stakes were ultimately the same: "Because hallucination is closely tied to the political and religious life of peoples,"[146] the leading role could not be played by a lunatic maid or God-sent messenger on a divine mission that was impossible to prove scientifically. Instead, the history of the nation, and the nation's liberation from the English occupier, hinged on the actions of a genius. Joan was the genius of the people, "the daughter of the people," who led to the regeneration of the French race after the nobles' decadence that was embodied by Gilles: "Thus it is from the people that emerged the French liberator, the woman who would free her land from the foreigner and raise the first flag of nationalities and the regeneration of races."[147] The medical distinction whereby a degenerate could also be a genius, and not a sick or criminal member of society, legitimated an exceptional historical figure as a symbol of the vigor and renewal of the French race, "the vital force" (Michelet, *Joan of Arc*, p. 9 [5:44]). Joan's symbolic status of "the daughter of the people" was useful in opposing the health of the "people" to the degeneracy of "aristocracy," yet her virginity and absence of menstruation were never brought to the fore of these discussions. For some, the absence of menstruation was proof of her insanity: "The partisans of this opinion [which declares Joan of Arc insane] have supposed that she had secret ecstasies which were no doubt favored by the frequent absence of any menstrual flow,"[148] whereas for others it was proof of her exceptionalism and intelligence, that "the physical curse of women never affected her. This was spared her, to the benefit of religious thought and inspiration" (Michelet, *Joan of Arc*, p. 9 [5:44]). Joan's genius was exceptional, and therefore a symbol, but not a model for young women to emulate, whose vitality was to be translated into fertility rather than sterile virginity. The vigor of its people at the waning of the Middle Ages symbolized the hope for the present regeneration and eugenics of the French nation.[149]

Gilles and Joan

Medical discourses provided a scientific and objective explanation of Joan's and Gilles's acts. They were in full possession of their mind and therefore entirely responsible for their actions. But medicine also reified the notion that they were an inextricably bound pathological pair: on one end of the spectrum, a degenerate criminal and, on the other, a degenerate genius. Their coupling became an undeniable scientific fact, used precisely to establish a contrast and raise the interest in the story, despite claims to the contrary: "We are not bringing together Gilles de Rays and Joan of Arc to establish a contrast between these two extraordinary figures from our history, nor to increase the interest of our story [*récit*], but because the fact is imposed to our conviction."[150] The two were inseparable sides of the same coin, two manifestations of a singular pathology of superior degeneracy. They formed an exceptional couple whose actions could not be explained by historical narrative, but only medical treatment. Joan's voices were hallucinations that made her a genius leader, while Gilles's exceptional monstrosity, "this exceptional creature,"[151] made him a sadist by imitation.

The pathological link between the genius and criminal legitimated and solidified the symbolical coupling of Joan and Gilles that writers had been creating throughout the nineteenth century. Gilles's image was double, one of rise and fall: on one end of the arc, he left "the memory of the great captain" and, on the other, "the memory of the brigand,"[152] "the *comrade*, Joan's pure knight, and the bandit, the monstrous murderer."[153] After "such a stunning role next to the young warrior," the most beautiful period in his life, "such great crimes."[154] Joan of Arc, in respect of whom he had been "the dignified emulator,"[155] had brought out the best in him. But with Joan's death,

> here seems to end the honorable and military career of Marshal of Retz. There remains no more than the painful task of offering a portrait of the extravagances, vices, and monstrous crimes that have contributed more to his unfortunate celebrity.[156]
>
> Thus, then, Gilles dedicated himself to the sustained study of magic only after the death of Joan of Arc, perhaps even only in 1432.[157]

In line with this tradition of the rise and fall of a shining example of French nobility, Michel Tournier's magnificent 1983 novel, *Gilles & Jeanne*, makes Joan's execution central to Gilles's transformation from angel to demon: "But if anyone had seen his face he would have realized that something had changed inside him: he had the face of a lying, pernicious, dissolute, blaspheming invoker of devils. . . . When the malign metamorphosis was complete, he emerged, an infernal angel, unfurling his wings."[158] Medicine gave the scientific approval to the historians' hypothesis of Gilles's infamy and made Gilles into Joan's explicit antithesis around the turn of the twentieth century, precisely at the time when there was a growing movement in France leading to Joan's beatification (1909) and elevation to sainthood (1920). Gilles's medical condemnation as a criminal degenerate was thus instrumental in elevating Joan to an inviolable virgin sainthood. In fact, genius talent could only come into the full light when projected against the blackness of the criminal sadistic degenerate that emerged due to the trauma of her especially violent death.

Most important, Joan was a glorified symbol of the regeneration and renewal of the French nation in its people. That French medieval nobility had fallen into decadence was best exemplified in the spectacular rise and fall of Gilles. Medicine purified the origin of the French race from the deviance of the dynastic, aristocratic privilege: the genius daughter of the people regenerated what the criminal noble threatened with degeneracy. Influential writers such as Paul Jacoby and Théodule Ribot, who wrote on hereditary degeneration, had otherwise concluded that privilege and power, "the exercise of absolute power, *Cesaritis*,"[159] "the delirium of power,"[160] lead to degeneracy and extinction: "The greater it [the power] is, the greater the evidence of the moral, intellectual and physical decadence."[161] The historical lesson of Gilles's rise and fall evinced the social role of medicine that surveyed and policed sexual behaviors, health, and the moral state of the nation:

> The possession of privileges and power seems to have had in every age the most disastrous effects on the intellectual and moral health of

> those invested with it. . . . No longer needing to restrain oneself, to vanquish one's desires, to resist one's passions, to restrain poor instincts, eventually makes the moral faculties suffer a profound disorganization. The conscience sleeps, the notion of good and evil dissipates, the base instincts are easily satisfied, the personality becomes exalted, a selfishness all the more powerful that it goes ignored suffocates all generous sentiment. . . . Thus we see the neuropathic vice in all its forms: neuroses, mental disequilibrium, psychic anomalies, crime, debauchery, mental ailments, chronic illness, infirmities, vices of conformation, and sterility successively attack the privileged races and make them die out with an often lightning rapidity.[162]

Degeneration of the privileged classes could only be cured "by the ennoblement of the commoners":[163] "Had it always been thus, France would perhaps not have had so to suffer under the administration of degenerate races, and pathological heredity is, in my opinion, one of the most powerful arguments against dynastic heredity."[164] Ultimately, Gilles's deplorable end served as the best argument against monarchy and in favor of the French Republic symbolized in Joan of Arc, "the daughter of the people": "We have just seen the difficulties that criticism, being based on experience, can raise against nobility considered to be a *natural fact.* What it was worth as an *institution* is not worth seeking here."[165] Joan of Arc's symbol of regeneration was the medical version of a prophetic promise for the Third French Republic proclaimed in 1870.

PART III

Sex and Love

Chapter 5

"Pornographic Archaeology": An *histoire des mœurs*

> The right of words—which is not that of the philologists—authorizes, therefore, the use of the term *archaeology*. . . . Archaeology describes discourses as practices specified in the element of the archive.
>
> —Michel Foucault, *The Archaeology of Knowledge*

> Literarity, a modern concept, cannot, without begging the question, found an archaeology.
>
> —Paul Zumthor, *Speaking of the Middle Ages*

In 1852, in his six-volume *Histoire de la prostitution*, Paul Lacroix included an essay on medieval urbanism and prostitution entitled "Les rues honteuses au Moyen-Âge" (Streets of Shame in the Middle Ages) and named this reconstitution of sexual topography of medieval Paris a "pornographic archaeology" (*archéologie pornographique*).[1] In calling this history "pornographic," Lacroix conformed to the etymology of the word: "The title of pornographic work is above all appropriate to books on prostitution."[2] But it is harder to intuit what Lacroix meant by "archaeology." Already a decade earlier, in 1842, in the wake of road works in Paris, "in this year of grace of expropriations, demolitions, reconstructions, and municipal transformations," Lacroix had called for an archaeology: "The old Paris already no longer exists: every day it disappears beneath the new. . . . The debris of the vessel, those are the names of the streets; the masts are the towers of Notre-Dame [de Paris]; and us, poor archaeologists, let us cling to the relics of this great wreck."[3] Lacroix's archaeology would not excavate objects in order to preserve and re-create. This kind of preservation was impossible because archaeology could no longer reconstitute the physical description of the places of prostitution, "the exterior façade of these places of ill-repute, . . . their names and their signs (at least for the majority), . . . the ordinary system of their shameless organization, . . . their interior physiognomy."[4] Lacroix's archaeology would instead be a cultural history of the name, a reconstitution not from monuments, but from documents. Lacroix therefore created an "archaeological inventory" of names of Parisian streets that, after demolition, were the only "relics" left for an archaeologist to read: "We must study their names, routinely modified, reformed by municipal decree, changed by events, like a dead language corrupted, which is lost day after day, and which will soon have no interpreter."[5] Lacroix took his readers on a "pornographic voyage," reading a "road map of the streets of shame in the capital." The history of prostitution was thus mapped onto a topography, "the topography of Parisian prostitution in the Middle Ages."[6]

But how could an archaeology of street names suffice to recreate history, as Lacroix claimed? His first point was that names had an obscure collective origin in the people, quite unlike now, when municipal authorities name the streets. The modern practice of naming imposes a history of individuals on the city: "We could summarize the annals of the arts, letters, sciences, crime and virtue, with the names of men inscribed at the intersection of streets, in as noble a fashion as on the bronze tablets of the Pantheon."[7] But since in the Middle Ages the origin of street names was in the people, their etymology could not be explained by proper names giving an intellectual history, but rather by diverse urban practices, sexual and other. The street Heuleu, "which led to the current Grand-Hurleur Street . . . exclusively reserved for Prostitution since its origins until the present day, . . . owed its name not . . . to a knight, named Hugo Lupus (*Heu-leu* in Old French), who lived in the twelfth century and made several donations to the Church of Saint-Magloire, but rather to the heckling [*huées*] which then nagged simple-minded or unsuspecting people that chance brought to this infamous place."[8] Thus Lacroix's second point was that in the Middle Ages street names did not come into existence because of individuals who may have lived there; individuals did not name streets but rather practices conducted in them. Indeed, street names were so marked by urban practices that they could become someone's proper name: "It is nonetheless certain that Pierre de Bordelles had been thus described in practice because he owned a house on this street, which was named *Bordelles*, *Bourdelle*, and *Bordel*, because of its primitive usage and the numerous *bordes* that the fortifications of Paris had enclosed in its walls."[9] To navigate the streets of the city, it was enough to know their names, "their nickname had to be the infallible sign of their birth, their natural appearance, their state, in a word, their entire physiological portrait": "Those streets there didn't fool anyone . . . a street name substituted for police and street lamps."[10] But when names are separated from the life of the city, from the practices of the people inhabiting it, then street names are no longer meaningful for cultural history: "Yet most of these street names, that the prefecture gives or consecrates every day, have no impact, no sympathy among the people, which adopts them with indifference and respects them by habit."[11] In the nineteenth century, the signifier and signified no longer correspond: "Go now then, on good faith in the names, to find shelter on *Gracieuse* [Gracious] Street…, to pick cherries on *Cerisaie* [Cherry Orchard] Street, to view the hour on *Cadran* [Dial] Street."[12] In contrast to the intellectual history written into modern street names, medieval street names could be read as cultural history because their origin was in the collective human practices of the city streets: "The moral and physical history of Paris is tied to that of its streets."[13] Streets of shame gave a most privileged access to mapping a cultural archaeology of the Middle Ages.

For Lacroix, sex was constitutive of the urban fabric, it was inscribed in street names, and therefore there was no history without a pornographic archaeology of the name. "Pornographic archaeology" also meant an archaeology of sexual practices that would paint a comprehensive picture of the medieval society through the lens of prostitution. Influenced by the context of policing and medicalization of prostitution that dominated the nineteenth century, Lacroix cast a wide sociological net in his *Histoire*: "This word *Prostitution* must be traced back to its etymology (*Prostitum*) and therefore understood as all types of obscene trafficking of the human body."[14] "Pornographic archaeology" was an archaeology of the sexuality of medieval society, anchored in the history of customs and morals [*histoire des mœurs*]. Lacroix's history of prostitution thus developed into a broadly conceived history of sexual practices of the Middle Ages: beyond laws and the policing of prostitution, he wrote on fashion, the role of clothes and shoes in the licentiousness of sexual mores, including the ecclesiastical fashions, sodomy, bestiality, urban prostitution and the related question of urban development, sorcery and in particular the Sabbath, heresies and their relation to sexual behavior, the history of venereal disease (leprosy and syphilis), the culture (*mœurs*) of the French court throughout history, popular festivals such as the feast of fools (*la fête des fous*) during Saturnalia, theater and prostitution, literary history, and so on.

Figure 5. Paul Lacroix jeune (Bibliophile Jacob), from the *Journal des femmes* (1843), Bibliothèque nationale de France. "Mr. P.-L. Jacob is one of these half-Benedictine elderly figures that spend their lives gaining instruction and are in a position to do something when they die. . . . He's a tall gentleman, pale, dry, very wrinkled, half-parchemin, half-sheepskin" (M. P.-L. Jacob est un de ces vieillards à moitié bénédictins, qui passent leur vie à s'instruire et sont en état de faire quelque chose quand ils meurent. . . . C'est un grand monsieur, pâle, sec, très ridé, moitié parchemin, moitié basane) (Balzac, p. 654). Balzac found Lacroix to be "the summum of mediocrity" (le point culminant de médiocrité), and this portrait of him in 1830, when Lacroix was only twenty-four, is telling.

Paul Lacroix (1806–1884), also known as P.-L. Jacob, bibliophile, or just Bibliophile Jacob, was a superbly prolific writer (*polygraphe*), bibliographer, and early historian of culture (*historien des mœurs*). The catalogue of the Bibliothèque nationale de France contains no fewer than 1,325 entries, of which 781 were under his authorship or editorship (a number of them doubled in different editions, which equals about 500 works).[15] Lacroix belonged to the Romantic generation, known for its writerly productivity and social engagement. In the early years of the July Monarchy he held a salon at his residence in the rue d'Enfer with regulars such as Victor Hugo, Sainte-Beuve, Alexandre Dumas, Frédéric Soulié, and Gérard de Nerval.[16] He had a complicated relationship with Honoré de Balzac[17] and with historian Henri Martin (Figure 5).[18] Lacroix, Hugo, Sainte-Beuve, and Dumas were also regulars at Charles Nodier's salon at the Bibliothèque de l'Arsenal—Nodier was its chief librarian from 1823 to his death in 1844—where they gathered alongside Lamartine, Musset, Balzac, Vigny, Deschamps, Jules Janin, Delacroix, and Liszt.[19] Lacroix knew Nodier well since between 1838 and 1840 they prepared together the catalogue of the private library of Guilbert de Pixerécourt. Lacroix's career achievement was his appointment in 1855 as manuscripts conservator at the Bibliothèque de l'Arsenal, where his personal archive is housed today.

Although a great deal of Lacroix's voluminous publications were dedicated to the Middle Ages, Lacroix, unlike some other early medievalists of the Romantic generation, such as Francisque Michel, Prosper Mérimée, Antoine Le Roux de Lincy, and Paulin Paris, has not been consecrated by the history of French medieval studies. For instance, the only mention of him in the most recent complete history of medievalism by Charles Ridoux, *Évolution des études médiévales en France de 1860 à 1914*, is as one of the first modern editors of François Villon in 1854.[20] The purpose of reading Paul Lacroix here is nevertheless not an effort to rehabilitate him or to present the contestable quality and tone of his publications as a loss to medieval studies.[21] Rather, I wish to discuss the posterity of his works within the larger framework of the ambivalent relationship of Romance philology and medieval studies to sexuality. Before 1860, there was a proliferation of discourses on medieval sexuality, especially in the domain of the *histoire des mœurs*, to which Lacroix's work belongs, and in particular in the study of the medieval literary genre of the fabliaux. The branch of history of the *histoire des mœurs* relied a great deal on archaeology for findings on the past practices of daily life, and sexuality, primarily in classical archaeology and philology, was one of its focal points. But by the 1890s, the new state-sanctioned profession of academic philologists and medievalists was promoting "courtly love" as the only possible expression of sexuality in the Middle Ages. This concept became dominant by the time André Gide wrote in 1934: "Mystic love: we only hear of that love, it is the only one brought to the fore. . . . This problem should be studied in secret chronicles, in criminal trials, in monastic documents, without taking literature into account."[22] How and why did the academic discipline of medieval studies successfully neglect and diverge from its own sexual "pre-history" of "pornographic archaeology"? Is it possible to advance the theory of nineteenth-century Victorian repression? The fathers of medieval studies would thus have been victims of their own cultural and historical moment, purifying love of sex and exalting this pure "courtly" or "mystic" love into a civilizational art form. I contend, however, that the national stakes were much larger and inscribed in the context of the French civilizing mission at home and abroad. This archaeology of sexuality was precisely at cross-purposes with the incipient Romance philology and the emerging French national literary history. Put simply, pornographic archaeology was incompatible with the identity of the French nation-state and its imperial civilizing mission, past and present. This chapter retraces first the kinship of archaeology and the *histoire des mœurs*, then analyzes the separation of Romance philology from the discipline of archaeology, and finally looks at the medievalists' philological exclusion of archaeology and the *histoire des mœurs*, embodied in the rejection of the fabliaux in favor of the national French epic and chivalric literature of courtly love.

Histoire des mœurs and Archaeology

Histoire de la prostitution was part of Lacroix's historical project that he himself characterized as a "history of customs and morals" (*histoire des mœurs*): "The history of Prostitution . . . this subject, delicate and suspicious all at once, ties on all sides to the history of religions, laws, and customs."[23] *Grand dictionnaire universel du XIXe siècle* likewise placed Lacroix's work in the domain of the *histoire des mœurs*, "beautiful studies of customs that he had long ago begun on French society."[24] *The Encyclopaedia Britannica* in its seminal eleventh edition in 1911 characterized Lacroix's works as "monographs on phases of the history of culture" and called his 1847–1852 edition of *Le Moyen Âge et la Renaissance* "a standard work on the manners, customs, and dress of those times."[25] *Histoire des mœurs*, understood here as a history of culture, was an early and somewhat distant precursor of the history of civilization (German *Kulturgeschichte*), which itself evolved into the history of mentalities (*histoire des mentalités*) of the Annales School and then into cultural history. While history of civilization had its beginnings in Germany in the 1860s with Jacob Burckhardt, *histoire des mœurs* can be traced back in France to Voltaire's 1756 *Essai sur les mœurs et l'esprit des nations.*[26] Voltaire's *Essai* had an immediate conceptual and philosophical effect on the writing of history, even if the written product did not always match the stated project, as Voltaire was still attached to events and political history. Nevertheless, in his preface, Voltaire replaced "the disgust that modern History causes you," in which aspiring scholars learn "in what year a prince unworthy of being known succeeded after a barbarian prince in a crude nation," by "what is worth being known by you; the spirit, the morals, the customs of chief nations."[27] First, instead of a political history of events, Voltaire offered up an *histoire des mœurs* defined by notions of foreignness, importation, and imitation. From the outset, he situated the West in relation to the East, "the Orient, cradle of all the arts, and which gave everything to the West."[28] No Western civilization, and no record of it, would exist had it not been for its Eastern victors who brought culture and writing to them; Gauls, Germans, Spaniards, Celts of Brittany, and Sarmatians were plunged into such deep darkness that it took eighteen centuries "before that we might learn enough to know that we were barbarian."[29] Second, the Middle Ages occupied one half of the *Essai* in the effort both to illustrate their barbarism and uncanniness and to domesticate it. Despite the fact that "one half was called the language from *Oui* or *Oil*, and the other the language from *Oc*," Voltaire nevertheless considered the heterogenous nation of France to be "a country."[30] Medieval France was "an unhappy country lacking set laws, troubled by civil wars, without commerce, without police, without written customs, and governed by a thousand different customs," where "the most ridiculous and barbaric practices were then established," such as *le droit de cuissage.*[31] Still, in comparison with other Western nations of Germany, Spain, and southern Lombardy, French nobility could at least boast "a more polished façade than the other nations."[32]

Voltaire's refutation of political and institutional history in favor of an *histoire des mœurs* set the tone, and between 1756 and 1860, and especially between 1840 and 1860, there was a proliferation of books that can be roughly divided into two categories: one interested in the foreign and the other in the national customs and morals. The first trend was an *histoire des mœurs* linked to travel narratives.[33] The second trend was synonymous with an effort at reappropriation of the French historical self. These national histories were structured along two axes, either tracing the history of the French civilization from the Roman Gaul to the present (often stopping at the Revolution), or integrating France into a European (hence Western) historical perspective beginning at the dawn of Christianity.[34] This Westernized *histoire des mœurs* by definition included the Middle Ages, and works that were devoted almost entirely if not exclusively to the Middle Ages started to appear, such as the very early La Curne de Sainte-Palaye's *Mémoires de l'ancienne chevalerie* (1759) or Francisque Michel's *Mœurs du Moyen-Âge* (1832). Paul Lacroix, with Ferdinand Seré, edited between 1847 and 1852 a five-volume *Le Moyen Âge et la Renaissance: Histoire et description des mœurs et usages, du commerce et de l'industrie, des sciences, des arts, des littératures et des beaux-arts en Europe*, divided into four sections: Culture (*Mœurs*), Sciences, Letters, and Arts.[35] In the late 1860s, Lacroix undertook alone the rewriting and reediting of the original work, now divided into four volumes: *Les Arts au Moyen Âge et à l'époque de la Renaissance* (1869), *Mœurs, usages et costumes au Moyen Âge et à l'époque de la Renaissance* (1871), *Vie militaire et religieuse au Moyen Âge et à l'époque de la Renaissance* (1873), and *Sciences et lettres au Moyen Âge et à l'époque de la Renaissance* (1877). The fragmentation into an episodic history of culture can be felt in the studies of other medievalists, such as A. Lecoy de la Marche and L. Petit de Julleville, who continued the research of the early generation of Romantic medievalists, represented in Francisque Michel and Achille Jubinal, and who produced titles such as *Recherches sur le commerce, la fabrication et l'usage des étoffes de soie, d'or et d'argent et autres tissus précieux en Occident* (Research on the commerce, making and uses of silk, gold and silver material, and other precious fabrics in the West), or *Histoire des hôtelleries, cabarets, hôtels garnis, restaurants et cafés* (History of hotels, cabarets, bed and breakfasts, restaurants and cafés), *La société au XIIIe siècle* (Society in the thirteenth century), *La Comédie et les mœurs en France au Moyen Âge* (Comedy and customs in Medieval France), or *L'esprit de nos aïeux* (The spirit of our ancestors).[36] The effervescence of the *histoire des mœurs* died down in appearance around the 1860s. However, the interest did not wane but was modified by the history of civilization. Only the name of "histoire des mœurs" disappeared because the term "mœurs" was no longer thought of as an object of historical research but primarily as a contemporary issue of morality. The term "mœurs" was thus frequently used in religious discourse, medical discourse (as in "offense against public decency" [*attentats aux mœurs*]), in the discourse of the police (as in "vice squad" [*la police des mœurs*]), or sociological discourse.[37]

Already in 1842, Honoré de Balzac pointed to the predicament of the *histoire des mœurs* as the history forgotten by historians: "In compiling the inventory of the vices and virtues, in bringing together the principal outcomes of the passions, in painting characters, in choosing the main events of Society, in forming types from a combination of the traits from several homogeneous characters, perhaps I could manage to write the history forgotten by so many historians, the history of customs and morals."[38] But Balzac's fictional rendering of customs and morals in *La comédie humaine* also underscored the complicated status of the *histoire des mœurs* and the validity of its methodology. What Balzac may have meant was that "real" historians, those dealing with political and institutional history, were not interested in a history of national traditions. There was but one. In the 1869 "Préface" to his *Histoire de France,* Jules Michelet described it as an "integral resurrection of life" (*résurrection de la vie intégrale*).[39] The comprehensiveness of his history depended on adding everything that traditional histories of political events and institutions lacked completely: materiality (soil, climate, food, physical and physiological circumstances) and culture (ideas, customs, progressive, interior movement of the national soul) (Michelet, 1:viii). Throughout the nineteenth century, Michelet's work retained a unique status of an initiator without disciples, whose "whims, . . . fantasy, . . . unruly imagination wandering on an adventure through nature and history, vigorously seizing onto such or such a thing in passing and as though at random, without establishing any guidelines or proposing any goal," were often criticized.[40]

The *histoire des mœurs* faced an important obstacle when compared to history proper. Lacroix's *Le Moyen Âge et la Renaissance,* with its division into "arts," "traditions, customs, and dress," "military and religious life," and "sciences and letters," and numerous other histories of "luxury" [luxe], "fashion" [mode], and "footwear" [chaussure], is a perfect example of the tendency to produce "a history in compartments, the 'civilization' or 'culture' or 'arts and sciences' following after, and often at the end, the 'political,' 'economic,' 'social' compartments."[41] The *Grand dictionnaire universel* criticized Lacroix's execution because "political and religious history is rejected in an insufficient introduction. . . . We don't have a full picture of any century. We have indeed the history of each item in particular; but it is an abstract history robbed of its relationships which, logically uniting all of the diverse human manifestations of an age, come together to specify the physiognomy of that same age."[42] The stated objective of the work, "as complete a whole as possible," resulted in its opposite, "a splintering which prevents the reader from developing an idea of the whole, and . . . introduces a great confusion in the general impression."[43] The failure to incorporate the smaller heterogeneous cultural elements into a coherent political-religious historical narrative, and ostensibly to ground culture in the national political, religious, and institutional history, was the major flaw of the *histoire des mœurs*. For how could one understand the history of the Middle Ages only through individual histories of, for instance, "Tapisseries" (Tapestries), "Horlogerie" (Clockwork), "Cartes à jouer" (Playing Cards), "Vie privée dans les châteaux, les villes et les campagnes" (Private Life in Castles, Cities and Countrysides), "Nourriture et cuisine" (Food and Cooking), "Chasse" (Hunting), "Pénalité" (Penal Law), or "Funérailles et sépultures" (Funerals and sepulchers)? Histories of culture appeared defective because they were written outside of the master narrative of national political and institutional history. Therein lay the disciplinary problem for the cultural historian of the nineteenth century: the *histoire des mœurs* was simply not on the same footing as history. The article in the *Grand dictionnaire universel* was less a qualification of Lacroix's work and more a criticism of the inability inherent to the *histoire des mœurs* to write also political and institutional history as a part of the total history of the kind Jules Michelet had imagined and practiced. Anticipating perhaps such criticism, Émile de la Bédollière, for whose history Paul Lacroix prepared the bibliography, gave a subtitle to his three-volume *Histoire des mœurs et de la vie privée* of "a supplement to all the Histories of France."[44] Once political history established the course of events, "the genealogy of kings, the revolutions of peoples, the changing of dynasties," the remaining question of interest was: "How did our ancestors live?"[45] Only an *histoire des mœurs* could give the answer to how we became who we are and explain the origin of our present identities—"Where is the source of so many living traditions that we have accepted?"—but such an answer could be given only as a complement to the political-institutional history.[46]

The narrative of the *histoire des mœurs* was a more circumscribed view of the history of culture that relied primarily on literary monuments (i.e., documents) rather than physical monuments, but it partook of the approach to civilization practiced by archaeology.[47] Like the *histoire des mœurs*, archaeology was a disparate but complementary discipline to history. Archaeology was defined as a "science that is concerned with all that relates to the Ancients, especially their arts and monuments."[48] This science had several "objects and categories": "literary archaeology," "archaeology of art," and "archaeology of customs and utensils." In this order, archaeology collected and decoded documents, monuments, and customs and tools, but archaeology's ultimate importance lay in the reconstruction of daily life from the artifacts collected: "While historians tell us the general facts regarding politics, war, revolutions, pay heed to remarkable events that broadly outline the profile of peoples—in a word, the dramatic and external life of societies—*archaeology* makes us penetrate the inside, into the depths of that life; it teaches us the mores, traditions, customs, and beliefs of the ancients, makes them visible to us by carrying our imagination to the midst of antiquity."[49] When archaeology digs "into the depths of that life," it produces a history of a culture: it reveals the internal workings of a society. Archaeology infers human behaviors from material remains. And because it studies this human behavior and internal life, it opposes itself to history in the strict sense of the term. Political history is a series of great events, steered by the cause and effect principle and performed by great actors. On the historical stage the "people" play a minor part, though they play a major part in the "practice of everyday life" of a society, *le quotidien*, made visible by archaeology.[50] The opposition of events versus *le quotidien* sums up the opposition of history to archaeology. Both archaeology and the *histoire des mœurs* thus stood in opposition to history proper: based on daily practices, they provided narratives of the inner life and character of the people.

Romantic Archaeology of the Middle Ages

Lacroix's pornographic archaeology of the Middle Ages partook of the same Romantic enthusiasm as Alexandre Lenoir's *Musée des monuments français* that opened in 1795: they both followed the same desire to display and describe artifacts as facts and sources of historical truth, not merely as aesthetic objects that one likes or dislikes.[51] The eighteenth century had developed systematic archaeological excavations of Roman classical sites, such as Pompeii, and, as a consequence, in the Enlightenment "the idea of antiques had already evolved, and their study began to be approached with a new perspective that was, in a way, that of 'culture.'"[52] In the aftermath of the Revolution, this notion of culture superseded the contingencies of the present; patrimony was slowly extricated from its immediate religious, monarchic, and aristocratic context: "The French nineteenth century will be able to be deemed the *century of history*"[53] because "we began to paint the past, whereas beforehand we had only judged it from the viewpoint of contemporary opinion."[54] What was previously classified as "antiques" (*antiquités*) or "curiosities" (*curiosités*) now provided prime material access to the history of the nation. Classical archaeology of the eighteenth century—known as antiquarianism—had been interested in the common history of European nations and humanity. The Romantic interest, on the other hand, lay in "primitive" societies; it therefore favored the local and the culturally specific. In exploring the past of the specific people, Romanticism was linked to nationalism.[55] Displacing antiquity as a privileged archaeological site, the Middle Ages became for the Romantic age the matrix of national consciousness and a historical way to validate its identity. Medieval history was the grounds on which to construct the present of the nation: "The Middle Ages are therefore, in this research, a special field of study. Then is the time when the matrix of French civilization was constituted through successive ties, when the systems and instruments of government were forged; then, too, in this long period, when the middle classes, the bourgeoisie, the people were formed and emerged, these forces which are today the substance of the nation."[56] Archaeology was inextricably bound to the nation's present needs for history, but also to the French imperial enterprise, in the aftermath of the Napoleonic military expedition to Egypt (1798–1801) and the colonization of Algeria (1830–1962) and, much later, of Indochina (1859–1954) (see Chapter 3). The nation was being constructed in relation to the Orient and its past; both archaeology at home and archaeology abroad were invested in the making of the national identity.[57] The universalist classical archaeology that explored the common Greco-Roman origins was sidelined in the rivalry of specific claims of imperial nation-states that each argued for their own piece of the shared classical past. The realignment of national origins with the medieval period, during which the French had colonial contact with the Orient as leaders of the Crusades and also carried out their first civilizing mission (*mission civilisatrice*), gave a parallel chronology to medievalism and Orientalism. The medieval and the Oriental past submerged the universalizing classical past common to all European nations—the uniqueness of the French nation-state could be affirmed in its relation to its medieval metropolitan and colonial past without recourse to antiquity. The cultural politics of the past could not be separated from the historical-cultural present of the metropole and the colony.

Romantic archaeology of the Middle Ages was consecrated under the July Monarchy, when the restoration of national monuments became an institutional and state priority, no longer the effort of individuals, chiefly expert amateurs (*savants-amateurs*) and antiquarians. A number of different commissions for the restoration, conservation, and cataloging of French monuments were set up. In 1830, historian and newly appointed Minister of the Interior François Guizot (1787–1874) successfully advocated for the creation of the position of the General Inspector of Historical Monuments (Inspecteur général des Monuments historiques), occupied by Ludovic Vitet. In October 1832, Guizot became the Minister of Public Instruction and the portfolios of the Collège de France, the École des chartes, and the libraries (in particular the Bibliothèque royale and the Institut de France) were granted to him. He subsequently participated in the foundation of the Société de l'histoire de France in 1833, and the Comité historique des Arts et des Monuments and Comité des Travaux historiques et scientifiques were created in 1834. In 1837, the Commission des Monuments historiques was charged with compiling the inventory of buildings to be protected and restored by the state. A number of societies in the provinces were founded at that time as well.[58] In 1834, Prosper Mérimée took over Ludovic Vitet's job as General Inspector of Historical Monuments, which he held until 1860. Many of Paul Lacroix's friends, including Victor Hugo on the Historical Committee of Arts and Monuments and Guglielmo Libri as the secretary to the Commission des bibliothèques, became involved. Lacroix himself served until 1851 on "several historical committees of the minister of public instruction,"[59] such as the Commission des monuments historiques and the Comité des documents inédits de l'Histoire de France.

With the creation of the Société de l'histoire de France in 1833, dedicated to the publication "of original documents relative to our national history," papers, in addition to monuments, were now officially designated as the object of archaeology, sponsored by the state and its efforts to "popularize the study of and taste for our national history in a healthy critical perspective, and above all through research and the use of original documents."[60] Guizot's project was to "find and publish all documents able to contribute to knowledge of the national history," working especially on "the search for and . . . publication of unknown documents in literature, philosophy, the sciences and the arts assessed according to their relationships to the general history of France."[61] For that purpose, Guizot in particular solicited the help of the *archivistes-paléographes* trained at the École des chartes, founded in 1821.[62] This led to efforts to integrate medieval literature into the national history and literary canon: "There has been much talk . . . of national greatness. Well, nothing contributes more to this greatness than the monuments that are erected, and, among these monuments, literary undertakings occupy the highest rank."[63]

La Revue des deux mondes, a major outlet for literary and scholarly activity, started regularly publishing articles on the literature of the Middle Ages in the 1830s. Coinciding with the colonization of Algeria, these articles portrayed medieval language and literature as forerunners of the French civilizing mission. Claude Fauriel (1772–1844), professor of foreign literature at the Sorbonne, as well as Jean-Jacques Ampère (1800–1864), his replacement at the Sorbonne and later a professor at the Collège de France, both published in the journal extensively, a sign of the interest and accessibility of medieval literature for the Romantics. In 1832, Fauriel published his Sorbonne lectures on the origins of the "chivalric epic" that he attributed to medieval France: "A fact about this genre, which is neither contestable nor contested, is that, among all the literatures from the Middle Ages, French literature (within which I include the Anglo-Norman) is by far the richest in chivalric epics. It is also certain, also recognized, that the majority of these epics were translated or imitated from French into the other languages of Europe."[64] For Fauriel, the chivalric epic genre consisted both of "the novels of Charlemagne and of the Round Table" (*les romans de Charlemagne et de la Table Ronde*), that is, the Carolingian *chanson de geste* and Arthurian romances. In a 1833 article "De la littérature française dans ses rapports avec les littératures étrangères au Moyen Âge," Ampère propagated Fauriel's theory by arguing that "chivalric genius" (*génie chevaleresque*) extended from France to other northern countries.[65] In 1839, Ampère divided French medieval literature by "chivalric inspiration" (*inspiration chevaleresque*), "religious inspiration" (*inspiration religieuse*), "independence of thought" (*l'indépendance de la pensée*), and "satirical power" (*puissance satirique*), all four intertwined within the matrix of the greatness that French civilization would confirm in the following centuries: "I desired to show the already thriving tendencies that will form, through their combinations and struggles, in very large part, the complex life of modern centuries. In returning to these more known, or at least more studied centuries . . . we will perhaps better understand the development of the modern spirit, after having come across its seed in the vigorous womb of the Middle Ages."[66] Thirty years later, Larousse's *Grand dictionnaire universel du XIXe siècle* called for the making of "the genealogy of our national idiom."[67] Following the theory of three linguistic "races," Larousse promoted the idea that the French language was already French in the Middle Ages, even if it had been a somewhat impure mixture of three heterogeneous components, Latin, Celtic, and German: "From this fermenting mixture will surface a new form, imperfect and crude, but at least national."[68] From the thirteenth century onward, the medieval writers shaped the language with "simplicity, clarity, and concision."[69] French literature spread the language across the European continent: "These works . . . had . . . earlier played a great role and exerted a wide influence. They didn't stay encircled by narrow limits: all of Europe was their vast theater. . . . And it is there . . . that begins the European use of the French language."[70] From Ampère to Larousse, there was agreement that the literature of the medieval period was "the celebrated agent of world civilization," because it embodied chivalric ideals of honor, rules of military engagement, and respect for women and "our language has not exerted any lesser influence in Europe than our customs, our ideas, and our arms."[71] Under the French influence, Europe turned chivalric. Since the Middle Ages, France had been fulfilling its civilizing mission: "France is in and of herself an entire civilization. Her elected mission is to precede others and guide them toward new destinies."[72] Its "civilizing force"[73] had proven the supremacy of French culture well before the classical age, the age of Louis XIV, *le grand siècle*: "French civilization has thrice been placed at the head of Europe: in the Middle Ages, by our literature, our Crusades and our chivalry; in the seventeenth century, by the genius of our writers and the reign of Louis XIV; in the eighteenth century, by the ascendancy of our philosophy and the triumphs of our glorious revolution."[74] Romantic medievalism thus privileged a holistic approach to the Middle Ages, where language, literature, and civilizing mission were all a part of the medieval French spirit (*le génie national*): "It is not art that must be sought in these great compositions, but precious details on history, on the geography of traditions."[75] The French Middle Ages has been and will continue to be the fertile matrix of the European civilization: "Let us keep step, gentlemen, and for the fourth time, let us reclaim our standing at the head of movement in Europe. It is watching us and waiting."[76]

Thus, medievalism and archaeology were synonymous in the Romantic period. But the goal of popularizing medieval literature resulted in Romantic medievalism being prone to auto-didacticism, subjective criticism, and bibliophilia. The use of medieval literature as one of the sources for the history of national customs and morals, rather than a self-standing literary object, also meant that the Romantic study of medieval literature never rose to the level of literary history. These characteristics of Romantic medievalism were objects of harsh criticism in the last quarter of the nineteenth century by the professional Romance philologists and medievalists. Figures such as Paul Lacroix, Francisque Michel, Prosper Mérimée, Antoine Le Roux de Lincy, Achille Jubinal, and Paulin Paris were medievalists *and* bibliophiles *and* popularizers. Romantics were committed to the popularization of medieval literature—"Great successes belong less to intellectuals than to popularizers"[77]—but this often translated into unscientific methods of text editing: "He followed a less critical than literary path. He wanted to popularize the romances of the Middle Ages."[78] In his edition of the *Romans de la Table Ronde*, published in five volumes between 1868 and 1877, Paulin Paris (1800–1881) scandalized because he mixed translation, adaptation, and commentary without distinction

(Ridoux, p. 36), putting them "'into new language":[79] "Imagination, which is the supreme virtue of the novelist or poet, is perilous for the scholar."[80] Like Paulin Paris, in his preface to *Les vieux conteurs français* (1841), Lacroix rejected the methods of the *chartistes*, who form "a small number of initiated for whom the *faithful* reproduction of ancient texts means serfdom of spelling; we have applied modern spelling to these tales from the fifteenth and sixteenth centuries."[81] Discovery and presentation of medieval manuscripts were guided by a form of subjective criticism that resembled mystical ecstasy: "He contemplated [these ancient volumes] with a surprised rapture."[82] Instead of judicious censure and scholarly restraint, Romantics would "print everything."[83] Along with "the lack of an established method or a philosophical link," auto-didacticism was perhaps the most embarrassing: "By this gap in his initial education, he was held back from being as good a philologist as a literary enthusiast, as he recognized himself."[84] As David Hult has remarked, Paulin Paris's achievements embodied "the unwelcome dilettantism of a previous generation, one that had gained access to the texts and their understanding without any formal instruction."[85] Bibliophilia was the final complaint against Romantic medievalists.[86] Paulin Paris was editor-in-chief of the *Bulletin du bibliophile* from 1836 to 1846, and as conservator of the Bibliothèque du roi "he didn't really distinguish between his own books and those in his keeping: a behavior that he had to forbid severely to the public, that he should perhaps also have refused himself, as a manuscript conservator."[87] But this personal relationship to manuscripts was to change with the Libri affair.

The Libri affair signaled the decisive break between Romantic archaeology and Romance philology. Guglielmo (Guillaume) Libri (1803–1869) was a historian of science and a passionate bibliophile, a naturalized French citizen of Italian birth, accused in 1848 of manuscript and rare book theft.[88] Following the overthrow of King Louis Philip that sent him and his prime minister François Guizot into exile in England on February 28, 1848, Libri, a close ally of Guizot, was accused first of having stolen prodigious quantities of manuscripts and rare books from the French public libraries while working as the inspector of provincial libraries between 1841 and 1846, and then of having put his library of 3,025 items on sale in 1847, from which he sold 1,923 items to the English Lord Ashburnham, thus allowing national heritage to fall into foreign hands.[89] The battle of Libri's accusers and defenders, which lasted from 1848 to 1888, was generational. It pitched bibliophiles against philologists, Romantic auto-didacts against the new positivist generation of the *chartistes*. On one side, along with Paul Lacroix, were Paulin Paris, Prosper Mérimée (accused in 1835 of stealing a manuscript by the *chartistes*),[90] the bibliographer Gustave Brunet, and Achille Jubinal. Less visibly, Sainte-Beuve, Victor Hugo, and Victor Cousin also supported his cause. Even Silvestre de Sacy and Jacques-Joseph Champollion-Figeac, Champollion's older brother and conservator of manuscripts at the Bibliothèque royale (also accused of pilphering some manuscripts), joined the ranks of Libri's defenders.[91] On the other side were the former students of the École des chartes: "They [Lalanne, Bordier, Bourquelot] were . . . convinced of Libri's guilt and regarded him not only as the enemy of the École des chartes, but of the French libraries and even of French culture as such."[92] They were supported in the background by Paul Meyer (appointed the school's secretary in 1872).[93] What this affair confirmed beyond any doubt was the mismanagement of French public libraries and the negligence of their staffs.[94] In essence, it was a fight for the restoration of national patrimony and reorganization of the libraries that were supposed to protect it. The nation needed to be straightened out after the turmoil of the post-revolutionary nineteenth century, and its literary patrimony needed to become not only a part of national history but also a self-standing object of literary analysis. Medieval literature would no longer be the province of amateur excavating bibliophiles who could not tell the difference between private and public good, and who saw in medieval literature an archaeological object of conservation by dubious methods. In the eyes of Romance philologists, the Romantic generation of medievalists *and* bibliophiles *and* popularizers had dispersed the national patrimony and, worse yet, in a certain way, prostituted the French patrimony to foreigners. The constitution of the philological discipline of medieval studies thus played out between the Romantic generation and the generation of the Third Republic, between Romantic medievalists and academic medievalists of the École des chartes, between archaeology and philology.

Archaeology and Philology

As seen above, both documents and monuments were the objects of archaeology. Linguistics and philology, "the study of languages and of ancient literatures," were essential to the study of documents as sister disciplines to archaeology:

> It is fair to say that *archaeology*, *linguistics,* and *philology* mutually support each other, and that archaeological, linguistic, and philological

> studies cannot separate from one another. Without detailed knowledge of ancient languages, literary archaeology (paleography, epigraphy) is impossible; on the other hand, the veritable archaeologist cannot forgo the support of ancient authors to explain monuments and, in return, the monuments enlighten a great number of difficulties in ancient authors' texts, [which are] insoluble without them.[95]

Philology here supports the history of culture because it serves as an "archaeology of language."[96] Literature, the other type of document, "owes no less than art to *archaeology*: without archaeological knowledge, there would be no local color in the works whose subjects are derived from ancient events."[97] Yet under the influence of positivist philosophy, by the mid-nineteenth century philology had become a master science, no longer assisting archaeology but replacing it. It performed the same kind of reconstruction of life in the past, no longer from monuments but from documents. Reassembly of lost words and lost works, rather than art and architecture, promised to bring history back to life.

The Romance philologists, led by Gaston Paris (1839–1903), today widely considered the founder of modern French academic medieval studies, rejected archaeology practiced by his own father, Paulin Paris, and declared that the findings of Romantic medievalists were just "bare bones" (*ossements*).[98] Gaston Paris's statements illustrate the effort to separate Romance philology from Romantic archaeology. The new philological project could be achieved only through a redefinition of the archaeological relationship between literature, history, and culture:

> We understand today a bit differently the study of the Middle Ages. We devote ourselves less to appreciating them and to making others appreciate them, than to knowing them and understanding them. What we seek above all is history. Of course the benefit that history finds in the study of literary works from the past also concerned my father: he published several texts of primarily historical interest, and he insisted on many occasions on the precious history of customs [*histoire des mœurs*], institutions, general civilization.[99]

In Gaston's definition, the study of literature as cultural history, "history of customs, institutions, general civilization," belongs to the aesthetic category of "appreciation," and not science. The cultural history of customs and morals is a history that never fleshes out bare bones. Instead, the new—positivist and philological—history pushes the limits of the definitions of history: "We look at poetic works themselves as being above all historical documents, as being part of history in its largest sense, as being made from the same history of the language, feelings, and thought."[100] Although history is defined "in its largest sense," history seems opposed to culture. It is language itself that is now viewed as a historical artifact and, moreover, as a mirror of sentiments and thought: "Literary works, and especially poetic works, are the direct product of souls and they reveal them to us, using the transparent signs of words, far more clearly than do the monuments of art, more clearly even and more completely than acts truly accomplished, of which we are moreover acquainted with such a small number."[101] Since language presumably mirrors its culture, this new historical approach precludes the need for further study of culture. The assumed transparency of language made the texts, extracted from their context, seem "interpretable in terms of an avowal or a reference to external nature, transcendent to the words whose supposed function was to manifest it."[102] Language was a reflection of "souls" of the people, that is, of its customs and morals: "The history of the literature of a people . . . is the history of its moral life, and particularly of its national consciousness."[103] Because "these are the original eras when nations spontaneously developed, without the trouble that the arbitrary intervention of personal will or external causes of pressure all too often provoke," philology insists on studying a pure state of language and text, outside of its cultural context.[104] While classical and Romantic archaeology was not independent from linguistics and philology (Larousse, "Archéologie," p. 571), Romance philology will read the medieval literary text as transparent, emancipating the document from the monument. And it will also exclude an archaeology of the Middle Ages: its history of culture.

Gaston Paris's Romance philology indeed was not an archaeology and it was a far cry from the classical philology of Pompeii, for example, which engaged with various types of documents. In this respect, Romance philologists followed the general evolution of comparative philology that had been inextricable from archaeology and a universal "science of man" (*science de l'homme*) to a philology that was a linguistic science of a national language.[105] Romance philologists imposed a very narrow concept of philology; since literary text was a transparent historical document, philology could forgo the cultural context of the document. The philology practiced by Gaston Paris and the *chartistes* was derived from Karl Lachmann's method of text editing, introduced in France through the intermediary of Gaston Paris.[106] With the arrival of philology, manuscripts were not only catalogued, the primary occupation of

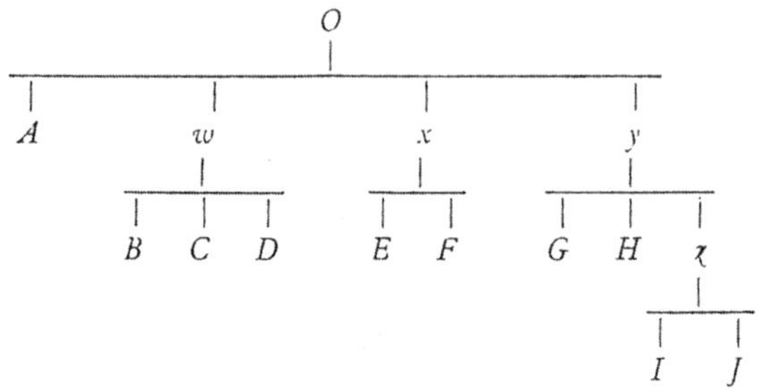

Figure 6. *Stemma codicum* (Lachmann method).

the previous Romantic period, but also organized in the order of their descent from the original text. This method proceeded by way of reconstructing the lineage of literary genres, and the filiation of manuscripts was supposed to lead back to the original, known as the MS O. The MS O was the reconstructed amalgam of extant manuscript copies that were less perfect than the lost original manuscript. The ultimate goal was to reconstruct and reproduce, from the union of the less perfect copies, the ideal original "archetype *Urtext*" (*Urtext archétypique*).[107] The philologists' diagrams resembled a genealogical tree: "It is the *stemma codicum*, the genealogical tree of manuscripts"[108] (Figure 6). The genealogical tree was intended to establish which of the extant manuscripts is closest to the lost original; these manuscripts were subsequently privileged in the reconstruction. In fact, manuscripts were understood to belong to an imaginary "family of manuscripts."

In constructing the genealogy of manuscripts, Romance philology also reflected the dominance of discourse on heredity in France. It interiorized the idea of degeneracy, the transmission of hereditary defects (transcription errors), and the notions of (original) purity. Philology could be described as a medicine of pathologies of manuscript families: "Philology is a bourgeois, paternalist, and hygienist system of thought about the family; it cherishes filiation, persecutes adultery, and fears contamination" (Cerquiglini, p. 49 [p. 76]). In drawing genealogical trees, the relatedness of manuscripts is traced through their shared anomalies, their heredity—it is the defect that allows the establishment of filiation: "The 'shared mistake' [*la 'faute commune'*] could only be an inherited defect, and consequently it even represented a form of filiation [of manuscripts]" (Cerquiglini, p. 49 [p. 77]). The deviance from the virtual original was corrected by the philologist—the "doctor"—composing the *Urtext*. Any autonomy, seen as a defect or error on the part of the scribe copying the manuscript, was undercut in favor of a healthy and pure origin. In fact, the autonomy of any given text had to be undermined if there was to be a genealogy: "This philology could not, without collapsing, think that a scribe, when confronted with an uncertain reading, for example, might be able to improve it or indeed even rediscover the 'original' reading. Otherwise, from that point on, the linear filiation, demonstrated by constant deterioration, would become unclear" (Cerquiglini, p. 49 [p. 76]). Deviance, perversion, or morbid pathology of manuscript lineage had to be corrected by the philologist himself. No transgressions in the ties of filiation interfering with the continuity of lineage were allowed.[109]

The genealogy of manuscripts was nevertheless a continuation of the earlier Romantic effort by Romance philologists to construct a genealogy of French literature that would include the Middle Ages: "The literature of the Middle Ages . . . is the first chapter of our family memoirs,"[110] and the study of literature was presented as a national concern: "Literature is the expression of national life."[111] Medieval literature was proof of the longevity and genealogy of the French race, since the French race was in essence the same from its origin to the nineteenth century: "The life of our forefathers from seven or eight centuries ago, who lived in our homeland, who handed their blood down to us, who spoke our language, and in whom we so often find our spirit and our heart. The French of the eleventh century were already true Frenchmen: they loved France as much as we do, and they cared for each other perhaps more than we do; their genius [*génie*] resembled ours, and we would understand them if only we were to give ourselves the trouble."[112] On the micro level, the philological-genealogical enterprise involves then the quest for the lost original; on the macro level, it implies the search for the origin of the nation. In short, the genealogy of the manuscript is the genealogy of the nation.[113] To find the original is to locate the origin of the nation, and to reconstitute the MS O is to purify the nation at the source. According to this mindset, the nation was not only identical to itself, but also pure, dating all the way back to its medieval beginnings—and if not absolutely pure, already in the Middle Ages France was undoubtedly ahead of any other European nation-state now competing with it for imperial supremacy.

Erotic Philology versus Chivalry: France's Civilizing Mission

Given the Romantic premise of the links between philology and archaeology, Paul Lacroix argued in his *Histoire de la prostitution* that "erotic philology" (*la philologie érotique*) was paramount to any archaeology. Such a philological study would be a major and innovative contribution to the history of language: "If erotic philology could start into the general history of Prostitution, we could dedicate to it several very new and very interesting chapters; for there exists no such curious work in which the origins of the language, or rather the jargon of places of ill-repute, might be studied in detail."[114] What made erotic philology necessary was the understanding of language as central to the study of medieval mores. In other words, erotic philology would study the deployment of erotic discourses in the Middle Ages. "Erotic language" (*la langue érotique*) was omnipresent, because it was the most innovative and the most flexible: "Erotic language in French, which already emerges very obviously in the thirteenth century, usually proceeds by pleonasms and redundancy, translates words for its purposes from foreign languages, or appropriates them as they are with their indigenous consonance, seeks out images with a figurative style, triumphs in the realm of equivocation, and ceaselessly averts the monotony of discourse by means of the most extraordinary philological combinations."[115] This erotic language covered whole areas of culture and civilization, including literature, to the point that no single text written in Old French was safe from its unstoppable contamination: from Guillaume IX of Aquitaine, the authors of *Partenopeu de Blois*, *Garin le Loherain*, *Ami et Amile*, the *Miracles de Nostre-Dame* to the *Romance of the Rose*, this "gospel of French gallantry," Rutebeuf, Christine de Pisan, and, finally, François Villon, "*Master of Arts* in the schools of slang and Prostitution."[116] "The prostitution of literary language" (*la prostitution du langage littéraire*) was a common and widespread phenomenon. Erotic language infected and spread through literature, especially in the poetry of the trouvères, the poets of northern France: "The trouvères, most of whom were commoners, conserved original sin in their works and applied to compositions full of verve, gaiety and mischief, the crude and vulgar language that they had learned in their parents' homes; they called each thing by its name and preferred to use the most popular expression, which was always the most picturesque."[117] Lacroix judges jongleurs and minstrels even more harshly because of their nomadic lifestyle. They lived outside of any social law, practiced all kinds of vice, and "they were the ones who, traveling the country, corrupted both its language and mores."[118] Taken to the city streets and country roads, this literature contaminated an entire society: "The trouvères of the thirteenth century . . . had been the poets of Prostitution."[119] The fabliaux represent the ultimate stage of this linguistic degradation, the popular farce dethroning the noble epic (Lacroix, *Histoire*, 4:300). When the literary "fashion" changed from public recitation to silent reading, in order to attract the listener, the trouvères and the jongleurs abandoned and "entirely forgot the *chansons de geste* and honest poetry" in favor of the fabliaux.[120]

In Lacroix's view, erotic philology was thus vital to understanding the linguistic evolution of the crude French vernacular from the fusion of three heterogeneous elements—Latin, Celtic, and Germanic—to its linguistic maturity in the thirteenth century: "The language had to become supple, amplified, enriched, in order to lend itself to these riotous and biting acts; it drew especially on the everyday and even trivial language of the commoners in the town and countryside; it became more lively, more alert, more subtle, more defined, and more robust."[121] Erotic language was the primary source for any philological study because it was the richest (Lacroix, *Histoire*, 5:93). Yet the most inventive part of the French language was also its most ostracized element: "The fabliau is the most perfect and most scandalous thing there is, alas! among the trouvères' and the jongleurs' wares in the northern French language. These are sometimes masterpieces . . . and not the least immoral."[122] The most perfect and the most scandalous, the entire dilemma of the philological study of the French language and its literature is contained in that expression.

Despite the repetitive moralizing of the history of French literature seen through the restrictive lens of sexual practices, Lacroix was able to point out what Romance philology would choose to neglect: erotic language was everywhere, and a study of the deployment of erotic discourses must be part and parcel of the study of literature of the Middle Ages. Moreover, an erotic philology could retrieve the origins of the French language because erotic language was universal in the era of nearly infinite linguistic dispersion that occurred at the dawn of vernacular speech: "It could be easily proven in a philological study on the jargon of Prostitution, that this jargon is contemporary to vernacular language and that it was created from a confused mixture of all of the idioms and dialects, as though it held the pretension of representing a universal language."[123] Not only was there "a slew of words that didn't bother dropping their national character and made their way into French while remaining Greek, Latin, Italian, German, or Spanish," but this was "a language of convention that was spoken and heard also in the different provinces of France, at a time when two neighboring towns were often foreign to each other because of their local dialects."[124] Lacroix saw medieval France as an agglomerate of distinct regions with different dialects that communicated via a transdialectical erotic language that transcended all political and regional boundaries. The erotic transdialect needed no translation and no national domestication—it was the first national language. This element would prove hard to digest for French national Romance philology.

There are several conclusions to be drawn from Lacroix's erotic philology. First, Lacroix's compilation of literary knowledge of the Middle Ages identified the fabliaux as the first mature expression of the French language in the thirteenth century. Indeed, since the second half of the eighteenth century, when they became the object of erudite interest, the fabliaux were considered to be the first full-fledged literary expressions from France and, through the first half of the nineteenth century, they constituted a major site of inquiry for the *histoire des mœurs*. Moreover, Lacroix argued, philology cannot be divorced from the fabliaux. They belong to a genre that Lacroix defines expansively, for it includes any number of works that deal with overt sexual activity, and philology should adopt them as its privileged object of study. This is not because language expresses itself better in the fabliaux, but because the erotic language that they employ made possible the emergence of a national literature. Put differently, Lacroix explains the development of the French language and literature as rooted in an erotic transdialect. Finally, erotic language knows no regional bounds. Through literature, it spreads to infect all of society, crossing all political boundaries and identities. If language cannot be identified with a particular group, then Lacroix's erotic philology brings home the fact that the undisciplined sexuality of the nation is what is at stake in its literary genealogy.

Although erotic philology was Lacroix's original contribution to the pornographic archaeology, he was indebted to the fabliaux studies of the eighteenth century. As Lionel Gossman pointed out, many of the earliest medievalists in the eighteenth century, who belonged to the Modern movement, put forward arguments for the study of the national rather than classical past.[125] The identification and classification of medieval literature was one of its main objectives, and medieval literary monuments were to be treated in the same spirit as traditional historical sources. For the first time, this approach brought the study of literary texts of the Middle Ages under the purview of serious historical study, including almost every material and moral aspect of medieval life. The fabliaux were thus a major source for the study of the national past and for the *histoire des mœurs*.

In July 1746, the count of Caylus argued at the Académie des Inscriptions that "there is no part necessary to the perfection of a Work of this nature that has yet to be rendered in the Fabliaux of our ancient Poets in such a way as to serve as a model."[126] In 1756, Barbazan published three volumes of fabliaux; La Curne de Sainte-Palaye, who published the *Mémoires de l'ancienne chevalerie*

(1759), likewise amassed and collated a large number of fabliaux manuscripts, though he did not publish them.[127] That fabliaux were the single most important source of information on these distant and foreign times, with special insight into "the morals and customs of our forefathers,"[128] "our ancestors" (*nos aïeux*), can be seen from the work of Pierre Jean-Baptiste Legrand d'Aussy (1737–1800), La Curne de Sainte-Palaye's disciple, the first to make a sustained and successful attempt at popularizing medieval French literature. In 1782, he published *Histoire de la vie privée des Français depuis l'origine de la nation jusqu'à nos jours*, after having printed *Fabliaux ou contes du XIIe et du XIIIe siècle* between 1779 and 1781. The history of private life was thus born from the publication of several volumes of fabliaux:

> It is by this so interesting portrayal of the customs and dress of their time above all . . . that the fabliaux may please. . . . Made by their nature . . . to depict the ordinary actions of private life, they show the Nation undressed, if it may be permitted to speak thus. Opinions, prejudices, superstitions, customs, tones of conversation, ways of making love, all is found there; and many things are found exclusively there. I dare even to believe that once we have read them, we will know the French of the thirteenth century better than if we were to read all of our modern histories.[129]

Legrand d'Aussy began the *Fabliaux ou contes* in 1779 in defense of the "centuries of ignorance" and of the "ancient rhymers," furnishing evidence of precedence in French literary superiority even before *le grand siècle*.[130] Legrand d'Aussy thought the fabliaux to be the first unadulterated products in the language because they corresponded to the rise of the French vernacular in the thirteenth century. Namely, in the period following the Albigensian crusade (1209–1229), the distinction between the *langue d'oil* (northern French) and the *langue d'oc* (Occitan or Provençal, southern French) is seen as abolished. Since then, the French language has corresponded to a certain degree to the geography of contemporary France, and the fabliaux are the initial display of the preeminence of the *langue d'oil* in the thirteenth century, "the beginning of our poetry."[131]

What is of interest for the *histoire des mœurs* is their "triviality," "domestic events."[132] However, the study of domesticity and mores had one unwelcome aspect: obscenity. Nevertheless, writers argued, the vulgarity of the fabliaux "incriminate[s] less the liberty of the era than its naïveté. The Frenchman then *braved honesty in words*; he knew not periphrases or circumlocutions imposed by modern rigorism, and called all things by their name, without ill intention."[133] They were true to their nature, without the refinements of the later ages that learned how to dissimulate with words: "The fabliaux writers call a spade a spade, and nothing is named in their writing other than by its name"; "one was only scandalized by the poor usage made of them [words]."[134] "We cannot blame an Author for employing terms that usage or politeness have not banished from language," since such defects are "rather those of their century than their own."[135] Instead, in the interest of the *histoire des mœurs*, "however strange may be the mores in the fabliaux, it is my duty to represent them as they are, since they illustrate their century."[136] The fabliaux had been "neglected and even disdained because some were persuaded that they were vulgar, without creativity, imagination or etiquette," but in the interest of national history, their richness is proof of the vigor of the French race and the nascent French language: "We find a delightful simplicity, interesting storylines, vivacious images, subtle ideas, accurate reflections, energetic expressions, a pleasant variety of etiquette and order."[137] Such statements, from Barbazan's 1756 edition throughout the nineteenth century and continuing into the twentieth, characterize the study of the fabliaux and accompany their popularizing editions: an evaluation of their "natural" language and customs "true" to nature, "neither so barbarian, nor so obscure," "a simple, clear and naïve way of narrating," without "any affectation,"[138] an apologia of the "good old days" (*le bon vieux temps*),[139] with some self-censorship, as, for example, "There are some lewd tales that I will remove in entirety."[140] The fabliaux are simply of unmatched interest for the *histoire des mœurs*, and even scholarly editions, which concentrate primarily on questions of origin, principles of edition, and the historical record of French literature, occasionally allowed room for such arguments: "These texts are not only useful, but even necessary for the history of the language and for literary history. . . . It is not up to us to judge them."[141]

Informed by the *histoire des mœurs*, writers and historians saw the Middle Ages through the lens of the fabliaux, but the fabliaux simply could not represent the best of the French language and literature because of their "salacious repertoire. It teems with salty interjections, droll contrivances, features that have a contagious gaiety; but it is often of a disgusting obscenity."[142] They only reinforced the general perception of the Middle Ages created by Voltaire as one of the reign of immorality and decadence, an age "lacking set laws, troubled by civil wars, without commerce, without police, without written customs, and governed by a thousand different customs" (Voltaire, 16:437).

Since Voltaire, the idea of "jargon," which is considered primitive, haunts the earliest literary expressions of the Middle Ages. By 1879, the general perception, especially among the defenders of the *belles lettres* had not changed: if the language is "barbarian, coarse like manners and crass like an appetite," it is such because of "this impotence of spirit and . . . this sterility of thought."[143] If they had been useful for the study of the national past and its mores, they nonetheless had no literary value. The fabliaux had given a barbarian image of medieval language and culture: "Most of our fabliaux are literally illegible . . . what interest could there be, not even for literary history, but for the history of the language, to pull from the obscurity that until now concealed them from us, these hideous models of crudeness in manners, tactlessness in humor, and artlessness in impropriety."[144]

What saved the Middle Ages from the image of "complete lack of scruples [*dévergondage*]"[145] was chivalry, the quintessential export of French civilization: "There remains at least one fact that it would be difficult to deny, which is that Chivalry was born on French soil. . . . We know nothing in the annals of France that is more essentially French" (see Chapter 6).[146] Chivalry was "a rampart against violence." This perception of chivalry as a civilizational safeguard attenuated and rescued the Middle Ages in the eyes of writers, since "the nation had forever been brave."[147] The preeminence of the French chivalry and chivalric values came into force around the time of the Crusades: "It takes but little study of literature from the Middle Ages to be stupefied by the intellectual movement that followed the Crusades. . . . Our ancestors, whose memory we have denigrated for far too long, would appear surrounded by a splendid halo before us; we would see fertility, light, the civilizing force, where we had only perceived sterility, shadows, and barbarism."[148] Authors associated the civilizing refinement of chivalry with the French North that carried its banner across the world: "Came the Crusades, where the French appeared in the front ranks, for it is not without reason that the name Frank was given to all of these adventurous warriors who went to fight for the cause of civilization, believing only to follow the banner of faith."[149] One of the civilizational aspects of chivalry was the refinement of medieval customs and morals, especially in the treatment of women. This was known as "chivalric love" (*amour chevaleresque*), an ideal of courtly behavior that the French also gave to European civilization. Thus, since the Middle Ages, France has been fulfilling its civilizing mission by spreading chivalry and chivalric love through military might and literary production: "France is in and of herself an entire civilization. Her elected mission is to precede others and guide them toward new destinies" (Larousse, "Moyen, -enne," p. 657). Civilizing force had asserted the hegemony of French culture and language; Europe turned chivalric and "the national language of France spread throughout all of Europe."[150]

To the philologists and academic medievalists the tradition of the *histoire des mœurs*, in which they had been raised and trained, gave two parallel narratives: a civilizing French mission (chivalry) that co-existed with the crude customs and unrefined language (the fabliaux). Chivalry and chivalric love had become the characteristic civilizational marker identified with the French nation, its original civilizing mission and export. In contrast, the sexuality that permeated the fabliaux did not lend itself to nationalization. To begin with, it was impossible to contain sexuality and its erotic language within the confines of a nation. Paul Lacroix exemplified this notion when he showed how erotic language crossed regional lines in the Middle Ages. Sexuality rises above political and national boundaries and cannot be contained within them. In addition, the sexuality of the fabliaux, unlike that of chivalric love, was not a model of universal value that the nation could export across Europe. The fabliaux were not the best representative of French imperialism, despite the fact that the eighteenth- and early nineteenth-century writers established that the emergence of this literary genre coincided with the rise of the French vernacular. The sexuality of the fabliaux was thus central to the choices that were made in the construction of the literary genealogy of France and its civilizing mission. This incompatibility between France's civilizing mission and sexuality played out in the marginalization of the fabliaux as non-French, which were replaced with chivalry and chivalric love.

For Romance philologists in the 1880s, the fabliaux were of Oriental origin, rather than French. In the age of Romantic archaeology, the French appropriations of these stories, thought to be Oriental and above all universal, did not prove to be embarrassing but rather served as proof of the superiority of the French language and literature over all other European nations, which imported even the fabliaux from the French: "In the Middle Ages, we gave much and received quite little. . . . We received many tales from the Orient, ourselves like all the other peoples of Europe, perhaps more than any other, and, in addition, it is very often through us that the transfer took place. . . . We were constantly the means through which these Oriental tales, which we transformed into fabliaux, were disseminated in the rest of Europe."[151] Even if their origin was Oriental, their translation, transformation, and dissemination were done in France by the French. What ultimately mattered was that the French Middle Ages were ahead of all others: "In this history of the French spirit, there are two things to note: precedence and fame."[152] But it was perhaps the ever-expanding French empire (as in Indochina in 1859) and the subsequent intensification in colonial contact that contributed to the negative connotations of fabliaux's Oriental origins and the rejection of the idea that they were a domestic product, indigenous to the French spirit.

Gaston Paris, the leader of the French academic medievalists, did not contribute anything new to the understanding of the fabliaux from this tradition that dated back to the eighteenth century; rather, unlike the earlier scholars, he rejected them because of their Oriental, Indian, origin: "Thus gradually formed this literature of the *fableaux* [*sic*], which . . . ended up being the most truly popular of our ancient poetic genres, although it had its source and roots at the edge of the Orient."[153] Although the authors tried "to adapt [*approprier*] the Oriental tales to European ways of life,"[154] the fabliaux were shocking even in earlier times due to their abusive portrayal and treatment of women. To nineteenth-century readers, they seemed all the more foreign "when we approach this literature with common ideas about tactful and passionate gallantry and the cult of the woman that are attributed to chivalric times."[155] Because chivalry came to be expected as the French civilizational norm, the foreignness of the fabliaux was all the more striking. The fabliaux could not contribute anything to the study of the French national literature, since the customs they portrayed were Oriental and not French: "As for the innumerable tales, almost always lighthearted, too often disrespectful, which address the ruses and perfidies of women, they were not spontaneously born of society in the Middle Ages: they originate in India and they have their underlying *raison d'être* in the milieu that produced them."[156]

This categorical confirmation of fabliaux's Oriental origins was accompanied by Gaston Paris's affirmation of the epic as the truly national genre in the aftermath of the Franco-Prussian War. Even before the War of 1870, authors acknowledged the French epic (*chanson de geste*) as the earliest stage of French literature that preceded the fabliaux, but did so mainly in the effort to divorce it from Frankish influence, and even to suggest it as a model for the Germanic epic: "Our chivalric epics, French and Provençal, were the model for chivalric epics in England and Germany, which are generally only translations of them or, at the very most, slightly modified reproductions; and it was thus not only for our national hero, Charlemagne, but even for heroes who do not belong to us by birthright, such as Arthur or Tristan."[157] But after 1870, *La Chanson de Roland* assumed the role of "the most ancient of our French romances" and to this day the epic is considered to be quintessentially Northern French.[158]

After 1870, the French civilizing mission is described as shared between the Carolingian epic and Arthurian chivalric literature: "For, afterwards, French literature was the queen and the initiator of neighboring literatures; no work of our classical era was translated into more languages, or exerted around us a larger and more durable influence."[159] Moreover, the epic represented better than any other genre of medieval literature the classical norms of literary value. Since philology had transformed the study of medieval literary monument from an object of history into literature as an autonomous object, it had also to palliate the obscenity and the irregularity of form in the fabliaux with a genre that could dignify French national literature at its source. Alongside heroism, domesticity of the fabliaux had no place, and it was then the language of the epic that assumed the qualities of a natural and naive language: "When we return to listen to this naïve language . . . we believe to hear the babblings of childhood. . . . This child who is already the great French people also speaks the great French language."[160] This problem of national origins to language and literature, whereby the French epic replaces what could be Oriental sexuality, also confines the site of medieval and national sexuality and ultimately curbs its importance. Thus, in 1883, Gaston Paris redubs "chivalric love," naming it instead "courtly love," and defines it as a French civilizational phenomenon, an art form developed in the northern Arthurian romances of chivalry (see Chapter 6). By the turn of the century, "courtly love" (*amour courtois*) is a part of the school curriculum with Gustave Lanson's *Histoire de la littérature française* (1895) and Ernest Lavisse's *Histoire de France* (1901).[161] Thus, the French epic is the quintessential national genre and the courtly love of Arthurian romances is the first French civilizational export. By the time Gaston Paris's disciple Joseph Bédier publishes his doctoral study on the fabliaux in 1893, in which he definitively refutes their Oriental origin and concludes that their origin is multiple (thus not national) and the province of oral tradition, courtly love will be established as the French civilizational phenomenon and the *chanson de geste* as a national epic.[162] It will be too late to rehabilitate the fabliaux; sexuality remained Oriental, while chivalry and courtly love were French. Sexuality had lost its place in the study of French medieval literature.

The Philological Exclusion

With the definitive orientalization of the fabliaux, Romance philologists excluded pornographic archaeology and unruly sexuality from the literary study of the French Middle Ages. This exclusion occurred at the time when medical concerns about the colonial experience were central to the regulation of sexuality in the metropole. And it occurred also at the time when imperial medicine was focused on the prevention and eradication of sexual vices and diseases threatening the metropole with importation from the colonies (see Chapter 3). Philology would not nationalize sexuality, but it could purify the nation's genealogy and its literary heredity.

Philology had always been a historical discourse, one of the three forms that historical knowledge could take: "the one [form] which is focused on nationalities, and which will essentially find itself in continuity, at the outset, with linguistic developments, and, as a result, with philology," the second one centered on social class and fundamental to political economy, and the third, centered on race and, thus, relating historical discourse to biology: "Philology, political economy, biology. Speak, work, live."[163] Philology and biology were equal parts of the historical-political discourse, as well as the apparatus of sexuality. As seen in Chapters 1 and 2, sexuality arose from the new way of deploying historical discourse within the tactic of modern power over life (bio-power), which exercised control over heredity and genealogy of the nation. Not only were the antiquity of ancestry and alliances important, but so were the health and hygiene of the family (lineage) and nation (race). The strength and improvement of the nation depended on the protection and improvement of its biological capital, that is, on prophylaxis and eugenics. Philology constructed the literary and linguistic genealogy of the French nation, from whose lineage all defects were to be expunged. The nation's genealogy depended upon the regularization and discipline of the medieval body of literature.

For both doctors and medievalists the heredity of defects and vices began in the Middle Ages. Steeped in the medical discourses on heredity, the philological discourses of academic medievalists converged around national identity—pure origin, flawless manuscript—constructed on the prevention and exclusion of the sexual, especially in its "deviant," pathological forms: "courtly love" was still preferable to the decadence of the fabliaux; heteronormative adultery was preferable to the unruliness of historical passions and lifestyles (*mœurs*). The academic discipline of medieval studies, born out of the complicity of philology and biology, purged all "deviance" from the study of history, and most effectively so by exiling the *histoire des mœurs* and its pornographic archaeology. It is in this context that Gide, in the letter with which we began this *Pornographic Archaeology*, spoke of *mystic love*—the disembodied, platonic, contrived and highly codified love—and criticized medieval studies that promoted *amour mystique* as the only possible expression of sexuality in the Middle Ages, urging scholars to probe whether there was more behind "the charming décor" of literature. Gide upbraided medievalists precisely for not taking up courtly love as one of the medieval discourses of sexuality.[164]

While Romantic archaeology had studied and catalogued sex in all of its variety of practices and pathologies, beyond national and European boundaries, Romance philology was profoundly anti-archaeological and pro-genealogical when it dealt with the origins and husbandry of national heredity and normativity. If chivalry was France's original civilizational export, sex on the contrary corrupted a civilization by importation and imitation of foreign Oriental lifestyles. Sexuality eluded both the control of philology and the metropole; it escaped the grasp of normative genealogy, the discipline of "errors," and the medical discipline of the nation. In sum, sexuality—whether as pornography or courtliness—was at the heart of the formation of French medieval studies as a discipline, where philology integrated it as a silenced presence in their very foundations.

Chapter 6

Courtly Love, Courtly Marriage, and Republican Divorce

In the previous chapter, I traced the philological exclusion of sexuality in the last quarter of the nineteenth century. I argued that the civilizing mission of chivalry and chivalric love was preferable to the unbridled and contaminating influence of sexuality of the fabliaux that could not adequately represent the French nation. The preceding chapter, however, brings up the question of why chivalric or courtly love, widely perceived to advocate heterosexual adultery, was not ostracized by academic medievalists and doctors as a form of deviant behavior. In the tradition of the *histoire des mœurs* elaborated between ca. 1760 and 1855, chivalric love was considered a civilizational phenomenon riddled with paradox: what made the Middle Ages civilized, because it refined a brutal and violent feudal society, also provoked sexual excesses outside marriage. There was something inherently decadent about this "religion of love, having reached its supreme exaltation":[1] the "laws of love" (*lois d'amour*) were "a theory very widespread and very harmful to conjugal fidelity"[2] that worked against its protagonists by provoking extreme acts of violence, "hence the cases of adultery and bloody vengeance."[3] In 1883, thirty years after the *histoire des mœurs* reached the peak of its production, Gaston Paris coined the term "courtly love" (*amour courtois*).[4] Many doctors were writing at the same time about chivalric love (*amour chevaleresque*) and the courts of love (*cours d'amour*): while "pretty pages, elegant squires, proud knights, languorous troubadours, jongleurs, singers and minstrels" were "animated by the most chivalric of sentiments," the women were presiding over "the *courts of love* where principles that would make the hair of honest modern women stand on end were professed under the pretext of a 'purified love.'"[5] The answer as to why doctors and medievalists alike embraced courtly love could be as simple as follows: adultery is the lesser of all evils among "deviant" behaviors and it is less immoral to speak of love than of sexuality. But another historical-cultural reason exists. The early 1880s, when academic medievalists defined the concept of "courtly love," was also the height of the legislative and public battle for the legalization of divorce, made into law in July 1884, which had begun eight years earlier, on June 6, 1876. An ever-increasing pressure on the institution of marriage to sustain itself, not by denying the right to divorce, but by the very nature of the husband-wife relationship, resulted in the assimilation of chivalric love into marriage. That is, doctors aimed to transform love outside of wedlock into marriage based on love; the force of law prohibiting divorce that had kept marriage indissoluble would become irrelevant if marriage were held together by the bonds of love. In this context, chivalric love continued to have a wide cultural resonance when it underwent a series of transformations that made it productive to the social engineering of the French nation. Both professional groups, doctors and medievalists, inherited the term from the *histoire des mœurs* and in both cases seized on the paradox of decadent love to improve radically the morals of the nation. While medievalists applied it to a carefully contained and crafted literary phenomenon that further elevated French medieval literature as the paragon of civilizational refinement, doctors, in trying to discipline the population, used it to redefine marriage, transform conjugal love and marital sexuality, and improve national health and birth rates. Consequently, this concept continued to be socially productive when, in the attempt to transform chivalric love into courtly marriage, doctors wrote popular marriage manuals in the last decade of the nineteenth century and the first decade of the twentieth century, in the middle of the French demographic crisis and natalist debate.

Chivalric Love: Another *histoire des mœurs*

By the early 1880s, medieval chivalry and its "code of love" (*code d'amour*) had been around for over 100 years, following the initial publication of Jean-Baptiste de La Curne de Sainte-Palaye's *Mémoires sur l'ancienne chevalerie* in 1759.[6] In this earliest study of chivalry, La Curne de Sainte-Palaye repeatedly used "*courtoisie*, meaning civility or honesty," "a gentleness, a modesty, a politeness that the name *courtoisie* expressed perfectly, concerning which no other rules are as formal as those of Chivalry."[7] Chivalry prescribed a set of rules of *courtoisie* that made possible, in the eleventh century, that "the French government emerged from the chaos where it had been plunged by the troubles following the extinction of the second race of our kings [the Carolingians]. Already, royal authority began to be respected; everything took on a new appearance; laws formed; communes and townships were established; fiefs acquired a more consistent form and discipline."[8] The refinement of the *mœurs* was also felt in the culture of the *cours d'amour*: "A dispute judge who reported to what was called a *Prince of Love*, or *Prince de Puy in the courts of love*, jurisdictions established in some regions, in order to understand these important matters; a judge, I say, pronounced sentences that were almost always equivocal, unclear, and often enigmatic, to which the parties submitted with a respectful docility."[9] Continuing La Curne de Sainte-Palaye's work on chivalry and the courts of love as institutions of the feudal age, François Raynouard wanted to show in 1816–1817 that the southern *langue d'oc* was the archetype of all Romance languages, "an intermediary idiom which, by the decomposition of the language of the Romans and the establishment of a new grammatical system, provided the common model from which the diverse idioms of Latin Europe successively emerged."[10] This original idiom, consecrated by the troubadours (Raynouard, 2:i), would have sprung from exemplary moral uprightness. For Raynouard, social behavior was mirrored in literature. The literary production of the troubadours, "a new literature," made to celebrate love, was born out of "the strict frankness of their moral and political opinions, their enthusiasm for honorable exploits and for the illustrious figures who accomplished them, their just and courageous indignation against the errors and mistakes of their contemporaries."[11] The literature and success of troubadours were linked to the institution of the courts of love, a "remarkable institution," "one of the most interesting points in the *histoire des mœurs*, in the customs and spirit of the Middle Ages."[12] Raynouard thought that the courts of love pronounced judgments in the matters of love—their main distinctive trait was that they were presided over by women, of whom at least three were identifiable historical figures: Eleanor of Aquitaine, Marie of Champagne, and Ermengarde of Narbonne (as well as an unnamed countess of Flanders). Informed by the works of Andreas Capellanus (André le Chapelain), Raynouard argued that the courts of love existed before 1200 in both the south and north of France (Raynouard, 2:lxxx; also 2:xcvi, cxxi–cxxii).[13] In his view, this institution "was not the work of the legislator, but the result of civilization, mores, customs, and discrimination [*des préjugés*] of chivalry."[14] A result of civilizing forces of chivalry, the courts of love transformed medieval civilization: they were a veritable institution that "purified and ennobled, to the benefit of civilization, morals, and chivalric enthusiasm, this impetuous and tender feeling"[15] that is called love. This literature was one of cultural and linguistic purity: it owed nothing to the ancients and was completely original and independent (Raynouard, 2:ii). Raynouard's work thus reinforced La Curne de Sainte-Palaye's argument that the civilizational phenomenon of *galanterie* was a specifically French invention: "One of the traits that most characterized the Frenchman in the time of Chivalry, of which he still holds traces that distinguish him from all other nations."[16]

The terminology of the courts of love, *courtoisie*, gallantry, and chivalry dominated different historical narratives of France and European civilization through the 1860s. Eighteenth- and nineteenth-century authors conflated medieval *courtoisie* and classical, seventeenth-century *galanterie* as essentially the same phenomenon.[17] La Curne de Sainte-Palaye declared the Middle Ages "the Golden Age of gallantry."[18] Raynouard exalted "the gallant ways of chivalry" and spoke of "chivalric manners," "chivalric spirit," "chivalric passion," and of the "strict, chivalric, and gallant mood of the troubadours."[19] On numerous occasions Raynouard referred to "the code of love" or the "love code."[20] Historian Edgar Quinet in his early writings before 1830 wrote of "chivalric love" (*amour chevaleresque*).[21] During the 1830s, Claude Fauriel, in his lectures at the Sorbonne, published posthumously in the *Histoire de la poésie provençale* (1846), spoke of "chivalric gallantry" (*galanterie chevaleresque*) and "chivalric love" (*amour chevaleresque*).[22] Émile de la Bédollière, in the *Histoire des mœurs et de la vie privée* (1847–1849), affirmed that "the rules which [Master André] provides are those which still bear in current times description as chivalric, and so that they not be infringed upon with impunity, special courts, the *courts of love*, oversaw their application."[23] And in his *Histoire de France*, Henri Martin wrote about the revolution of "the ideal of pure love" north and south of the river Loire: "It is a general and concurrent revolution [on both sides of the Loire] that introduces, in the practices as well as in the ideas of medieval society, a new element, almost immediately dominant, beside the religious and purely military elements."[24] In the second half of the twelfth century, says Martin, "chivalry does not limit itself to a morality taught by poetry and spread through opinion: it creates an institution that focuses the force of opinion and gives sanction to this morality, which is as different from ecclesiastical teaching as from feudal maxims."[25] The courts of love were this institution that chivalry purportedly created, an embodiment of the high civilization of the Middle Ages, a juridically sanctioned essence of civilized manners and morals in medieval French society.

In the late eighteenth and first half of the nineteenth century, chivalry, the courts of love, and gallantry were systematically identified as lofty ideals transformed into quintessentially French institutions: "There remains at least one fact that it would be difficult to deny, which is that Chivalry was born on French soil. . . . We know nothing in the annals of France that is more essentially French."[26] A medievalist and professor at the École des chartes, Vallet de Viriville wrote in 1853 that "France is the classic land of gallantry. . . . The mere words *galanterie* and *culte des dames* [worship of ladies] suffice to evoke a more or less vague idea of romantic sentiments that are refined yet real, and this idea is intimately linked to our historical memories and national traditions . . . [and] confirmed by the agreement of foreigners" (Viriville, qtd. in Hepp, p. 234 [p. 3678]). The ongoing speculation as to the origins of medieval literature played a large role in such claims. Although authors differed on the way they interpreted the early medieval fusion into a new French nation of three racial elements—Gallic, Roman, and Frankish—they all agreed that by the eleventh century French civilization had become dominant in Europe.[27] Raynouard and Fauriel believed in the autonomous emergence of southern French literature and culture, independent of other influences. Consistent with her views on chivalry as a European institution, "a sort of European patriotism," Mme de Staël wrote that chivalry was born in the north, but she acknowledged that it was "embellished by the charm of poetry and love" in the south of France.[28] Much like Charles Nodier, who rejected the need to identify external influences on national literature as a misplaced "rage of genealogies"[29] and who believed in the spontaneous emergence of the chivalric institution as a product of its time, Edgar Quinet also believed in the spontaneity of the French "genius" (*génie*). In the late Middle Ages, France led all other nations in literary history and it expressed the essence of European civilization, a view to which Jules Michelet and Henri Martin also subscribed (Glencross, pp. 100, 102). They too considered France a superior mixture of the three races, a tripartite division that France had transcended. Chivalry's longevity and influence were due to "morality . . . based on the national character of the French," "one of the great phases in the development of the human spirit," "a field open to all of these sentiments, all of these principles of civilization."[30] But whatever their beliefs about chivalry's origins, all authors agreed that the Northern French spread it throughout the world. Likewise, the courts of love were "one of the most original institutions . . . the most frankly civilizing in medieval life."[31] Henri Martin, in his *Histoire de France*, went so far as to affirm that the influence of these medieval values extends to modernity: "What still exists of moral delicacy in the modern world comes from it [chivalric poetry]."[32] The connection between chivalry and civilization, society and literature, was articulated most clearly by Claude Fauriel: "In order to promote good social actions, a motivation was needed that was more immediate, more tangible and, to be fully honest, less lofty than religion. Chivalry finds this motivation in love."[33] Taking precedence over religion, love became the essence of morality and the symbol of civilization: "Another sign that these mores were far advanced on the road to true civilization is that—hardly past the horrors of the Middle Ages, and the feudal system where force was everything—we see the weaker sex less tyrannized than it is *legally* today."[34] Such arguments were a veritable ideological *tour de force*: in replacing religion with love, the writers secularized the Middle Ages, which they then enlisted to the benefit of Republican, anti-clerical ideals. The profound morality of the French, encoded in the laws of the courts of love, meant that, since the earliest constitution of European civilization in the Middle Ages, France had been performing a civilizing mission by spreading chivalry and chivalric love through its military might and literary production of chivalric literature, epic and Arthurian. And this civilizing mission was secular. All of Europe followed the lead of the "gallant and great lords, whose urban superiority was unanimously recognized by the European nations."[35]

The idea of a civilizing force of chivalry and the courts of love, thanks to which "we would see fertility, light, the civilizing force, where we have only seen sterility, shadows, and barbarism,"[36] was crucial to the Romantic historians and writers of the *histoire des mœurs* who were perplexed by two parallel and contradictory features of the medieval period: "Reading all that we have said in honor of Chivalry, we recall that the centuries during which it was most flourishing, were those of debauchery, pillaging, barbarism, and horror; and that often all of these vices and these crimes united within the same knights, which were thus acclaimed as heroes. . . . Nevertheless, these two truths, so contrary in appearance, are equally observed."[37] La Curne de Sainte-Palaye opposed what he believed were two antithetical and irreconcilable institutions: chivalry and feudalism. In 1813, Mme de Staël was among the first to provide a corrective to La Curne de Sainte-Palaye. She described chivalry as an ideal within the feudal institution rather than an incongruous institution: "The feudal regime, this sad and severe political institution, but which consolidated, in some respects, the spirit of chivalry by transforming it into laws."[38] The brutality of the institution gave birth to an ideal. The connection between the "bandits of the eleventh century" and the "galant and great lords" who were leaders among their peers "of European nations" was thus not one of succession; there was no progressive and successive transformation of the medieval mores, only a relative equilibrium between immorality, violence, and refinement. As Henri Martin explained, that was precisely the difference between feudalism and chivalry: "What a contrast between this crown of virtues that chivalry would set upon its brow, and the brutality, avidity, tyranny, disloyal volatility that the feudal world inherited from barbarians."[39] Love and its codes counterbalanced the brutality and lawlessness of the Middle Ages with civilized and good morals: chivalry was "a safeguard against violence," against "the ill-fortunes of the state and the anarchy of the government."[40] Chivalry was not a sign of progress; it did not eradicate but rather attenuated this violence. It was born from the necessity of its age: "It could only live in the social state where it was born. The confusion of powers, the absence of justice . . . , the rigidity of feudal customs."[41] It was a useful and purifying social remedy and an enriching cultural medium, and entirely a product of this difficult time that it aimed to improve. Chivalry was thus the age of both splendor and decadence (Nodier, p. xxii), rise and decline: "Our chivalry,"[42] the French "civilizing force," did not supersede "barbarism"; rather, they co-existed. Barbarism and civilization, violence and love, were two inextricable and mutually dependent features of the period between the eleventh and the fifteenth centuries.

This tension between disorder and restraint ultimately tore at the institution of chivalry. Starting with La Curne de Sainte-Palaye onward, authors in general blamed the decline of the Middle Ages on the decline of chivalric discipline: "The primitive discipline of the old Chivalry had fallen into abandon as early as that period" and pointed to the difference between rules and behavior, for "the wisest regulations were not capable of halting the advance of corruption."[43] Copying La Curne de Sainte-Palaye, authors dated the decline of chivalry to "the divisions between the princes of royal blood during the bouts of Charles VI's illness, [which] caused in all parts of the government infinite disorder; and those which occurred in Chivalry were not the least pernicious."[44] The knights sank into "ignorance and barbarism . . . principally after having abandoned the glorious functions of justice," they abandoned themselves to "idleness" and "reading works of their trouvères and jongleurs, uncouth and libertine people."[45] While Stendhal wrote in 1822 that "thus we owe what is noble in our customs to these Crusades and the Moors in Spain,"[46] authors writing after 1830 and the conquest of Algeria considered that the Crusades bore fatal consequences, for their *métissage* of cultures, religions, and peoples resulted in "the most despicable depravity," "the most shameful of all prostitutions": "A horrible libertinage had slipped into all of these classes of society since the Crusades, and the sin against nature, that the French travels to Palestine had acclimated to France, again threatened, despite chivalry, to infect the mores and corrupt the entire population."[47] Even the Church did not remain impervious to such corruption, falling "into the same degree of decadence as chivalry, and society in its entirety seemed to head toward its dissolution," to the point that "it would be impossible to say whether the Church demoralized chivalry, or chivalry demoralized the Church."[48] This decadence, that began with the Crusades, spread by "the prostitution of literary language," culminated in Gilles de Rais, who exemplified the "degree of perversity reached by some noblemen, some great lords, who gave in to all of the aberrations of debauchery" (see Chapter 4).[49]

A very similar tension also undermined chivalric love. Just as chivalry was the product of violence, so chivalric love was the product of onerous marriage bonds: "If anything can help to conceive the exaggerated pretensions, the refinements, the subtleties of this love, then it is the self-interested and precarious reasons behind feudal marriage."[50] Chivalric love came into existence because of the duress of feudal marriage, as it revealed marriage in "its most direct and most primitive necessity."[51] The central role in chivalry was thus reserved for women: "Chivalric ideas attributed a veritable supremacy to women over men," "her service was a ritual [*un culte*]."[52] The woman was placed on "a pedestal of tender admiration and a throne of honor":

> This is obviously the principle of chivalry: it liberates a sex that Prostitution had subjected to the most degrading servility. Prior, the woman was a slave and humiliated in an undignified role; now, she is queen, and her sovereignty still relies upon love; but it is no longer carnal love, whose guilty pleasures oppress good instinct and predispose the heart to all of the vices; it is perfect love, heroic love, which draws its source in the most beautiful sentiments and which is exalted by the imagination in breaking away from the encumbrances of physical nature.[53]

The *histoire des mœurs* identified all the hallmarks of courtly love—woman's sovereignty and superiority, the idea of a "gift" (*le don*), and the chivalric prowess of men:

> It was in order to gain distinction in her eyes and, also, to be loved that he proved himself heroic [*preu*] and valiant, honest and courteous, loyal

> and virtuous. The name and the colors of this woman acted for him like a talisman under the most trying circumstances in his life. . . . Nothing less resembled physical love than this deep and subtle amorous devotion in honor of a lone woman, who often gave not so much as a chaste kiss in recompense for such an exalted sentiment; but this sentiment, pure and ardent all at once, found within itself an invincible force that grew ceaselessly from an obsession [*idée fixe*] and ecstasy.[54]

Relations between men and women became more virtuous, thereby liberating the woman to some extent. Nevertheless, compensating for the absence of love outside marriage only undermined the codes of love. All authors believed in what Fauriel summed up as "the incompatibility of love and marriage."[55] Many but not all followed in the path of Paul Lacroix, who found a middle ground by proposing that chivalric love did not exclude physical relations, but it did not emphasize them, either. Sexual appetite and "physical love" were elevated by "the most beautiful sentiments" and imagination: "One would have said that the lovers took pleasure in overexciting their desires, so as to prove to what extent they could then combat and defeat them; seeking out peril and exposing themselves to it with a sort of pride, it can be supposed that they sometimes succumbed to it."[56] "Carnal love," "physical love" was replaced by "this almost mystic love," "metaphysical love,"[57] "heroic love," "perfect love." Other authors followed La Curne de Sainte-Palaye's condemnation of the "metaphysics of love, this vast field where the best minds practiced, shining among our most respectful servants of women."[58] These literary and social rituals made a mockery of true love because they hid the difference between being (*être*) and seeming (*paraître*), between "the essence and character of perfect and true love" and "false subtleties"; they were nothing more than "vain ceremonials of reverence, genuflection, prostration down to the ground."[59] The lovers' antics and mimicry were misleading, "the supposed innocence of their amusement with the ladies and ladies in waiting, their conversations, their stories," for in fact "never did we see manners more corrupt than in the time of our knights, never the reign of debauchery more universal."[60] Voltaire's condemnation of medieval chivalry as merely a veneer over a barbarous and superstitious epoch held some water: "French nobility had only the advantage of a more polished façade than the other nations."[61]

The sociohistorical analysis practiced by the *histoire des mœurs* consistently produced an interpretation of the cyclical decline of civilizations. The interpretations of medieval chivalry and chivalric love were firmly inscribed in the nineteenth-century context of degeneration, following the curve of civilizational rise and fall, rising to counter "vulgar and demeaning tyrannies of physical love," but ultimately falling into decadence: "It was under the reign of Charles VI that this decadence of chivalrous manners began."[62] They represented the paradox of social institutions and phenomena that corrupted precisely by purifying the morals of the nation. Archaeology as method and cultural history as object could not transcend the pessimism and decadence of modernity: chivalry and chivalric love were seen as yet another example in a series of evidence proving the decline of civilization(s) and degeneration of races to which the nineteenth century was witness. Yet, given the central role of *courtoisie* and gallantry in modern French culture, as well as the increasing focus of medicine and social science on marriage, fin-de-siècle writers, academic medievalists and doctors alike, were invested in resolving the paradox of chivalric love in order to improve the morals of the nation drastically.

Courtly Love and Republican Divorce

In his 1883 article, well known to medievalists, Gaston Paris spoke of both "chivalric and courtly love" (*l'amour chevaleresque et courtois*) (G. Paris, "*Lancelot du Lac*," pp. 520, 521). His use of "chivalric" was a throwback to the lexical heritage of the *histoire des mœurs*, just like his expression "exalted and almost mystic love."[63] Paris's coinage of "courtly love" (*amour courtois*) struck the imagination of twentieth-century medievalists and has since provoked numerous disputes over its accuracy, application, and usefulness.[64] But his description and definition of courtly love were not novel; rather they surfaced from the *histoire des mœurs*. Just as he did fundamentally little to change interpretations of the fabliaux (see Chapter 5), Gaston Paris was not either a groundbreaking innovator in regard to other forms of medieval literature, but rather an heir to eighteenth-century thought. He added "courtly" to the name of chivalric love and narrowed down its scope to a code of love at work only in Arthurian romances. But he still rehearsed the themes of the *histoire des mœurs*: courtly love was "illegitimate, furtive," and extra-marital; the man was always in an inferior position to the woman who is superior; to be worthy of "the tenderness that he desires or has already received," he obeyed her even in her contradictions and was capable of any act of prowess; and, finally, "love is an art, a science, a virtue, which has its rules just like chivalry and *courtoisie*, rules that are better mastered and applied as one progresses, and which one must not disobey under penalty of being judged unworthy."[65] The terminology chosen—art, science, rules—underscores the fact that the emphasis was not on physical love but on rules of behavior, "the rules of a refined art," "a science, . . . a code."[66] Paris evoked the carnality of love with one word, "tenderness," and conceded, in a similar vein to Lacroix, that it was "nevertheless in no way platonic, and founded upon full possession, but attributing only a secondary role to the senses."[67] The reason Paris took interest in this phenomenon was principally its structure and formal rules that enabled "the creation of a courtly society by the establishment in the aristocracy . . . of the rules of a subtle and often strange etiquette whose rigorous observation constituted a science."[68] The idea that courtly love was performing important social work also came directly out of the long tradition of the *histoire des mœurs*: "an effort to create and make men accept an ideal and refined love . . . closely linked to the practice and growth of social virtues."[69]

The historical existence of the courts of love is one interesting point of departure from the tradition of the *histoire des mœurs* in Paris's analysis and definition of courtly love. La Curne de Sainte-Palaye, Raynouard, Martin, Bédollière, and Lacroix had all believed in the historical existence of the courts of love as real tribunals: "The form of the judgments and sentences was the same as in the tribunals of royal and seigniorial justice."[70] Gaston Paris repeatedly denied the historical existence of the courts of love and underlined that they existed only as "games of wit, . . . social amusements" and the word itself was "unknown in the Middle Ages."[71] This divergence from the *histoire des mœurs* opened up a whole new dimension in the exploration of the invention of courtly love. By taking out the dimension of an institutional-historical reality, Paris disengaged courtly love from the question of *mœurs* and sexuality; subsequently, the analysis of the phenomenon was freed from the perplexing problem of simultaneous feudal decadence and chivalric refinement, as well as that of inevitable civilizational decline.[72] It thus could be analyzed as a self-standing literary and linguistic phenomenon. The second major effect of challenging the view that the courts of love corresponded to historical reality was a change in the role that the rhetoric of courtly love played in efforts to counter decadence and degeneration in the nineteenth century. In the context of the divorce debate, for academic medievalists this will mean a quest for the origins of pure love.

The debate on the legalization of divorce lasted for eight years, between June 1876 and July 1884. For eight years, a newly elected representative, Alfred Naquet, fought a sustained public and legislative campaign, lobbying public opinion in its favor. When his first and second legislative proposals on June 6, 1876, and May 21, 1878, failed, he simply requested the reestablishment of Title 6 of the old Civil Code in 1879. Two years later, on November 11, 1881, he submitted a third legislative proposal that the parliament would pass into law on July 27, 1884.[73] This law did not allow divorce by mutual consent, but allowed for a court trial that weighed merits of causes.

The Naquet law was the end point of a process that had begun with the first legalization of divorce on September 20, 1792, and confirmed on March 21, 1803, in a more restrictive version by the Napoleonean Civil Code, making divorce more difficult to obtain, turning it into a trial and reestablishing "legal separation" (*séparation de corps*).[74] Divorce was abolished on May 8, 1816, by a decree proclaiming the indissolubility of marriage, part of the ultra-royalist and Catholic backlash in the wake of Napoleon's defeat and exile: "The only great argument of the legislators in 1816 was the religious argument."[75] Between 1816 and 1884, the so-called "legal separation" was the only option for unhappy spouses. Still, divorce was debated regularly at each change of government. In 1830 and 1848, the debate stirred a great deal of public interest along with a flurry of publications, books, pamphlets, and newspaper articles (Ronsin, *Les divorciaires*, pp. 12, 57, 125, 164; Naquet, *Le divorce*, p. 30). At each turn, the key question was the "restitution" and not the "creation" of divorce, for two reasons: first, because of the precedent set between 1792 and 1816, and, second, because historical research disproved the notion of a transhistorical indissolubility of marriage, since it had been introduced by the Catholic Church a relatively short time prior (Ronsin, *Les divorciaires*, p. 11; also Naquet, *Le divorce*). The discipline of anthropology, which came into being with the study of kinship, had been studying "primitive marriage" as a sub-province of the study of kinship since the 1860s and had formed the view that patriarchal family and marriage in its modern form were not at the origin of society. Some authors tried to prove an early form of primitive promiscuity, but most were proposing matriarchy as the early organization of family, challenging implicitly the self-evidence of Western types of marriage.[76] In short, the social and political battles over divorce were an expression of two opposing views on the role of the Church in the State.

The long divorce debate fueled the proliferation of discourses on adultery. Already Romantic literature from the 1830s to 1850s, with George Sand in the lead, was considered to have moved from the condemnation of marriage to the glorification of adultery, creating a veritable "theory of adultery."[77] According to this theory, adultery is natural for men, and some famous doctors, Virey,

Cabanis, and Forel, encouraged it. In the 1860s, public opinion even encouraged female adultery.[78] In reaction to this theory, as early as the late 1840s, doctors intervened by openly promoting divorce over adultery: "Whether committed by the man or woman, adultery is always a crime with enormous consequences, which solid legislation should at least make increasingly rare, if not eliminate, by using divorce to moralize marriage."[79] Adultery was immoral and divorce was its moral cure. By the 1870s, a small group of socialists and radical feminists inspired by Enfantin and Fourier grew (Ronsin, *Les divorciaires*, p. 91), encouraged by the ever-increasing numbers of legal separations, doubling for instance under the Second Empire, between 1850 and 1870 (Ronsin, *Les divorciaires*, p. 319). The advocates of divorce were also motivated by the accompanying number of trials that called for public testimony in the *séparation de corps* procedure. In 1872, the affair Leroy-Dubourg of a husband on trial for the murder of his adulterous wife saw Alexandre Dumas *fils*, despite his conservative leanings, write an editorial in favor of divorce that provoked a violent public debate: "It is obvious that in contemporary society, divorce, in a multitude of cases, has become indispensable, so to speak, and that it is better to reestablish divorce by law than to be forced to accept murder in social conduct."[80] Dumas's opinion was echoed by a number of doctors in the early 1870s: "Better divorce than fornication, prostitution, adultery, which have become the most detrimental elements of marriage and family; better divorce than legal separation which is a lie and has become the most radical expression of the freedom to bachelorhood within marriage [*la liberté du célibat dans le mariage*]."[81]

Over the life of the debate, adultery had thus entered the public discourse and divorce became the legitimate solution to the problem; the promise of a necessary, and overdue, regeneration of marriage now depended on the freedom to divorce. While for its detractors such discourse was proof of society's moral dissolution and civilizational decline and its advocates were "enemies of family,"[82] by the 1880s it was possible for authors to affirm, without scandalizing the public, that divorce would rehabilitate the soundness and moral foundations of marriage, thus reinvigorating society: "Far from harming marriage, the reestablishment of divorce would, on the contrary, moralize it, and would reduce the so very considerable number of bachelors. Today, as we know, the majority of marriages are a commerce of deception; people wear a mask before being married, they disguise their flaws, they hide, for some time, their vices and they let them have free rein after the forever binding sermon has been pronounced. Divorce would put an end to this shameful ploy."[83] By this view, the restitution of divorce will salvage the institution of marriage, guarantee the happiness of spouses, and ensure the equality of women: "In asking for divorce, reduced to rightly limited cases, we ask for an encouragement to marry, a counter-insurance of conjugal happiness: we plead above all the case of the weaker sex, the woman for whom divorce would become the firmest guarantee, the surest protection, the most honorable shelter."[84] After its restitution, some furthermore observed that divorce became an encouragement to remarry: "Since the institution of divorce and its annexation in our code, marriage has been brought within the grasp of all, there is no merit to shying away from it."[85] Marriage guarantees "the endurence and happiness of human societies," while bachelorhood destroys them.[86] Indeed, soon after its legalization, except for a small minority, its opponents ceased to contest the legality of divorce, and instead a new discourse emerged on France's demographic crisis and the urgent need to stimulate birth rates.

The long nineteenth-century history of the divorce and the intensity of the debate that led to passing the law of 1884 shed new light on Gaston Paris's renaming of chivalric love. As we saw, Paris published his first thoughts on courtly love in 1881, then consolidated its definition in 1883. Throughout the thick of the debate, he also held his Decadent literary salon: "the Salon at 11, rue de Varenne . . . was held between 1880 and 1885, and . . . was attended by Renan, Taine, Paul Bouget, Sully Prudhomme, Hérédia, Alexandre Dumas *fils*, Sorel, Bethelot, Pasteur, and the viscount of Vogüé. Moreover, the beginning of the 1880s is the period when Gaston Paris worked on *courtly love* [and] when the Decadent movement flourished."[87] The basic premise of the Decadent movement was that to write about Evil is to inspire Good. If prostitution and adultery had been "a necessary evil" to save marriage when there was no divorce, divorce was now a necessary evil to reinvigorate marriage; by extension, courtly love was "a necessary evil" in the revitalization of marriage and pure love.[88] The idea that decadence is a source of regeneration must therefore be understood in the overall context of the nineteenth century. Cyclical convulsions of political and social crises at that time culminated in the military disaster of the Franco-Prussian War (1870) and the bloodshed of the Paris Commune (1871). In the aftermath of these upheavals and following the proclamation of the Third Republic, national decadence was an opportunity for a new departure. Gaston Paris's scientific thoughts on decadence were indebted to the theory of cyclical linguistic and literary decadence: the supremacy of "Gallic" literary production in the twelfth and thirteenth centuries fell into decadence in the fourteenth century, when the Italian and

Spanish languages and their literary production burst onto the scene, but the French language was able to regain its vigor and reassert its supremacy in the seventeenth century (see Chapter 5). Many fin-de-siècle writers sought their inspiration for yet another literary renewal in the imagination of the Middle Ages, hence the quest for the bizarre, the obscure, and the refined, stylized violence of the Decadent literary movements, as opposed to the pessimism of the Neo-Gothic. The Middle Ages were the source that could bring a desired and much-needed fresh vitality to the exhausted century: "[Our oldest poetry] teaches us to turn into children again to taste the naïve joys of childhood; it reopens the treasures of our fathers' imagination, and makes surge up again, in our withered minds, the sparkling fountains of young and joyous poetry."[89] Decadence was not the expression of the darkness of an age, but rather a quest for the new, an optimism and re-sourcing of literature in the new topoi: "We have in our day seen literatures create nations, meaning that national consciousness, almost completely faded . . . recovered . . . the fullness of its strength and life."[90] In 1872, Larousse's *Grand dictionnaire universel du XIXe siècle* expressed this general optimism of renewal: "All again predicts long centuries of prosperity [for our literature] because it seems to cross into a new phase only to be enriched by a new perfection, new resources."[91]

Thus the renewed interest in "courtly love" was made possible by the divorce debate, and writing about adultery was no longer a problem for a philologist. Gaston Paris carefully circumscribed the fact that courtly love could only take place outside of marriage: "It is illegitimate, furtive. We do not conceive such ties between man and wife." Paris also carefully assigned courtly love to the realm of Arthurian romances and, even more specifically, associated it with the love of Lancelot and Guinevere, who was married to King Arthur. In contrast, he described the equally adulterous love of Tristan and Yseut (queen by marriage to Tristan's uncle, King Mark) as "a simple, ardent, natural passion."[92] The difference between these two loves is not in the opposition between marriage and adultery, but rather in "the details [*les subtilités*] and refinements" of the codified love between Lancelot and Guinevere,[93] compared to the natural and pure emotion that placed Tristan and Yseut's relationship outside the purview of courtly love.[94] In contrast to pure love, courtly love is already a decadent form. It is for this reason that "courtly love" is an art and a science, that it is a convention of *courtoisie*, a code. The decadence of courtly love prompts the desire to return to the pure love-passion shared by Tristan and Yseut, which serves, in turn, as the source of rejuvenation. Courtly love is thus the point of departure for this process which proves necessary for regeneration in the future. Just as a philologist can remedy degeneration in manuscripts by reconstructing the manuscript family in order to arrive at the ideal manuscript original, MS O, he can inspire a return to the simple, original, natural love that preceded codes. The philological project to reinvigorate the nation lies then in the reconstitution of "courtly love." In the end, pure love will "moralize marriage," just as divorce will renew it.

Gaston Paris and the French philologists inherited from the *histoire des mœurs* the idea of "chivalric love" as the greatest civilizational export of France. But the civilizational decadence and degeneration that had prevailed in the *histoire des mœurs* made the code of love outdated and fatalistic because they described an inevitable cycle of rise and fall. By contrast, the philological revival of courtly love, viewed in the larger context of the divorce debate and the Decadence movement, can be viewed as the desire to transform decadence into a moral and intellectual revitalization of France at the end of the nineteenth century. Courtly love was an opportunity for the renewal of not only national literature, but also national vigor and health. Interestingly, courtly love was just as topical for *scientia sexualis* that also instrumentalized it in order to improve marriage and *mœurs*, though in a different way. Unlike philologists, for whom the decadence of courtly love paled in comparison to pure love-passion, doctors attempted to purify courtly love and transform it into conjugal love, thus creating a new kind of marriage: courtly marriage.

Courtly Marriage

The idea that chivalric love can renew marriage guided the doctors who inherited from the *histoire des mœurs* the idea that "starting in the eleventh century, a fortunate improvement was felt in public and private mores."[95] A spontaneous outpouring of love almost instantly made the Middle Ages into a civilization: "With the twelfth century, mores suddenly change character. The most purified sentiments of divine and earthly love awaken in every heart and take on the form of an incredible exaltation; man rushes to the Orient to defend God's cause, he crosses cities and kingdoms to offer aid to widows, orphans, beautiful, innocent and persecuted women."[96] As certain doctors point out, "Be that as it may, [in chivalry] there was a commendable attempt to reform and discipline the two great motives behind human action: erotic passion and the love of combat."[97] The doctors remembered well the lesson from the *histoire des mœurs* that "the Middle Ages had made love into the principle of chivalry. It even made it into a sort of institution against marriage. The courts of love did not recognize the possibility of love within marriage."[98] If the inequalities and the brutality of feudal marriage gave rise in the Middle Ages to a "theory of chivalric love" (Fauriel) and to conjugal infidelity, they nevertheless improved the medieval civilization and culture in general. Doctors seized on the notion that the principles of chivalric love could improve the foundations of marriage and equality in male-female relations and, if imported into present-day France, they could invalidate the need for divorce. Following its legalization, numerous marriage manuals spurred the population to value conjugal courtliness and marital love—in short, courtly marriage. Thus, courtly love played an important role in enacting profound social changes in the institution of marriage in the last two decades of the nineteenth and the first decade of the twentieth century.

Throughout the nineteenth century, medical discourse upheld marriage as the foundation of human society. The first medical writings on marriage in the 1850s placed it at the root of social structure: "Whatever viewpoint is adopted, it is impossible to conceive a social state whose fundamental base would not be marriage. . . . Yes, without marriage, no society is possible; family is a word empty of meaning and property an uncertain thing without attachments anywhere."[99] In the wake of the anthropological debates on kinship and marriage of the 1860s and 1870s, marriage in fin-de-siècle France was no longer considered the basis of society since time immemorial, but the doctors could still defend the position that it was the basis of family, which was the institutional groundwork of the modern nineteenth-century society: "Marriage and its laws constitute the foundation for the entire social edifice,"[100] and "Marriage is family, and family comprises the support, the *raison d'être* of our society."[101] Marriage had a supreme biological and social function: "Marriage, we have said, is the legal union of man and wife instituted for the satisfaction of physical needs, to perpetuate the species, for the education of children and to help each other, by mutual assistance, throughout the course of life," "the union of man and wife commanded by nature to perpetuate the species."[102] Marriage is a union "which best suits the state of man,"[103] because it "decreases the chances of mortality, insanity and, above all, criminality; it is favorable to fertility and still more to the vitality of newborns."[104] What benefits individuals also benefits the whole of society: "The more you will have married men, the less there will be crimes. . . . A father does not want to blush before his children."[105] Marriage is "the most natural and moral solution, while also being the most beneficial to society and to the individual,"[106] and it guarantees the physical and moral health of individuals ("marriage becomes also the most precious guarantee of health for the body, assuring the vitality and longevity of the spouses")[107] and nation ("This legal union has a marked influence on the physical and moral health of populations").[108]

The question dominating the medical discourse on marriage before the 1870s involved the perfectibility and health of the French race: "What is the goal of marriage? To perpetuate society, not only by the number of individuals, but even more so by the beauty, vigor, and the perfect organization, in so much as possible, of the races."[109] Doctors promoted "the mysteries of conjugal love and the hygiene of marriage, as means to perfect the human species,"[110] efforts that sparked a discussion over consanguine marriages (see Chapters 1 and 2). Marriage was a matter of social engineering of the French race: "It is nonetheless always necessary to protect, and at certain times more than ever, the family and the race."[111] By the 1880s, marriage policy and hygiene were no longer concerned only with eugenics, but also with population numbers: "The extreme mortality rates of children in France, and our weak *birth rates* put into peril the existence itself of our nation."[112] This demographic crisis came in the early years of the French Third Republic following the 1870 Franco-Prussian War. In its immediate aftermath, Armand Hayem wrote in 1872: "If the German armies had only vanquished upon us military victories, the fortune of France would be quickly restored. But it is the victory of one society over another, one civilization over another, one great human family over another great human family more divided, less numerous, less hard-working. The wrong is not only in the ranks of soldiers, it is in the ranks of society itself, even in the family. . . . In France, the population is just about stagnant; in Germany, it doubles in less than 60 years."[113] France suffered a humiliating social and civilizational defeat because its population had declined. If population numbers had dwindled prior to the war, well before the legalization of divorce, the main culprit was the constraint of marriage: "But it is incontestable that the fear of being confined by indissoluble ties must stop many young people despite their liking for marriage; and, thus, the regime of legal separation is perhaps one of the most serious obstacles to a greater number of marriages."[114] For its proponents, divorce would foster greater fertility. Moreover, the newly perfected discipline of statistics supported the idea that the downward tendency of depopulation was not a consequence of divorce, but rather tensions within the institution of marriage, namely, the unhealthy nature of the male-female relationship.[115]

The dire consequences of depopulation—"denationalization"—were observed with increasing alarm starting in the early 1880s. Already in 1882, economist Paul Leroy-Beaulieu began teaching a course at the Collège de France later published as *La question de la population*.[116] "Denationalization" was a growing trend because of the technological advancements and democratization of contraceptive methods, as well as the influence of neo-Malthusianism, which attacked marriage in its primary social function of procreation: "This excessive precaution of medical Malthusianism, against the danger of excessive births, is the very destruction of marriage and its negation."[117] Neo-Malthusians were seen as enemies of the nation: "Malthus's theory is an economic absurdity, and its practice, a crime of treason against the nation [*un crime de lèse-nation*]."[118] In a related development, sexual pleasure without the fear of pregnancy was becoming more widely accessible, thus further undermining the procreative function of marriage. Dr. Brennus's 1893 *Amour et sécurité*, which would reach its 108th edition by 1908, was put on trial for pornography in 1895 because of openly advocating "unnatural practices" to achieve pleasure and avoid procreation.[119] The publisher was acquitted in 1896. Medical manuals for family households dispensed serious medical advice, but some, like Dr. Riolan's medical collection, also contained paid advertisements for books such as *L'école de la séduction* and *L'art de plaire*: "No more hesitation or disappointments in love, thanks to the secrets revealed by Doctors Jaf and Raphaël Saldo. This is the Guide for young men and young ladies, the Quintessential Breviary for Lovers."[120] Another volume from the same series advertised Dr. Eynon's *Manuel de l'amour conjugal* not as "a medical work," but as "a guide, a spousal manual, as much from a moral as from a hygienic point of view. . . . Contrary to widely accepted doctrine, a wise and intelligent husband's essential duty is to seek his happiness in his wife's happiness."[121] Two more volumes from the same series contained graphic ads for "Condoms for men," "the Chaste, the Cutie, the Favorite," and "Condoms for Women," "the Simplex, the Discrete."[122] Faced with the theories of degeneration and decline, armed with demographic facts and statistics, doctors feared that France was at risk of losing its civilizational and imperial dominion. After the original concern with eugenics, they confronted another conundrum: How could they encourage procreation despite the quest for pleasure? How could they make marriage appealing and combat the decline of the nation? In the wake of the divorce debate, when it was philosophically justifiable and technologically possible to use contraception, it was no longer feasible to insist simply on marital procreation as a duty to the nation and individual health.

Medieval courtly love provided tools to reinvest modern marriage with new emotional capital. At the height of the Romantic age, in 1823, J. J. Virey had declared that medicine did not recognize in modern society such a phenomenon as chivalric love:

> We no longer know the rapture of love in our centuries; they are done, these times of chivalry and *courts of love*, when women doled out glory, became the arbiters of *courtoisie* and prowess among wayfaring knights; they reigned just by looks, and their sweet dominion [*empire*] was perpetuated by the purest virtue and the most faithful bond. Such were also these *galois* and these *galoises* [*sic*], a sort of brotherhood in the Middle Ages, who took oath to suffer both the heat of summer and the cold of winter, and every torment, if need be, for a beloved.[123]

Instead, modern love was equated with procreation and sexual instinct: "What we call *love* is only the external sign of this vital movement that tends to spread into other beings, to transmit life to them."[124] "Marital love" (*l'amour conjugal*) meant sexual intercourse and procreation: "Loving is living for one's species,"[125] and "The primordial cause of love is without a doubt in the instinct of reproduction."[126] Throughout the nineteenth century, love continued to be conflated with the principles of making life: "Love is the principle of life, life is characterized by love, that is, reproductive vigor."[127] But once divorce was legal and procreation within wedlock was no longer a given, marriage had to be made into a source of happiness to give spouses reason to choose their union: "[The] intimate sharing of two people seemingly made for one another and which will create for each of them the conditions of a most complete happiness."[128] In order to encourage procreation and the stability of families at the foundation of society, doctors began to associate love and marriage, substituting love within wedlock for love outside of wedlock. Courtly love became courtly marriage: "Love in marriage is not merely a condition of domestic happiness that all should seek, in preference over all of the other elements that ordinarily enter into matrimonial combinations."[129] Love in marriage was touted as the foundation of familial happiness, in contrast to past practices that encouraged normal heredity, strong education, or a substantial dowry as essentials in the creation of the healthy family lineage. Divorce thus made a happy marriage, no longer an ideal, but a desirable norm and even a necessity. Robert Nye pointed out that "initially, at least, these positive valuations of bourgeois 'amour conjugal' may have depended on models of aristocratic sexual gallantry."[130] But doctors in fact rejected gallantry, for what it had to offer was superficial: "Many moralists seem to have taken up the task of conflating gallantry with love. Thus we owe to this confusion the disagreement that reigns in a fair number of writings on this last passion. And yet what a difference! Less vibrant, less serious, but more clairvoyant and more sensual than love, gallantry searches for physical rather than moral beauty. . . . Love commits us uniquely, generously and without reserve to the object of our affection."[131] Instead, the doctors proceeded to codify conjugal love. Love was a cult and followers of this cult had to observe a certain number of rituals. In the manuals, the complete ritual of courtly marriage and conjugal love—from wooing, first night, honeymoon, and so on to marital hygiene—is subject to a code of love. Sex education thus became the necessary condition of social medicine: "In the interest of bodily and intellectual hygiene of humanity, it is necessary to push young men to marry and give them the anatomical precepts of love."[132]

Doctors were popularizing courtly love as a marriage model, a model of courtship and conjugal love, but they emphasized first of all that it was not passionate love, "love-passion" (*amour-passion*), "fiery ardor" (*fougue amoureuse*):[133] "When love, whatever its vehemence may be, has no other foundation than the physical charms of youth and beauty, it is rare that passion and, above all, that the abuse of pleasures not end up leading bit by bit to indifference and even distaste. So it has rightly been said when speaking about unions of this nature that the nuptial bed is love's grave."[134] Love-passion was a transient kind of love, and that subsequently could not form a firm base for a durable union: "It can therefore not be love that is the principle of an enduring union between the man and woman. Love determines the union, but it is impotent to maintain it, and, as a cherub [*un dieu enfant*] who knows not how to wield his power, it does and undoes its work."[135] Nor was love to be mistaken for the other form of passion, that is voluptuousness and lust: "Love is the reason for all things: we speak here of that tender, honest love that the heart inspires and reason guides, of that calm love that one would take for *friendship* [*amitié*], not of that fiery love, born from the ardor of the senses, which disappears as soon as desires are satisfied."[136] This careful moderation corresponded to the ideals of the courtly lover's self-mastery before the lady—the marital process became a well-ordered set of necessary and well-measured steps toward happiness.

The ritual starts with courtship (Adler, pp. 15–30) and continues with the first night. This is the most critical moment in the rite of passage to future happiness: "Often the happiness and future of a household depends on that first night."[137] The longevity of marriage depends on this initial experience: a man's brutality and inability to satisfy his wife while expecting her to achieve orgasm at his own pace is "the most common cause of a great number of conjugal separations":[138]

> What pain the unfortunate woman must endure when, without preliminaries, without preparation, under the thrusts of the one from whom she hoped the most delectable caresses, she feels painful bruises of the flesh and the senses. Some words of love and affection must, on the contrary, precede and prepare; first, upon entry into the bed, then, at the first bodily contact. The man will therefore proceed with caution, will attempt some preliminary tries prior to full penetration, which will arouse the external organs and will make the woman enter in advance into the spasms of pleasure.[139]

In short, "The man attends carefully not just to the organs, but also to the emotional sensitivity of the woman."[140] Young women also need advice, proffered by Dr. Fauconney in "L'ami des jeunes femmes ou Les devoirs du mariage et de la maternité" (The friend of young women, or The duties of marriage and maternity).[141] Happiness on the first night combats successfully against frigidity and, ultimately, sterility. The honeymoon is the perfect occasion to understand each other: "It is a period of euphoria with no bounds, granted by nature to the spouses for them to understand each other and reach agreement on the normal and complete satisfaction of the sexual needs and senses."[142] But spouses must learn to moderate their passion and desire: "It must be used, and not abused. . . . Moderation in the pleasures is a wise economy that will give a charm always new."[143] Moderation in sexual activity was also an ancient medical notion, affirmed by Cabanis in 1802, but framed here by a very clear set of marital rules and codes of behavior that were to inspire a happy union over time. In the most direct borrowing from chivalric love, in the code of courtly marriage, conjugal love is a cult:

> Marriage, so they say, is the grave of . . . entirely physical love, which, according to the proverb, dies out in marriage. . . . True love begins at this death, *the change from gallantry to cult* is the necessary effect of marriage, every husband, whatever face he may outwardly project, is, inwardly, an *idolater*; if there is ostensibly a conspiracy among men to shake off the oppression of the [female] sex [*le joug du sexe*], there is also

> an unspoken convention *to adore* it; only the weakness of the woman occasionally forces the man to take back control; barring such rare exceptions, the woman is *sovereign*, and therein resides the principle of affection and conjugal harmony.[144]

Ultimately, courtly love sheds its ties to adulterous passion outside of wedlock and is thus transformed into a medical model of conjugal love, "a marriage of inclination."[145] In 1886, the French translation of Paolo Mantegazza's *Gli amori degli uomini* (Sexual relations of mankind), was published: "Free choice for the two sexes; no contract imposed by the parents and endured by the children."[146] One of the most successful medical popularizers, Dr. Garnier, encouraged this: "All the more so for those who chose each other, freely loved, and that a mutual love unites. Free and joined, they have only to tell and to prove to each other reciprocally all of the happiness that they derive from it."[147] By 1905, this idea will be firmly lodged in the spirit of doctors writing conjugal manuals: "It matters not to hamper the natural freedom of choice of a couple, and the parents who impose a tyrannical influence on the will of their children are guilty," since "marriage . . . is first of all a free union of two hearts later sanctioned by the laws in order to found a family."[148]

Courtly marriage thus unites procreation and pleasure: "It is even authorized to affirm that the act is always imperfect without this union of hearts with the cooperation of the senses."[149] "True love" occurs within marriage in large part because pleasure is now supposedly connected to it: "In true love, the pleasure of the senses plays a large part."[150] "In the coupling of lovers, there should in reality exist only a shared and reciprocal pleasure."[151] Pleasure, true love, and marriage are the new recipe for hygiene, health, and happiness, that is, "the *perfect marriage*. Enlightened and impassioned by love—born of it or before it—such a union gives to the two *friends* [*amis*] that it unites for all time a moral tranquility, a spiritual balance, an energy in the will that multiplies their worth and their strength. For us, in a word, the *perfect marriage*, soars above social and purely conventional form, and exudes an emotion that, for lack of a better word, we call *love*."[152] The language of medieval, adulterous courtly love—*amour, ami(s), amitié, culte*—was transformed into a ritualized, that is, regulatory model of conjugal love and courtly marriage. It was a well-suited technology for the social engineering of family. Given its heteronormativity, when brought within the bounds of marriage and procreation, it could cure society of social ills such as adultery, prostitution, and homosexuality and repopulate the nation. Converging with the normativity of medicine, its heteronormativity created a lasting myth of courtship and "romantic" love in courtly marriage, while excluding from this mix the damaging paradox of medieval courtly love that improved the violent and brutal feudal mores only by encouraging adultery, as examined above in the medieval *histoire des mœurs* of the first half of the nineteenth century.

Nowhere is this nineteenth-century technology of courtly marriage better exposed than in Jacques Lacan's Seminar VII, *The Ethics of Psychoanalysis*. By the time Lacan lectured on the "Courtly Love as Anamorphosis" in 1959–1960, the term "courtly love" made immensely popular by Denis de Rougemont's *L'amour et l'Occident* (1939) had gained widespread currency. Lacan was not interested in the historicity of the cultural phenomenon. Did courts of love exist or not? Was courtly love an actual social practice in male-female relations? "Personally, I don't believe a word of it," he declared, and these ideals "couldn't have any real concrete equivalent."[153] Rather, Lacan read courtly love as a poetic practice that had existed between the beginning of the eleventh and early thirteenth century in the south and north of France and the Germanic realm. This poetic practice "finds its point of origin in a certain systematic and deliberate use of the signifier as such" (p. 148 [p. 178]). Its use of the signifier is purposeful in that it makes stand out the absence of the Real at the center of signifiers. It brings into light precisely their signifying function as they reach the limit of symbolization and open a gap within and beyond signification. This symbolic limit against which is felt the pressure of the Real is the Thing (*das Ding*, *la Chose*). Because the Real resists the symbolic order, primary symbolization, having missed it, is followed by an ensuing chain of signifiers, each in turn falling short of the target. The pleasure principle leads the subject in the endless pursuit of the next signifier that promises yet withholds it. But in the poetic practice of courtly love, man's adoration of an inaccessible Lady and his submission to a cult of love reveal this dispossession from the Real, which man craves and sublimates: "That final demand to be deprived of something real is essentially linked to the primary symbolization which is wholly contained in the signification of the gift of love [*le don d'amour*]" (p. 150 [p. 179]). The Lady is emptied of all real substance and transformed into a symbolic function of the subject's ideal; the distance from her is the measure of the impossibility to represent the Real. Courtly love insists on this inaccessibility and thus pursues the pleasure principle in its very unpleasure or dispossession. The "gift of love" is a fantasy because the gift of the Real is endlessly deferred in the chain of signifiers; in courtly love, the "systematic and deliberate use of the signifier as such" causes the Real, which remains in reality lacking, to emerge precisely as fantasy, illusion. This illusion is not emptiness, from which it differs (p. 140 [p. 169]); rather, the illusion transcends itself by revealing its signifying function. Courtly love is therefore a rare anamorphosis where "what the function of the signifier means" (p. 153 [p. 183]), otherwise hidden in the pursuit of the infinite chain of signifiers, becomes visible. This poetic practice interested Lacan from a functional point of view because of its "determining influence on manners [*les mœurs*] . . . and subsequently in its historical consequences" (p. 149 [p. 178]). If its actual historical effects—had it improved medieval male-female relations?—were of lesser consequence for Lacan, courtly love nevertheless had an effect on psychic organization, by means of its "signifying system or scheme [*organisation signifiante*]" (p. 172 [p. 143]): "These ideals, first among which is that of the Lady, are to be found in subsequent periods, down to our own. The influence of these ideals is a highly concrete one in the organization of contemporary man's sentimental attachments, and it continues its forward march" (p. 148 [p. 178]). We can infer from Lacan's comments that the assimilation of courtly love within the structures of marriage entailed a major psychic reorganization of the nation's mores. When doctors appropriated medieval courtly love to use it as a technology regulating the nation's population, they transformed it from an anamorphosis of illusion into an attainable reality by fantasy. That is, the "gift of love" precisely as illusion, fantasy, was harnessed in order to draw together the population into desiring a healthy and modern nation-state, a Real that could be attained, the doctors claimed, through a series of rituals. By loving and procreating in the medically approved manner, the population could satisfy its obligation to the nation, perform its individual contribution to the species, and at the same time achieve its personal psychic fulfillment. The assimilation of love and marriage, combined into a new creation of courtly marriage, in addition to the reification of heterosexuality and the nuclear family as transhistorical truths, were the crowning achievements, in the eyes of nineteenth-century doctors, of medical medievalism.

Epilogue

From Pornography to Archaeology: Priapus at the Cluny Museum

The investigation of *Pornographic Archaeology* started with the premise of double transference: sexuality and nation formed the nineteenth-century notions of the medieval world and its sexual practices and, at the same time, these ideas shaped the moral and social views in contemporary France and, more locally, disciplinary knowledge. While the first half of the nineteenth century teemed with open, even if at times phantasmagoric, explorations of the nation's past medieval sexuality, the second half saw the waning of pornographic archaeology that had been practiced in the *histoire des mœurs*. Academic disciplines preferred the notion of courtly love that was never taken up as a discourse of sexuality, while examination of sexuality—sex, sexual practices, and desires—was limited to medical literature.[1] What were some of the forces at work in this gradual closing off to the eroticism and sexuality present within the nation's past?

In January 1860, sigillographer Arthur Forgeais (1822–1878) put up for acquisition his collection of "historiated pewters" (*plombs historiés*) in a letter to Edmond du Sommerard, conservator and director of the Musée national du Moyen Âge in Paris. Forgeais's collection of pilgrim and secular lead badges (*enseignes de pèlerinage, enseignes profanes*) had been dredged out of the riverbed during the works begun in 1848 to make the Seine navigable. In his final letter addressed to du Sommerard to negotiate the transaction, dated March 5 (or 6), 1861, Forgeais spoke with enthusiasm of "this unique and unrivaled combination of small monuments from the Middle Ages destined to embellish one of the museums in the capital," a collection that would also "shed a new light on this shadowy and distant era."[2] Acquired by the Musée national du Moyen Âge-Thermes de Cluny in the course of 1861–1862, Arthur Forgeais's collection is indeed one of the most important in the world, containing more than 3,000 lead badges, principally from the late Middle Ages. Forgeais published the full inventory of his collection between 1862 and 1866 in five volumes.[3] A small part of this collection included secular badges of overt sexual content, called "erotic (or sexual) badges" (*enseignes érotiques*). But, as the editors of the 2006 catalogue of new acquisitions at the Cluny Museum note, "Because of their content and despite their presence since the beginning within the collection of the museum, they [erotic badges] had hardly been examined or published in the past."[4]

Anticipating this attitude of propriety and secrecy, Forgeais published, sometime before his death in 1878, a book on his collection of erotic badges, in a volume that was separate from the five-volume inventory of his collection. His first concern, from the opening pages of this sixteen-page book entitled *Priapées*, was to justify the motivation for its publication: "There exist some turpitudes that must not be brought together without a serious reason, and I have this excuse that my work will probably not be glimpsed outside the circle of earnest people. The form is moreover always crude enough so as not to inflame the imagination, which knows how to abstain when the heart does not give over willingly to the inebriation of the senses."[5] Despite the crude representation of sexual organs that, Forgeais hoped, was not sufficiently sophisticated to arouse its viewers, this publication, which proved unusual for its time, was warranted in the interest of erudition. But Forgeais remained perplexed as to the possible use of these badges: "As for the purpose that they may have served . . . that's what is difficult to determine."[6] The collection of these "little priapic monuments"[7] was sizable enough, and if this many badges were found in just one locale, they were not only widespread, Forgeais concluded, but also of some practical significance beyond ornamentation that must have been clear to all their users. Although Forgeais understood that these medieval artifacts were not made with the intent to moralize, he nevertheless felt strongly that he had to offer an interpretation for his contemporaries that was indeed moral: "Let us draw at least from these erotic clowneries what they can offer as deeper morality, although the purpose of the artist who drew them may have been entirely different; for it does not seem that one had the pretension of moralizing with such crass forms, since the senses are so easily prepared to triumph if one absolutely disrobes them of their veils."[8] He thus described a winged phallus (*phallus ailé*) as "mounted on two legs, as though the animal were thus complete, except for the wings (which can nonetheless designate impatience and anxious ardor). It is not a fully absurd portrait of

Figure 7. Erotic badge: Winged phallus (ca. 1375–1425). CL 3739, Musée national du Moyen Âge-Thermes de Cluny, Paris, France (collection of Arthur Forgeais). Réunion des musées nationaux / Art Resource, NY.

certain natures where lewdness [*lubricité*] has been inflamed by habit, until it becomes constant monomania"[9] (Figure 7).

In his description, Forgeais deployed not only the language of moralistic condemnation of "lewdness," but he also displayed his familiarity with the medical notion of "monomania," probably in the sense of "erotomania." In other words, Forgeais used the medical language to moralize an otherwise crude, yet potentially arousing, representation. Two other items in the collection are described in a similarly moralizing tone: "Thus the woman straddling a *phallus* can solicit this idea: *humanity that abandons herself to the bestial inclination ends up no longer in control of herself*"[10] (Figure 8). In another example, "We would again come across a similar lesson in the figure of the man who bears an organ larger than the rest of his body. To use the expression from Balzar . . . this man has become a temperament"[11] (Figure 9).

Along with these attempts to moralize the medieval artifacts and render them palatable to his contemporaries, Forgeais also historicized them as representative of medieval sexuality, an interpretation he derived from the *histoire*

Figure 8. Erotic badge: Woman astride a phallus (ca. 1400–1425). CL 3737, Musée national du Moyen Âge-Thermes de Cluny, Paris, France (collection of Arthur Forgeais). Réunion des musées nationaux / Art Resource, NY.

des mœurs. Thus, as if borrowing a page from Paul Lacroix's *Histoire de la prostitution*, Forgeais hypothesized that these erotic badges could only serve as a form of an unspoken language of prostitution, as well as a sign of infamy in the attempt by the medieval urban authorities to promulgate public hygiene: "But why would they not be either a sort of *badge* adopted by women of loose morals in order to be understood without discussion, or a token of infamy imposed upon them in order to classify them clearly, or, perhaps, some sign of subscription to a place of ill-repute?"[12] Forgeais resorted to prostitution as an explanation of the historical artifact that he believed his contemporaries would see as potentially arousing, that is, pornographic. His interpretation thus epitomizes nineteenth-century concerns with the definition and function

Figure 9. Erotic badge: Ithyphallic man (ca. 1400–1425). CL 23469, Musée national du Moyen Âge-Thermes de Cluny, Paris, France (collection of Arthur Forgeais). Réunion des musées nationaux / Art Resource, NY.

of pornography, signaling the development of an autonomous visual pornographic genre, while highlighting the ancestry of pornography in the past practices and customs of prostitution.

The original meaning of "pornography" was "prostitution": "Consistent with the Greek etymology of the word, the title of pornographic work belongs above all to books on prostitution. Such is the book by Restif de la Bretonne entitled: *Le pornographe ou Idées d'un honnête homme sur un projet de règlement pour les prostituées, propre à prévenir les malheurs qu'occasionne le publicisme des femmes* (The pornographer or Ideas of an honest man about a project to regulate prostitutes, proper to prevent the evils that cause the publicness of women)."[13] But Alexandre-Jean-Baptiste Parent-Duchâtelet's 1836 *De la prostitution dans la ville de Paris, considérée sous le rapport de l'hygiène publique, de la morale et de l'administration* (On prostitution in the city of Paris, considered in relationship to public hygiene, morality, and administrative government), a work of serious research, hygienic and regulatory intention, and scientific language, could no longer bear the name of "pornography," as Rétif de la Bretonne's did: because of "modifications that occurred in the meaning of certain words . . . we would no longer dare to give the name 'pornographic' to the book by Parent-Duchâtelet."[14] By the mid-nineteenth century, the meaning of "pornography" as prostitution had definitively shifted to "pornography" in its modern meaning. Pornography had evolved into a self-standing genre, a legal and artistic category, understood to have a single aim of sexual arousal and pleasure: works "of which immorality constitutes the basis and the purpose" and that have "a more thorough intention of immorality, are more completely made in the aim of abetting lewd appetites, awakening the indifferent [*blasés*] senses."[15] The development of pornography into an autonomous genre of representation of sexual organs and sexual activities with the aim to arouse was made possible by the spread of print culture and rise of literacy. The wide accessibility of text and image required regularization, both legal and moral: "Pornography as a regulatory category was invented in response to the perceived menace of the democratization of culture,"[16] though what Forgeais's example shows us is that laws were best enforced when interiorized morally: "In actuality, what is propriety [*pudeur*], if not the habitual care not to awaken, neither in ourselves, nor in others, the perilous instincts always ready to break out like the ferocious beast who only dreams of prey?"[17]

Articles on "Pornographie" and "Prostitution" in Pierre Larousse's *Grand dictionnaire universel du XIXe siècle* are helpful not only in tracing the linguistic but also this epistemic shift that was complete by the mid-nineteenth century. For pornography was not just a textual-visual genre that needed to be named and categorized, but also a concept whose contours were shifting and fluid, and difficult to define. The major problem faced by authors was how to avoid incitement to imitation, by text and/or image; the worst-case scenario was to provide an example such as Suetonius did for Gilles de Rais, and Gilles de Rais for Baron Fersen (see Chapter 4). When writing about prostitution, Larousse himself promised to impose "on our quill pen, all the greater and simpler a reserve given that such images offer a sufficient moral lesson by view alone. On the other hand, it is with severity, it is with the most extreme rigor that we will choose our quotations, that we will recall the historical facts."[18] If "reserve" was not exercised, description could have an *effet pervers*. Text and images intended to provide historical examples or medical diagnoses could instead arouse and incite to sexual activity by virtue of imitation; "reserve," "rigor," and "severity" were thus intended to restrain the imagination and render it impotent. This reserve, or propriety, however, seriously limited the ability of writers, and especially doctors, to provide scientific accounts of sexual diseases and practices:

> Because of a misplaced commitment to propriety [*pudeur*], these questions [of pederasty] have been only addressed from a scientific standpoint with a certain timidity. The authors, even those who believe themselves the most emancipated from any religiosity, fear to be suspected of pornography or scientific inappropriateness [*inconvenance*] and spew [*se répandent*] with the fiery spirit of a predicator various epithets about a *vice* that they classify as an abomination, a monstrosity, an infamy, etc., etc., as though the truly extraordinary character of these facts, in all societies, at all eras in history, should not attract the attention of the psychologist and the doctor of legal medicine in order to elucidate these complexes and to know whether . . . a treatment is required.[19]

Writers worried that even scientific language could have the *effet pervers* of pornographic stimulation, but the by-product of their restraint was that diseases and behaviors could not receive adequate treatment—whether medical or in writing—and scientific exchange was hampered.

This problem of arousal as a side effect of description or representation was common to all studies of prostitution (and, later in the nineteenth century, of inversion). The combination of image and text was the most dangerous, as the review of Auguste Rabutaux's *De la Prostitution en Europe depuis l'antiquité jusqu'à la fin du XVIe siècle* (On prostitution in Europe from antiquity to the end of the 16th century) (1851) showed: "It is hardly appropriate, we believe, to excite the people, through the attraction of engravings of dubious taste, to buy a book which, by the care taken in its composition and by the very nature of the subject that it treats, must remain only in the hands of moralists and learned scholars."[20] Another book on prostitution judged most severely was Paul Lacroix's *Histoire de la prostitution, chez tous les peuples du monde depuis l'antiquité la plus reculée jusqu'à nos jours* (History of prostitution among all of the peoples of the world from the earliest antiquity to the current day) (1851–1853), written under the pseudonym of Pierre Dufour (see Chapter 5). Compared to Rabutaux or Parent-Duchâtelet, Lacroix's project was the most "popularizing," and thus the most threatening to public hygiene and morals: "The editor spared nothing to give the book by M. Dufour all of the seductive airs designed to attract the attention of idle [*désœuvrés*] and libertine students and old men. By acting in this manner, perhaps, the success from scandal was obtained, but this work no longer has the dignity to warrant a place in the libraries next to the serious authors that we have just named."[21] The way Lacroix popularized prostitution and its history meant, in the interpretation of Larousse's *Grand dictionnaire universel du XIXe siècle*, to participate in the degradation of the moral and physical health of the reading public. The goal of the publication should be physical and moral hygiene, not the commercial ends of appealing to the public.

Lacroix was, of course, aware of this and in fact played on the double meaning of "pornographic" in his *Histoire*. While "pornographic archaeology" (*archéologie pornographique*) indicated prostitution as the subject of the book, it was an intentional throwback to an earlier etymology that was no longer current. At the same time, Lacroix's use of the word "pornographic" was a play on the potential of his topic to arouse. This was not exclusive to Lacroix's *Histoire*, for all archaeology dealing with prostitution had pornographic potential. In fact, the term "pornography" emerged out of classical archaeology, when it became clear to eighteenth-century archaeologists that Romans of Pompeii lived indiscriminately among images of Priapus.[22] Lacroix asserted that in fact only archaeologists had dared write about the history of prostitution, the subject that is "delicate and suspicious at the same time."[23] But they were unable to do so with restraint and reserve and archaeology was incapable of being moral because of its methodology:

> We will never lose sight of the fact that this book, prepared slowly for the benefit of science, must serve morality and that its principal objective is to make vice hateful by revealing its turpitudes. . . . It is above all in this manner that we will distinguish ourselves from archaeologists and intellectuals properly speaking, who are not concerned by the morality of the facts and who are not worried about deriving from them philosophical consequences. They discuss at length, . . . they divulge its monstrosities, they retrace its infamies, but they forget thereafter to purify our thinking and to calm our spirit, by opposing to these impure and degrading images the chaste lessons of philosophy and the healing [*bienfaisante*] action of Christianity.[24]

In their commitment to erudition, archaeologists forgot their debt to the moral fortification of the public. Before an archaeologist can be a scientist, he must be an educator, a moralist: "As for ourselves, as much an archaeologist as we also are, we will not forget that . . . we are addressing a French public that wants to be instructed, but that wants at the same time to be respected."[25] The principal objective of a history of prostitution—a "pornography"—must be to educate, the task at which archaeology had failed.

But unlike Greco-Roman artifacts, tributes of a lost civilization that perished because of its moral degeneracy, medieval artifacts were an unfortunate testament to the obscenity of the nation's origins. The fabliaux especially, a worthy companion to the erotic badges, were "often of a disgusting obscenity. The dirtiest words of the French language seem prodigious in them as much as could be desired; the most vulgar functions of the human machine are the subject of crude jokes, the most secret parts of the body are named in terms that would make prostitutes today blush."[26] The display of medieval archaeological artifacts contained not only the risk of inciting and arousing the public but also the risk of shaming the nation. The Art of Sex could circulate only in the form of Greco-Roman artifacts, whereas the Middle Ages had to be content with forging an Art of Love. It will be a long road for academic

Figure 10. Erotic badge: Phallus (ca. 1400–1425). CL 23470, Musée national du Moyen Âge-Thermes de Cluny, Paris, France (collection of Arthur Forgeais). Réunion des musées nationaux / Art Resource, NY. (There seems to be some confusion as to whether this badge belonged to the original Forgeais collection. While the acquisition archives confirm this, Bruna classifies it in his inventory [D.B. #609] as of "unknown origin" [*provenance inconnue*] [Bruna, p. 318].)

Figure 11. Erotic badge: Three phalluses and a vagina (ca. 1375–1425). CL 23653, Musée national du Moyen Âge-Thermes de Cluny, Paris, France. Réunion des musées nationaux / Art Resource, NY.

disciplines to establish "a clear, unmistakable, scientific language understood and accepted by all"[27] and for the medieval artifacts, and the fabliaux, to return to the public eye as archaeological artifacts, stripped of their power to incite. This return to archaeology took place at the Cluny Museum only in July 1998, when the first erotic badge from the Forgeais collection acquired in 1861–1862 was finally exhibited, following the 1996 publication of Denis Bruna's catalogue and inventory of pilgrim badges held there (Figure 10).

This badge, representing a phallus on two legs wearing a bell, remains the only artifact of its kind exhibited from Forgeais's collection, thus countering Forgeais's wish for a new light to be shed on the Middle Ages, "this shadowy and distant era." But have the floodgates opened? This first move was followed by the museum's acquisition of a rare pilgrim badge immediately thereafter in 1999, which has also been on exhibit (Figure 11).

Lacroix's pornographic archaeology belongs to the prehistory of the medieval discipline. Since national roots of philology were called into question by new philology and new medievalism in the 1990s, the relationship of the academic discipline of medieval studies to sexuality also needed to be reconsidered and historicized.[28] It seemed important to remember those who were excluded by philology that rejected the obscenity—pornography—of the fabliaux. But it remains to be seen if, after new medievalism and new philology, this warrants a claim to the new pornographic archaeology.

Notes

INTRODUCTION

1. "L'étude approfondie des romans de chevalerie et de la littérature que les érudits sont seuls à connaître, et que les curieux n'effleurent même pas, mènerait à bien des découvertes intéressantes. Espérons que ce champ fertile ne sera pas longtemps négligé; à mesure que le nombre des travailleurs érudits augmente, les demandes qu'on a à leur faire s'accroissent aussi et se révèlent. Et l'histoire de l'unisexualité réclame cette étude"; Marc André Raffalovich, *Uranisme et unisexualité: Étude sur différentes manifestations de l'instinct sexuel* (Lyon: A. Storck; Paris: Masson, 1896), p. 36.

2. André Gide (1869–1951) published his defense of inversion (homosexuality) in *Corydon* (1924). Georges Hérelle (1848–1934) was a prolific and erudite historian and ethnographer of Basque pastoral poetry, and professor of philosophy. He is the translator (under the pseudonym L. R. de Pogey-Castries) of the *Histoire de l'amour grec* by M. H. E. Meier (1837). The first translation, annotated, was published in 1930, and republished in 1952 and 1980. He is the author of an unpublished, three-volume manuscript "Nouvelles études sur l'amour grec" [New studies on Greek love], a work he began in 1884 and the reason for his correspondence with Gide; see "Georges Hérelle (1848–1935)," http://www.garae.fr/spip.php?article220, accessed October 11, 2011.

3. "Quelles furent, en réalité, les mœurs du Moyen-Âge? La littérature nous instruit-elle suffisamment sur les mœurs de cette époque? Cette littérature si tendre, si délicieuse, est très idéaliste et très artificielle; mais nous dit-elle la vérité? Qu'y a-t-il derrière ce charmant décor? . . . Où et comment se renseigner? Par exemple, quel était le rôle des pages, jeunes compagnons des chevaliers? Ceux-ci faisaient profession d'amour mystique: on ne parle que de cet amour-là, c'est le seul qu'on mette en avant. Mais comment supposer que tous ces gaillards restassent chastes? Et vers qui se portaient alors leurs désirs sensuels? À voir combien, aujourd'hui, la réalité diffère de l'apparence et combien le revêtement des mœurs est mensonger, il est permis de penser que ce mensonge n'est point particulier à notre époque, et qu'il était encore plus épais dans les temps où l'opinion, plus sévère, contraignait à plus de dissimulation. Il faudrait étudier ce problème, sans tenir compte de la littérature, dans les chroniques secrètes, dans les procès criminels, dans les documents relatifs aux cloîtres"; André Gide, Letter to Georges Hérelle, dated July 14, 1934, transcribed by G. Hérelle, Bibliothèque municipale de Troyes, MS 3188, fol. 359. I thank Monique Nemer for bringing this precious document to my attention.

4. Marc Bloch, *Les rois thaumaturges* (London: H. Milford, Oxford University Press, 1924). In a chapter on "Modes of Feeling and Thought," Bloch says, "A history more worthy of the name than the diffident speculations to which we are reduced by the paucity of our material would give space to the vicissitudes of the human body [*aux aventures du corps*]. It is very naive to claim to understand men without knowing what sort of health they enjoyed"; Marc Bloch, *Feudal Society*, trans. L. A. Manyon (London: Routledge & Kegan Paul, 1961), pp. 72–87, at p. 72, translation modified (*La société féodale* [Paris: Albin Michel, 1982 (1939)], pp. 115–35, at p. 115).

5. Marcel Mauss, "Body Techniques," in *Sociology and Psychology*, trans. Ben Brewster (London: Routledge & Kegan Paul, 1979), pp. 95–123, at p. 97 ("Les techniques du corps," in *Sociologie et anthropologie* [Paris: Presses universitaires de France, 1950 (1936)], pp. 365–86, at p. 365), presented to the Société de Psychologie on May 17, 1934, and published in the *Journal de Psychologie* 32: 3–4 (mars-avril 1936).

6. "The terms used in French to designate and evaluate the age (*moyennageux* [1865], *médiéval* [1874]) and to refer to those who make it a specific historical discipline (*médiéviste* [1867]) all date from the later nineteenth century"; Alain Boureau, *The Lord's First Night: The Myth of the* Droit de Cuissage, trans. Lydia G. Cochrane (Chicago: University of Chicago Press, 1998), p. 72 (*Le droit de cuissage: La fabrication d'un mythe [XIIIe–XXe siècle]* [Paris: Albin Michel, 1995], p. 86).

7. Michel Foucault, *The History of Sexuality, Volume I: An Introduction*, trans. Robert Hurley (New York: Vintage Books, 1990) (*Histoire de la sexualité I: La Volonté de savoir* [Paris: Gallimard/Tel, 2005 (1976)]).

8. While male, and female as its correlative, heterosexuality and male homosexuality figure prominently in these discussions, nineteenth-century doctors barely touch upon the medieval female homosexuality. The doctors consider that "tribadism" or "saphism" appears with regularity in the Renaissance. I was therefore not able to constitute a wide enough sample of writings on the subject. The popularizers, Dr. Hayès, Dr. Garnier, Dr. Fauconney, and Dr. Alibert, of the very end of the nineteenth and early twentieth century usually dismiss the topic with this summary: "Women, even nuns, gave in to orgies among themselves where the Roman *fascinum* reemerged and the art of fellatio had lost nothing from the lascivious lessons of antiquity. The term 'tribades' is reported in all of the penitentials. . . . Later [appears] . . . the famous flying squadron of the queen Catherine of Medici" (Les femmes, même les religieuses, se livraient entre elles à des orgies où reparaissait le *fascinum* romain et où l'art fellatoire n'avait rien oublié des leçons impudiques de l'antiquité. Le terme des tribades se trouve rapporté dans tous les penitentiels.... Plus tard [apparaît] . . . le fameux escadron volant de la reine Catherine de Médicis); Dr. Jaf (Dr. Fauconney), *Physiologie du vice, son histoire à travers les âges* (Paris: C. Offenstadt, 1903), p. 146.

9. "Il est plus facile . . . d'étudier les natures diverses dans leurs crises que dans leur normalité: l'état régulier ne laisse voir que la surface et dissimule les replis intimes; les troubles morbides font mieux ressortir ce qu'est la santé"; Dr. Cabanès, *L'histoire éclairée par la clinique* (Paris: Albin Michel, 1921), p. 100.

10. "l'engouement que montre aujourd'hui le public pour le *point de vue médical*"; Eugène Dally, *Des dangers attribués aux mariages entre consanguins.* Extrait de la *Gazette hebdomadaire de médecine et de chirurgie* 9 (1862): 1–45, at p. 21.

11. François Just Marie Raynouard, *Les Templiers, tragédie. Précédée d'un précis historique sur les Templiers* (Paris, 1805).

12. Pierre Dufour (Paul Lacroix), *Histoire de la prostitution, chez tous les peuples du monde depuis l'antiquité la plus reculée jusqu'à nos jours*, 6 vols. (Paris: Seré; Martinon, 1851–1853), vol. 4, p. 116.

13. "surtout aux livres relatifs à la prostitution"; Pierre Larousse, *Grand dictionnaire universel du XIXe siècle* (Paris: Administration du Grand dictionnaire universel, 1866–1877), "Pornographie," vol. 12 (1874), pp. 1438–39, at p. 1439. The best example of these early "pornographic" works is Rétif de La Bretonne, *Le pornographe ou Idées d'un honnête homme sur un projet de règlement pour les prostituées, propre à prévenir les malheurs qu'occasionne le publicisme des femmes* [The Pornographer or Ideas of an honest man about a project to regulate prostitutes, proper to prevent the evils that cause the publicness of women] (London; Paris, 1769). For the etymological history of "pornography," see Walter Kendrick, *The Secret Museum: Pornography in Modern Culture* (Berkeley: University of California Press, 1996 [1987]).

14. "l'*Encyclopédie* du XIXe siècle"; Pascal Ory, "Le 'Grand Dictionnnaire' de Pierre Larousse. Alphabet de la République," in *Les lieux de mémoire*, ed. Pierre Nora, 3 vols. (Paris: Gallimard/Quarto, 1997 [1986]), vol. 1, pp. 227–38, at p. 228. See also Alain Rey, "Les trésors de la langue," *Les lieux de mémoire*, ed. Nora, vol. 2, pp. 2189–205, at p. 2195: "Larousse . . . imparts to us the great Repubican didactic project" (Larousse . . . nous transmet le grand projet didactique républicain).

15. "nous fait pénétrer dans l'intérieur, dans le fond de cette vie [des sociétés]; elle nous apprend les mœurs, les usages, les coutumes, les croyances des anciens"; Larousse, *Grand dictionnaire universel*, "Archéologie," vol. 1 (1866), pp. 571–72, at p. 571.

16. "*le siècle de l'histoire*"; Larousse, *Grand dictionnaire universel*, "Histoire," vol. 9 (1873), pp. 300–314, at p. 301.

17. "les faits généraux de la politique, de la guerre, des révolutions, . . . [les] événements remarquables qui dessinent en grand les figures des peuples, en un mot [le] côté dramatique et extérieur de la vie des sociétés" (Larousse, "Archéologie," p. 571).

18. Michel de Certeau, *The Practice of Everyday Life*, trans. Steven Rendall (Berkeley: University of California Press, 1984) (*L'invention du quotidien* [Paris: Gallimard, 1990 (1980)]).

19. Gaston Paris, "*La Chanson de Roland* et la nationalité française. Leçon d'ouverture au Collège de France, le 8 décembre 1870," in *La poésie du Moyen Âge*, 2 vols. (Paris: Hachette, 1885), vol. 1, pp. 87–118. See R. Howard Bloch, "842: The Birth of Medieval Studies," in *A New History of French Literature*, ed. Denis Hollier (Cambridge, Mass.: Harvard University Press, 1989), pp. 6–13; David Hult, "Gaston Paris and the Invention of Courtly Love," in *Medievalism and the Modernist Temper*, ed. R. Howard Bloch and Stephen G. Nichols (Baltimore: Johns Hopkins University Press, 1996), pp. 192–224.

20. Stephen G. Nichols, "Introduction: Philology in a Manuscript Culture," in *The New Philology*, ed. Stephen G. Nichols, special issue of *Speculum* 65:1 (1990): 1–10, at p. 1.

21. Stephen G. Nichols, "The New Medievalism: Tradition and Discontinuity in Medieval Culture," in *The New Medievalism*, ed. Marina Brownlee, Kevin Brownlee, and Stephen G. Nichols (Baltimore: Johns Hopkins University Press, 1991), pp. 1–26, at p. 1.

22. Paul Zumthor, *Speaking of the Middle Ages*, trans. Sarah White (Lincoln: University of Nebraska Press, 1986), p. 54 (*Parler du Moyen Âge* [Paris: Éditions de Minuit, 1980], p. 62). See also Bernard Cerquiglini, *In Praise of the Variant: A Critical History of Philology*, trans. Betsy Wing (Baltimore: Johns Hopkins University Press, 1999) (*Éloge de la variante. Histoire critique de la philologie* [Paris: Seuil, 1989]).

23. R. Howard Bloch, "The Medieval Text—'Guigemar'—as a Provocation to the Discipline of Medieval Studies," in *The New Medievalism*, ed. M. Brownlee, K. Brownlee, and S. Nichols, pp. 99–112, at p. 100.

24. R. Howard Bloch and Stephen G. Nichols, "Introduction," in *Medievalism and the Modernist Temper*, ed. Bloch and Nichols, pp. 1–22, at p. 4.

25. See, for example, Jeffrey Jerome Cohen, ed., *The Postcolonial Middle Ages* (New York: St. Martin's Press, 2000); John M. Ganim, *Medievalism and Orientalism* (New York: Palgrave Macmillan, 2005); Sharon Kinoshita, *Medieval Boundaries: Rethinking Difference in Old French Literature* (Philadelphia: University of Pennsylvania Press, 2006); S. Kinoshita, "Deprovincializing the Middle Ages," in *The Worlding Project: Doing Cultural Studies in the Era of Globalization*, ed. Rob Wilson and Christopher Leigh Connery (Santa Cruz, Calif.: New Pacific Press, 2007), pp. 75–89; Karla Mallette, *European Modernity and the Arab Mediterranean: Toward a New Philology and a Counter-Orientalism* (Philadelphia: University of Pennsylvania Press, 2010); Kathleen Davis and Nadia Altschul, eds., *Medievalisms in the Postcolonial World: The Idea of "the Middle Ages" Outside Europe: Rethinking Theory* (Baltimore: Johns Hopkins University Press, 2010); Michelle R. Warren, *Creole Medievalism: Colonial France and Joseph Bédier's Middle Ages* (Minneapolis: University of Minnesota Press, 2011); Nadia R. Altschul, *Geographies of Philological Knowledge: Postcoloniality and the Transatlantic National Epic* (Chicago: University of Chicago Press, 2012).

26. *Desire and Discipline: Sex and Sexuality in the Premodern West*, ed. Jacqueline Murray and Konrad Eisenbichler (Toronto: University of Toronto Press, 1996); *Premodern Sexualities*, ed. Louise Fradenburg and Carla Freccero (New York: Routledge, 1996); *Handbook of Medieval Sexuality*, ed. Vern L. Bullough and James A. Brundage (New York: Garland, 1996); *Constructing Medieval Sexuality*, ed. Karma Lochrie, Peggy McCracken, and James A. Schultz (Minneapolis: University of Minnesota Press, 1997).

27. Kathleen Biddick, *The Shock of Medievalism* (Durham, N.C.: Duke University Press, 1998), p. 3.

28. With the notable exceptions of Biddick, *The Shock of Medievalism*; Karma Lochrie, *Heterosyncrasies: Female Sexuality When Normal Wasn't* (Minneapolis: University of Minnesota Press, 2005), especially pp. 1–25.

29. Ann Laura Stoler, *Carnal Knowledge and Imperial Power: Race and the Intimate in Colonial Rule* (Berkeley: University of California Press, 2002), p. 19.

30. This was successfully accomplished in a number of studies; see Albert Pauphilet, "Le mythe du Moyen Âge," in *Le legs du Moyen Âge. Études de littérature médiévale* (Melun: Librairie d'Argences, 1950), pp. 23–63; Janine R. Dakyns, *The Middle Ages in French Literature 1851–1900* (London: Oxford University Press, 1973); Barbara G. Keller, *The Middle Ages Reconsidered: Attitudes in France from the Eighteenth Century through the Romantic Movement* (New York: Peter Lang, 1994); Michael Glencross, *Reconstructing Camelot: French Romantic Medievalism and the Arthurian Tradition* (Cambridge: D. S. Brewer, 1995); Elizabeth Emery, *Romancing the Cathedral: Gothic Architecture in Fin-De-Siècle French Culture* (Albany: State University of New York Press, 2001); Elizabeth Emery and Laura Morowitz, *Consuming the Past: The Medieval Revival in Fin-De-Siècle France* (Aldershot, UK: Ashgate, 2003); Simone Bernard-Griffiths, Pierre Glaudes, and Bertrand Vibert, eds., *La fabrique du Moyen Âge au XIXe siècle: Représentations du Moyen Âge dans la culture et la littérature françaises du XIXe siècle* (Paris: Champion, 2006). None of these studies look at nineteenth-century medical sources. For the eighteenth century, see the classic by Lionel Gossman, *Medievalism and the Ideologies of the Enlightenment: The World and Work of LaCurne de Sainte-Palaye* (Baltimore: Johns Hopkins University Press, 1968).

31. Michel Foucault, *The Birth of the Clinic: An Archaeology of Medical Perception*, trans. A. M. Sheridan Smith (New York: Vintage Books, 1994) (*Naissance de la clinique* [Paris: Quadrige/Presses Universitaires de France, 1997 (1963)]). See also Elizabeth A. Williams, *The Physical and the Moral: Anthropology, Physiology, and Philosophical Medicine in France, 1750–1850* (Cambridge: Cambridge University Press, 1994); Jonathan Strauss, *Human Remains: Medicine, Death, and Desire in Nineteenth-Century Paris* (New York: Fordham University Press, 2011).

32. Jan Goldstein, *Console and Classify: The French Psychiatric Profession in the Nineteenth Century* (Cambridge: Cambridge University Press, 1987); Ian R. Dowbiggin, *Inheriting Madness: Professionalization and Psychiatric Knowledge in Nineteenth-Century France* (Berkeley: University of California Press, 1991); Laure Murat, *L'homme qui se prenait pour Napoléon: Pour une histoire politique de la folie* (Paris: Gallimard, 2011).

33. Jacques Léonard, *La médecine entre les savoirs et les pouvoirs: Histoire intellectuelle et politique de la médecine française au XIXe siècle* (Paris: Aubier Montaigne, 1981); Robert A. Nye, *Crime, Madness, and Politics in Modern France: The Medical Concept of National Decline* (Princeton. N.J.: Princeton University Press, 1984); Jacques Postel and Claude Quetel, eds., *Nouvelle histoire de la psychiatrie* (Paris: Dunod, 1994); Charles Coulston Gillispie, *Science and Polity in France: The Revolutionary and Napoleonic Years* (Princeton, N.J.: Princeton University Press, 2004), pp. 540–50.

34. David Arnold, ed., *Imperial Medicine and Indigenous Societies* (Manchester: Manchester University Press, 1988), p. 17. See also Roy MacLeod and Milton Lewis, eds., *Disease, Medicine, and Empire: Perspectives on Western Medicine and the Experience of European Expansion* (New York: Routledge, 1988); Mary P. Sutphen and Bridie Andrews, eds., *Medicine and Colonial Identity* (New York: Routledge, 2003).

35. "comprendre la Révolution, c'est en tirer toutes les conséquences positives pour en conjurer le renouvellement. Et pour la comprendre, il faut remonter aux origines"; Laurent

Theis, "Guizot et les institutions de mémoire," in *Les lieux de mémoire*, ed. Nora, vol. 1, pp. 1575–97, at p. 1582.

36. Émile Littré, "Un fragment de médecine rétrospective," in *Médecine et médecins*, 2nd ed. (Paris: Didier et Cie, 1872 [1869]), pp. 111–36.

37. "En suivant l'ordre chronologique des temps, nous arrivons à la grande période du moyen âge, qui, plus rapprochée de nous, va nous fournir des notions plus précises et nous donner des renseignemens [*sic*] de plus d'un genre"; Alexandre Brierre de Boismont, "De l'influence de la civilisation sur le développement de la folie," *Annales de l'hygiène publique et de médecine légale* 1:21 (1839): 242–97, at pp. 247–48.

38. "On vit . . . l'occident s'arracher de ses fondemens [*sic*] pour se précipiter sur l'orient" (Brierre de Boismont, p. 248).

39. "Une pareille exaltation était singulièrement favorable aux désordres de l'intelligence; aussi les annales des croisés sont-elles remplies d'apparitions d'anges, de saints, de révélations divines, d'exploits fabuleux. Les hallucinations de la vue et de l'ouïe sont alors très communes; les folies religieuses et guerrières appartiennent surtout aux deux premières croisades. Dans la troisième, les mœurs guerrières des Francs quittent le caractère éminemment épique qui les distingue pour prendre le caractère romanesque. C'est le règne des troubadours et des chevaliers, qui, tournant les imaginations vers l'amour et la gloire, fait éclater les folies amoureuses et chevaleresques. L'érotomanie, la nymphomanie, l'hystérie avec ses variétés, la manie des exploits sont les traits distinctifs de cette époque, dont Roland et le roi Arthur sont les deux types principaux" (Brierre de Boismont, pp. 248–49).

40. "L'esprit chevaleresque qui suivit les croisades multiplia la mélancolie érotique"; Étienne Esquirol, "Folie," in *Dictionnaire des sciences médicales*, ed. Charles Louis Fleury Pancoucke, vol. 16 (Paris, 1816), pp. 151–240, at p. 177.

41. "la santé de tel personnage avait exercé une action sérieuse sur les idées, les doctrines d'une époque"; Alfred Maury, "De l'hallucination envisagée au point de vue philosophique et historique, ou Examen critique de l'opinion émise par M. Brierre de Boismont, touchant les caractères auxquels on doit reconnaître l'hallucination chez certains personnages célèbres de l'histoire" [On hallucination considered from a philosophical and historical standpoint, or Critical examination of the opinion stated by M. Brierre de Boismont, concerning the characters which must be considered hallucinations in certain celebrities throughout history], *Annales médico-psychologiques. Journal de l'anatomie, de la physiologie et de la pathologie du système nerveux* 5 (mai 1845): 317–38, at p. 318.

42. Alexandre Lacassagne, "Préface," in Dr. Cabanès, *Les morts mystérieuses de l'Histoire. Souverains et princes français de Charlemagne à Louis XVII* (Paris: A. Maloine, 1901), pp. v-xi, at p. ix.

43. "avant et après la fistule"; Jules Michelet, *Histoire de France*, in *Œuvres complètes*, vol. 8 (Paris: Flammarion, 1893–1898 [1855]), p. 409.

44. "la philosophie de l'histoire, la médecine de l'histoire"; Maurice Beaujeu, *Psychologie des premiers Césars. Une étude de médecine légale dans l'histoire. Thèse pour obtenir le grade de docteur en médecine.* Faculté de médecine et de pharmacie de Lyon, no. 796 (Lyon: Imprimerie de A. Storck, 1893), p. 1.

45. Foucault, *History of Sexuality*, p. 108 [p. 143]. Foucault has been criticized for supersession of blood by sex, of sanguinity by sexuality, and its implicit periodization that reifies the pre-modern/modern divide; see Eve Kosofsky Sedgwick, *Epistemology of the Closet* (Berkeley: University of California Press, 1990), pp. 45–47 and chapter 4. Kathleen Biddick has argued that Foucault should have seen that "a discourse of blood *doubles* a genealogy of sexuality"; Kathleen Biddick, *The Typological Imaginary: Circumcision, Technology, History* (Philadelphia: University of Pennsylvania Press, 2003), p. 74; also "The Cut of Genealogy: Pedagogy in the Blood," *Journal of Medieval and Early Modern Studies* 30:3 (2000): 449–62.

46. "le thème de la pureté de la race se substitue à celui de la lutte des races"; Michel Foucault, *"Il faut défendre la société," Cours au Collège de France, 1976* (Paris: Gallimard/Seuil, 1997), p. 71.

47. Michel Foucault, "The Confession of the Flesh," in *Power/Knowledge. Selected Interviews and Other Writings, 1972–1977*, ed. Colin Gordon, trans. Colin Gordon, Leo Marshall, John Mepham, and Kate Soper (New York: Pantheon Books, 1980), pp. 194–228, at p. 223 ("Le Jeu de Michel Foucault," *Dits et écrits II, 1976–1988*, ed. Daniel Defert and François Ewald [Paris: Quarto Gallimard, 2001 (1977)], pp. 298–329, at p. 324).

48. Ann Laura Stoler, *Race and the Education of Desire: Foucault's History of Sexuality and the Colonial Order of Things* (Durham, N.C.: Duke University Press, 1995), p. 51. See also Homi K. Bhabha, *The Location of Culture* (New York: Routledge, 1994), pp. 248, 250: "Race and blood interfere with modern sexuality" because "retroversion of race or sanguinity haunts and doubles the contemporary analytic of power and sexuality and may be subversive of it."

49. "L'histoire des maladies éteintes, attenuées ou transformées est toujours captivante. Elle l'est surtout quand la maladie dont il s'agit, par l'horreur et la terreur qu'elle a inspirées aux populations, à certaines époques, leur a suggéré des mesures de prophylaxie et de défense qui ont fait partie des règlements et des institutions du passé. L'histoire de l'affection n'est plus alors un simple chapitre de l'histoire de la pathologie, elle devient un chapitre de l'histoire des mœurs, des réglements et des coutumes"; Gilbert Ballet, "Préface," in Dr. H.-M. Fay, *Histoire de la lèpre en France. Lépreux et cagots du sud-ouest* (Paris: Honoré Champion, 1910), pp. v–xvi, at p. vii.

CHAPTER 1

1. "La France est la terre classique de la valeur militaire. . . . Les Français du XIXe siècle n'avaient pas dégénéré, et, pour réaliser ce que l'audace a de plus éclatant, le sacrifice de plus noble, l'abnégation de plus sublime, ils n'ont eu qu'à suivre le penchant de leur nature généreuse, qu'à obéir aux suggestions de l'hérédité et de l'instinct"; G. Beaumont, *Discours de distribution des prix au Collège de Saint-Dié, le 12 juilliet 1919* (Saint-Dié: Imprimerie G. Freisz, 1919), p. 7. Beaumont, a professor of literature, is speaking in the aftermath of World War I, fought and won by men born in the nineteenth century. I am grateful to Virginie Greene for this reference.

2. "la tendance à la régulière transmission des caractères . . . un legs transmis plus ou moins fidèlement à travers la chaîne infiniment longue des générations"; Charles Letourneau, "Hérédité," in *Dictionnaire encyclopédique des sciences médicales*, ed. Amédée Dechambre (Paris: G. Masson, 1864–1889), 4e sèrie, t. 13 (1888), pp. 588–605, at p. 589. See also Louis-Marius Vialleton, *Principales théories de l'hérédité* (Lyon: A. Storck; Paris: G. Masson, 1893).

3. For heredity, see Staffan Müller-Wille and Hans-Jörg Rheinberger, "Heredity—The Formation of an Epistemic Space," and Carlos López-Beltrán, "The Medical Origins of Heredity," and for degeneration/degeneracy, see Laure Cartron, "Degeneration and 'Alienism' in Early Nineteenth-Century France," in *Heredity Produced: At the Crossroads of Biology, Politics, and Culture, 1500–1870*, ed. Staffan Müller-Wille and Hans-Jörg Rheinberger (Cambridge, Mass.: MIT Press, 2007), pp. 3–34, 105–32, and 155–74, respectively; Carlos López-Beltrán, "In the Cradle of Heredity: French Physicians and 'l'hérédité naturelle' in the Early Nineteenth Century," *Journal of the History of Biology* 37 (2004): 39–72; François Jacob, *The Logic of Life: A History of Heredity* (Princeton, N.J.: Princeton University Press, 1993 [1970]). For a history of hereditarianism and hereditary degeneration in psychiatry, see Ian R. Dowbiggin, *Inheriting Madness: Professionalization and Psychiatric Knowledge in Nineteenth-Century France* (Berkeley: University of California Press, 1991).

4. Michel Foucault, *The Birth of the Clinic: An Archaeology of Medical Perception*, trans. A. M. Sheridan Smith (New York: Vintage Books, 1994) (*Naissance de la clinique* [Paris: Quadrige/Presses Universitaires de France, 1997 (1963)]); Jan Goldstein, *Console and Classify: The French Psychiatric Profession in the Nineteenth Century* (Cambridge: Cambridge University Press, 1987); Elizabeth A. Williams, *The Physical and the Moral: Anthropology, Physiology, and Philosophical Medicine in France, 1750–1850* (Cambridge: Cambridge University Press, 1994).

5. George W. Stocking, *Race, Culture, and Evolution: Essays in the History of Anthropology* (New York: Free Press, 1968); William Schneider, "Toward the Improvement of the Human Race: The History of Eugenics in France," *Journal of Modern History* 54:2 (1982): 268–91.

6. Robert A. Nye, "Heredity or Milieu: The Foundations of Modern European Criminological Theory," *Isis* 67:3 (1976): 334–55; Robert A. Nye, *Crime, Madness, and Politics in Modern France: The Medical Concept of National Decline* (Princeton, N.J.: Princeton University Press, 1984); Eric T. Carlson, "Medicine and Degeneration: Theory and Praxis," in *Degeneration: The Dark Side of Progress*, ed. J. Edward Chamberlin and Sander L. Gilman (New York: Columbia University Press, 1985), pp. 120–44.

7. "La grande question de l'hérédité est loin d'être élucidée" (Letourneau, p. 592).

8. August Weismann, *Die Continuität des Keimplasmas als Grundlage einer Theorie der Vererbung* [The germ-plasm: A theory of heredity] (Jena: G. Fischer, 1885), translated into French and English in 1892 and 1893, respectively; Gregor Mendel, "Versuche über Pflanzen-Hybriden" [Experiments in plant hybridization] (1866). In 1913, the publication of Georges Genil-Perrin's thesis, *Histoire des origines et l'évolution de l'idée de dégénérescence en médecine mentale*, broke the dogmatic hold of these theories on the medical sciences

in France. See Charles Coulston Gillispie, *The Edge of Objectivity: An Essay in the History of Scientific Ideas* (Princeton, N.J.: Princeton University Press, 1960), pp. 303–51; Peter J. Bowler, *The Mendelian Revolution. The Emergence of Hereditarian Concepts in Modern Science and Society* (Baltimore: Johns Hopkins University Press, 1989); Ted Everson, *The Gene: A Historical Perspective* (Westport, Conn.: Greenwood Press, 2007), pp. 27–55.

9. "Transmission par voie de succession," "héritage, succession"; Pierre Larousse, *Grand dictionnaire universel du XIXe siècle* (Paris: Administration du Grand dictionnaire universel, 1866–1877), "Hérédité," vol. 9 (n.d.), pp. 217–18, at p. 217.

10. "Il est bon de remonter le plus haut qu'il sera possible, dans l'histoire médicale de la famille, et d'y rechercher, pour ainsi dire, le type originel de la santé des parents ascendants"; Dr. Francis Devay, *Hygiène des familles ou perfectionnement physique et moral de l'homme considéré particulièrement dans ses rapports avec l'éducation et les besoins de la civilisation moderne*, 2 vols. (Lyon: Dorier, 1846), vol. 1, pp. 155–56.

11. "*l'archéologie pathologique*"; Alexandre Lacassagne, "Préface," in Dr. Cabanès, *Les morts mystérieuses de l'Histoire. Souverains et princes français de Charlemagne à Louis XVII* (Paris: A. Maloine, 1901), pp. v–xi, at p. ix.

12. "L'émotion gagna le public; tous les journaux abordèrent le sujet; le ministre du commerce et des travaux publics en fit l'objet d'une circulaire aux préfets; des thèses furent soutenues devant les facultés de médecine; de nombreux mémoires publiés; une discussion eut lieu au Congrès médical de Lyon (1864)"; Alexandre Lacassagne, "Consanguinité," in *Dictionnaire encyclopédique des sciences médicales*, ed. Dechambre, 1ère série, t. 19 (1876), pp. 652–716, at p. 671.

13. David Warren Sabean and Simon Teuscher, "Kinship in Europe: A New Approach to Long Term Development," in *Kinship in Europe: Approaches to Long-Term Development (1300–1900)*, ed. David Warren Sabean, Simon Teuscher, and Jon Mathieu (New York: Berhahn Books, 2010 [2007]), pp. 1–32, esp. pp. 19–24; David Warren Sabean, "Kinship and Class Dynamics in Nineteenth-Century Europe," in *Kinship in Europe*, ed. Sabean, Teuscher, and Mathieu, pp. 301–13.

14. "de perpétuer la société, non seulement par le nombre des individus, mais plus encore par la beauté, par la vigueur et l'organisation parfaite, autant que possible, des races"; Dr. Serrurier, *Du mariage considéré dans ses rapports physiques et moraux. Inconvénients des unions entre des individus entachés d'un principe constitutif héréditaire* (Paris: J.-B. Baillière, 1845), p. 35.

15. "Si l'État a besoin de citoyens doués de toute la force, de toute la perfection d'une bonne organisation, la société revendique les mêmes droits. Les familles constituées sur des bases solides jouissent de l'espérance et de la certitude, autant physique que morale, d'une existence de plus longue durée" (Dr. Serrurier, *Du mariage*, p. 35).

16. "la cause des causes"; Ulysse Trélat, qtd. in Jules Dejerine, *L'hérédité dans les maladies du système nerveux* (Paris: Anselin et Huzeau, 1886), p. 28.

17. "On veut démontrer par les faits que la consanguinité des parents, *ipso facto*, en dehors de toute hérédité pathologique et de toute influence étrangère à l'innéité organique, augmente, dans une proportion notable, les chances d'infirmités et de maladies auxquelles

sont exposées les générations"; Eugène Dally, *Des dangers attribués aux mariages entre consanguins.* Extrait de la *Gazette hebdomadaire de médecine et de chirurgie* 9 (1862): 1–45, at p. 9.

18. "La consanguinité élève l'hérédité à sa plus haute puissance"; Émile Littré and Charles Robin, *Dictionnaire de médecine, de chirurgie, de pharmacie, de l'art vétérinaire et des sciences qui s'y rapportent*, 14th rev. ed. (Paris: J.-B. Baillière, 1878), "Consanguinité," p. 348.

19. "Les maladies les plus graves, les plus nombreuses, les plus irrémédiables" (Dally, p. 2).

20. "L'imbécillité, la surdi-mutité, la scrofule, le rachitisme, l'albinisme, les malformations tératologiques"; Yves Delage, *L'hérédité et les grands problèmes de la biologie générale*, 2nd ed. (Paris: Librairie C. Reinwald, 1903 [1895]), p. 268.

21. "On peut ranger désormais la consanguinité dans le cadre de l'étiologie morbide pour ce qui concerne l'espèce humaine"; Dr. Francis Devay, *Un mot sur le danger des mariages consanguins, réponse à une attaque. État de la question* (Paris: Victor Masson, 1863), p. 44.

22. "L'élection de *personne*," "l'élection des agents de la reproduction," "Les moyens de prévenir le transport séminal de la maladie dérivent nécessairement des lois et des formules de la génération"; Prosper Lucas, *Traité physiologique et philosophique de l'hérédité naturelle* (Paris, 1847), pp. 902, 903, 902. See the full quote on "the selection of *person*, the selection of *time*, the selection of *place*, and the selection of *state*, in the coming together of the sexes" (l'élection de *personne*, l'élection de *temps*, l'élection de *lieu*, et l'élection d'*état*, dans le rapprochement des sexes) (Lucas, p. 902).

23. "il doit ne rechercher, sous le rapport physique, dans une autre famille, que des conditions *semblables*" (Lucas, p. 907).

24. "ne donner comme conjoint à l'individu qui a des antécédents héréditaires morbides qu'un époux ou une épouse d'une constitution irréprochable, d'un état de santé parfait"; Dr. Émile Laurent, *Mariages consanguins et dégénérescences* (Paris: A. Maloine, 1895), p. 65. This was not originally advocated by Lucas; see p. 909.

25. "À quoi bon parler de mon orthodoxie religieuse, *bien connue*? . . . Je vous dirai que j'ai toujours vu et que je vois encore dans le Christianisme le seule remède aux misères physiques, morales, sociales, et j'ajouterai, . . . scientifiques de l'humanité" (Devay, *Un mot sur le danger des mariages consanguins*, pp. 8–9).

26. "aussi contraires à l'accroissement de la population qu'à sa validité"; Dr. Francis Devay, *Du danger des mariages consanguins au point de vue sanitaire* (Paris: Labé; Lyon: M. Savy, 1857), p. 23; revised and reprinted as Francis Devay, *Du danger des mariages consanguins sous le rapport sanitaire*, 2nd rev. ed. (Paris: Victor Masson, 1862).

27. Dr. Francis Devay, *Hygiène des familles*, p. 152; revised and reprinted as Francis Devay, *Traité spécial d'hygiène des familles, particulièrement dans ses rapports avec le mariage, au physique et au moral, et les maladies héréditaires* (Paris: Labé, 1858).

28. "Un second principe ressort de l'application de la science médicale au croisement des races: l'antagonisme des maladies et des prédispositions originelles" (Devay, *Hygiène des familles*, p. 157).

29. "Ainsi, nos mœurs, nos habitudes sociales portent généralement les pères de

famille, et souvent dans des vues honorables, à rechercher pour l'établissement de leurs enfants ce qu'on appelle généralement une *position identique* à la leur. . . . C'est là ce que nous appelerons une *consanguinité factice*" (Devay, *Du danger*, 1857, p. 62).

30. "Le sang étranger," "la race pure et le type individuel" (Dally, p. 21).

31. Lacassagne pointed out that anti-consanguinists were wrong to speak of mixing family lines as cross-breeding: "One makes a mistake in opposing the term *cross-breeding* to consanguinity. Cross-breeding is not a question of uniting one family to another, but one race to another, just as hybridity [in zootechnology] applies to the union of two different species" (on se trompe quand on fait de *croisement* un terme opposé à consanguinité. Le croisement ne s'entend pas de l'union de famille à famille, mais de race à race, de même que l'hybridité [en zootechnie] s'applique à l'union de deux espèces différentes) (Lacassagne, "Consanguinité," p. 677). Zootechnology had profound effects on theories of racial mixing and on eugenics, when the question of animal "races" was transformed into a question of human "races." Concurrently to the consanguinity debate, in his *Essai sur l'inégalité des races humaines* (1853–1855), Arthur de Gobineau developed racist theories by transposing the zootechnological term of "miscegenation" (*métissage*) to "ethnic cross-breeding" (*croisement ethnique*), to argue that racial mixing leads to degeneracy of race and nation, rather than their improvement (see Chapter 3).

32. "La consanguinité dans les mariages des rois devient, après plusieurs générations, funeste aux peuples: . . . héritage . . . se transmet sans altération, . . . il n'y a point de neutralisation"; Charles Girou de Buzareingues, *Philosophie physiologique, politique et morale* (Paris: Firmin Didot, 1828), pp. 312–13.

33. "les passions qui naissent de l'autorité, des résistances et de la flatterie, et le caractère qui en est le fruit, passent du père à la fille, de celle-ci au garçon, et sont enfin l'héritage commun de tous les enfants appelés à régner" (Buzareingues, p. 312).

34. "C'est par le choix de sa compagne, qu'on peut s'assurer de transmettre, avec la vie, des penchants heureux à ses enfants" (Buzareingues, p. 392).

35. "L'histoire témoigne . . . : les aristocraties, réduites à se recruter dans leur propre sein, s'éteignent . . . et souvent, en passant par la dégradation, la folie, la démence et l'imbécillité" (Lucas, pp. 905–6).

36. "Et cependant les historiens qui ont traité des causes du dépérissement des classes nobles en France ont passé sous silence précisément celle qui a eu le plus de portée, celle qui, littéralement, vicie le sang, atrophie la race, lorsque toutefois celle-ci peut éclore d'un monstrueux alliage. Et cependant rien n'était plus facile, car aucun siècle n'a fourni par ses nombreux mémoires, ses volumineux recueils de correspondance, un plus fidèle miroir de sa physionomie; dans aucun siècle ne furent plus étrangement méconnnues les obligations du lien conjugal; dans aucun siècle le mot mariage n'a été plus synonyme du mot *établissement*"; Dr. Francis Devay, *Nouvelles observations sur les dangers des mariages entre consanguins au point de vue sanitaire* (Paris: Victor Masson, 1860), p. 7.

37. "L'esprit s'éteint, les forces diminuent et la race s'abâtardit. . . . On sait ce qu'il advint à la fin du siècle suivant [XVIIIe], où les hautes classes, très affaiblies

numériquement, n'étant ni assez intelligentes pour améliorer, ni assez fortes pour pouvoir résister, s'anéantirent dans le tiers état" (Devay, *Nouvelles observations*, p. 8).

38. "L'histoire nous enseigne qu'elle [la consanguinité] fut une des causes de la dégénérescence de certaines races royales ou nobiliaires (famille de Charles-Quint, etc.), et elle nous montre en même temps l'heureux effet des croisements, qui concourent à la *régénérescence* en introduisant dans une lignée un sang meilleur"; Gilbert Ballet, ed., *Traité de pathologie mentale*, by D. Anglade, F.-L. Arnaud, H. Colin, E. Dupré, A. Dutil, J. Roubinovitch, J. Seglas, and Ch. Vallon (Paris: Octave Doin, 1903), p. 24.

39. "le défaut de croisement chez les familles nobles devait en amener l'affaiblissement, et par suite l'extinction. L'horreur des mésalliances, a-t-on dit, et le soin avec lequel les hautes classes se sont toujours unies entre elles, ont beaucoup contribué à leur dégénération"; Louis-François Benoiston de Châteauneuf, *Mémoire sur la durée des familles nobles de France* (Extrait du compte-rendu des séances et travaux de l'Académie des sciences morales et politiques, 1845), *Annales d'hygiène publique et de médecine légale* 35 (janvier 1846): 27–58, at p. 37. B. de Châteauneuf is the author of several works of statistics.

40. "D'après un relevé fait sur 300 familles nobles, j'ai trouvé que 2398 chefs de famille . . . ont eu 10490 enfans [*sic*], ou 4.3 chacun" (Châteauneuf, p. 32).

41. "La couche des nobles, au lieu d'être à-peu-près stérile comme on l'a dit, était, au contraire, très féconde" (Châteauneuf, p. 44).

42. "l'état militaire d'abord, et ensuite l'état ecclésiastique; de ces deux professions embrassées par la plus grande partie des nobles, l'une les obligeait au sacrifice de leur vie, l'autre les condamnait à ne pas la donner. J'ajouterais encore à ces deux causes une troisième: la grande quantité d'enfans [*sic*] qu'ils perdaient, ou qui, parvenus à l'âge d'homme, moururent sans avoir été mariés" (Châteauneuf, p. 58).

43. "Les mœurs du moyen âge n'étaient pas sans doute exemplaires," "dans une question où il s'agit bien plus de la vigueur du corps que de l'honnêteté des actions" (Châteauneuf, p. 42).

44. "La morale et la médecine, nobles sciences . . . ont depuis long-temps placé le dépérissement des hautes classes dans le dépérissement des mœurs, bien plus que dans le défaut de croisement" (Châteauneuf, p. 41).

45. "les évènemens [*sic*] remarquables qui sont les traits saillans [*sic*] d'un siècle n'appartiennent pas seulement à l'histoire, ils sont aussi du domaine de la médecine"; Alexandre Brierre de Boismont, "De l'influence de la civilisation sur le développement de la folie," *Annales de l'hygiène publique et de médecine légale* 1:21 (1839): 242–97, at p. 245.

46. "de quel droit l'historien s'érigerait-il en juge suprême et condamnerait-il sans avoir écouté le médecin?"; Émile Dusolier, *Psychologie des derniers Valois (François II, Charles IX, Henri III). Thèse présentée à la Faculté de médecine et de pharmacie de Lyon, et soutenue publiquement le mercredi 27 novembre 1895* (Lyon: Imprimerie de A.-H. Storck, 1895), p. 5.

47. There is yet another way that history and medicine, ideology and pathology, interact: recent historical events as causes of diseases, especially mental diseases; see Laure Murat, *L'homme qui se prenait pour Napoléon: Pour une histoire politique de la folie* (Paris: Gallimard, 2011).

48. Danielle Gourevitch, ed., *Médecins érudits de Coray à Sigerist. Actes du colloque de Saint-Julien-en-Beaujolais (juin 1994)* (Paris: De Broccard, 1995).

49. Danielle Gourevitch, "Charles Daremberg, His Friend Émile Littré, and Positivist Medical History," in *Locating Medical History: The Stories and Their Meanings*, ed. Frank Huisman and John Harley Warner (Baltimore: Johns Hopkins University Press, 2004), pp. 53–73.

50. "Ostracisme"; Charles Daremberg, "Préface," in *Histoire des sciences médicales*, vol. 1 (Paris: J.-B. Baillière, 1870), pp. vii–xxviii, at p. xv.

51. Isabelle Wohnlich-Despaigne, "Introduction," in *Les historiens français de la médecine au XIXe siècle* (Paris: J. Vrin, 1987), pp. 11–33, esp. p. 16; Charles Daremberg, "Préface" and Ch. Daremberg, "Introduction," in *Histoire des sciences médicales*, vol. 1 (Paris: J.-B. Baillière, 1870), 1–66, esp. pp. 2–7; Charles Daremberg, *Faculté de médecine de Paris. Cours sur l'histoire de la médecine et de la chirurgie, par le professeur Ch. Daremberg. Leçon d'ouverture, le 11 novembre 1871* (Paris: F. Malteste, 1871).

52. "n'était elle-même que de l'histoire" (Daremberg, "Introduction," p. 2).

53. "de tous côtés on fait une place à l'histoire de chaque science, tandis que l'histoire de la médecine n'est représentée officiellement, en France, que dans la chaire du Collége [*sic*] de France" (Daremberg, "Préface," p. xvi).

54. "un médecin qui prétendrait disserter sur le pouls sans connaître la circulation du sang"; José Miguel (Joseph-Michel) Guardia, *La médecine à travers les siècles* (Paris: J.-B. Baillière, 1865), p. xxviii.

55. "Pour apprécier les doctrines médicales, il est de rigueur d'en connaître les auteurs, l'origine précise, les modifications successives, les transformations diverses. . . . Et de fait, c'est de la réalité que naît la science positive, laquelle reste incomplète, si l'on ignore comment elle a été fondée, accrue et perfectionnée par une suite non interrompue d'efforts et de tentatives, comment les hommes ont procédé dans leur évolution intellectuelle" (Guardia, p. xxix).

56. See Émile Littré, *De la philosophie positive* (Paris: Ladrange, 1845), *Conservation, révolution et positivisme* (Paris: Ladrange, 1852), *Auguste Comte et la philosophie positive* (Paris: Librairie Hachette, 1863); Elme Caro, *M. Littré et le positivisme* (Paris: Libraire Hachette, 1883). See also Donald Geoffrey Charlton, *Positivist Thought in France during the Second Empire, 1852–1870* (Oxford: Clarendon Press, 1959); Walter Michael Simon, *European Positivism in the Nineteenth Century: An Essay in Intellectual History* (Ithaca, N.Y.: Cornell University Press, 1963).

57. "Biologie ou science des corps vivants, et sociologie ou science des sociétés"; Larousse, *Grand dictionnaire universel*, "Positivisme," vol. 12 (1874), pp. 1490–92, at p. 1490.

58. "d'un secours indispensable à la constitution de la sociologie"; Émile Littré, *Médecine et médecins*, 2nd ed. (Paris: Didier, 1872), p. viii.

59. "Ce mot [sociologie] ne figurerait pas ici si les hommes qui cultivent les sciences, et en particulier les médecins, n'avaient besoin d'une philosophie qui les guide"; Littré and Robin, *Dictionnaire de médecine*, "Sociologie," pp. 1456–57, at p. 1457.

60. "Guérir la dégénérescence en l'évitant et en s'opposant à sa propagation est une

question de sociologie pure," "indiquer le mal"; Valentin Magnan and Paul-Maurice Legrain, *Les dégénérés, état mental et syndromes épisodiques* (Paris: Rueff, 1895), p. 234.

61. "La science comprend la vie des sociétés, les lois de cette vie, le progrès entre autres" (Larousse, "Positivisme," p. 1491).

62. "en étudiant le passé politique des sociétés, vous apprendrez leur avenir, et non-seulement l'avenir comme probabilité, mais comme droit et devoir" (Larousse, "Positivisme," p. 1491).

63. "l'histoire, telle qu'elle a été faite jusqu'à présent, n'est pas une science, mais est simplement la connaissance des événements qui se sont passés au sein des sociétés" (Littré and Robin, "Sociologie," p. 1457).

64. "On a fait une part trop large au poison et au merveilleux, dans toutes les morts royales ou princières"; Dr. Cabanès, *Les indiscrétions de l'Histoire*, 6 vols. (Paris: Albin Michel, 1903), vol. 6, p. xix.

65. "Tant qu'on n'aura pas montré comment ces événements s'enchaînent les uns aux autres, dérivent les uns des autres, on n'aura, en fait d'histoire, que des matériaux d'érudition, mais point de théorie scientifique" (Littré and Robin, "Sociologie," p. 1457).

66. "L'hérédité intervient dans l'évolution de la civilisation, dans la *sociologie*. Sans elle, l'histoire manquerait d'une de ses conditions organiques ou biologiques essentielles"; Littré and Robin, *Dictionnaire de médecine*, "Hérédité," pp. 729–30, at p. 730.

67. "en présence d'un fait ou d'un personnage, qu'il faudra deviner comme une énigme, s'il ne possède pas le savoir qui permet au médecin des fous et au médecin légiste de porter un diagnostic motivé ou d'éclairer la conscience des juges" (José Miguel Guardia, qtd. in Dr. Cabanès, *Les indiscrétions*, 6:xviii).

68. "A-t-il le droit de dédaigner les menus faits, les anecdotes, les documents, pour ne considérer que les idées générales?"; Jean Carrère, qtd. in Dr. Cabanès, *Dans les coulisses de l'Histoire*, vol. 2 (Paris: Albin Michel, 1937), p. 7.

69. Lacassagne founded the *Archives de l'anthropologie criminelle et des sciences pénales* in 1886, having participated in 1885 at the First International Congress of Criminal Anthropology in Rome. See Laurent Mucchielli, "Criminologie, hygiénisme et eugénisme en France (1870–1914): Débats médicaux sur l'élimination des criminels réputés 'incorrigibles'," *Revue d'histoire des sciences humaines* 3 (2000): 57–88; *Crimino Corpus*, Dossier "Autour des *Archives d'Anthropologie Criminelle*," http://criminocorpus.revues.org/161, accessed November 6, 2011.

70. "l'importance de la médecine légale pour l'histoire. C'est en utilisant ses connaissances biologiques que le médecin légiste parvient à résoudre certains problèmes dont les historiens seuls ne pouvaient trouver la solution. . . . Nos connaissances de psycho-pathologie nous autorisent à comprendre et à expliquer quelques natures étranges qui sont restées comme des rébus historiques" (Lacassagne, "Préface," pp. v–vi).

71. "Les historiens sont semblables aux aveugles ou infirmes qui, de leur lit ou de leur fauteuil, grâce au théatrophone, entendent un drame ou un opera. . . . Il y a quelques éclaircies, parfois des auditions distinctes, souvent du brouhaha, des sons, des bruits. Mais

où sont les décors, les costumes, le jeu des acteurs, leurs attitudes, et leurs gestes, le mouvement scénique?" (Lacassagne, "Préface," pp. vii–viii).

72. "se sont conduits comme les experts devant la justice: ils ont rapporté, en leur honneur et conscience" (Lacassagne, "Préface," p. vii).

73. "de types morbides à mettre en évidence" (Lacassagne, "Préface," p. viii).

74. "Ne savons-nous pas des saints et des saintes, de grands mystiques, qui ont été des hystériques! De vaillants hommes de guerre, qui eurent l'anesthésie morale et le courage audacieux des épileptiques! Des monstres indiscutés, parce qu'ils ont été assez haut placés pour recueillir l'indignation générale et qu'il faut élever à la dignité de malades!" (Lacassagne, "Préface," p. viii).

75. "Combien de réhabilitations la médecine légale pourrait entreprendre; que de procès lui restent à réviser!" (Dr. Cabanès, *Les indiscrétions*, 6:xviii–xix).

76. "Depuis quelques années on voit s'affirmer un louable réveil du goût pour les études de *pathologie rétrospective*. On s'est mis de divers côtés à rechercher, dans les tableaux des musées et des églises et sur les sculptures appendues aux portiques ou aux piliers des cathédrales, les anomalies, les difformités, les contorsions nerveuses. . . . Si bien que l'examen attentif des œuvres d'art, fait du point de vue médical, a permis de recueillir un ensemble de documents, non seulement curieux mais quelquefois très instructifs touchant l'histoire de certaines affections"; Gilbert Ballet, "Préface," in Dr. H.-M. Fay, *Histoire de la lèpre en France. Lépreux et cagots du sud-ouest* (Paris: Honoré Champion, 1910), pp. v–xvi, at p. v, emphasis added. See Jean-Martin Charcot and Paul Richer, *Les démoniaques dans l'art* (Paris: A. Delahaye et E. Lecrosnier, 1887), and *Les difformes et les malades dans l'art* (Paris: Lecrosnier et Babé, 1889).

77. "Jusqu'alors il avait semblé qu'entre la constitution des sociétés et le tempérament des individus, il n'y avait que des influences bien lointaines; l'on n'eût osé avancer, sans crainte de ridicule, que la santé de tel personnage avait exercé une action sérieuse sur les idées, les doctrines d'une époque. . . . Il n'y a donc rien de déraisonnable de demander à la physiologie la solution de certains problèmes de la philosophie de l'histoire"; Alfred Maury, "De l'hallucination envisagée au point de vue philosophique et historique, ou Examen critique de l'opinion émise par M. Brierre de Boismont, touchant les caractères auxquels on doit reconnaître l'hallucination chez certains personnages célèbres de l'histoire" [On hallucination considered from a philosophical and historical standpoint, or Critical examination of the opinion stated by M. Brierre de Boismont, concerning the characters which must be considered hallucinations in certain celebrities throughout history], in *Annales médico-psychologiques. Journal de l'anatomie, de la physiologie et de la pathologie du système nerveux* 5 (mai 1845): 317–38, at p. 318.

78. "Fièvre intermittente"; Émile Littré, "De la toxicologie dans l'histoire et de la mort d'Alexandre," in *Médecine et médecins* (1853), pp. 393–428.

79. "Il est évident que la folie a exercé une grande influence sur les destinées des peuples"; Émile Littré, "Socrate et Pascal, pathologie mentale," in *Médecine et médecins* (1836–1846), pp. 82–110, at p. 88.

80. Auguste Brachet, *Pathologie mentale des rois de France. Louis XI et ses ascendants.*

Une vie humaine étudiée à travers six siècles d'hérédité, 852–1483 (Paris: Librairie Hachette, 1903 [1896]), p. xiii.

81. "Pour dissimuler une excroissance du pied," "à certaines époques, elles adoptèrent un ajustement qui les faisait paraître toutes enceintes; de la sorte, celles qui l'étaient réellement avaient un excellent moyen de cacher leur état"; Dr. Cabanès, "Comment une mode naît d'une infirmité," in *Les indiscrétions* 2:367–90, at pp. 377 and 382. Cabanès founded in 1893 *La Chronique médicale* and, according to Lacassagne, his first works on medical history date from 1885.

82. "À débrouiller des énigmes," "à expliquer certaines morts controversées de l'histoire" (Dr. Cabanès, *Les morts mystérieuses*, p. xvii).

83. "Qu'on nous entende bien: loin de nous la prétention de ramener l'histoire tout entière à une série de problèmes de psychologie morbide; . . . nous n'aspirons qu'à fournir une contribution, la contribution scientifique"; Dr. Cabanès, *L'histoire éclairée par la clinique* (Paris: Albin Michel, 1921), p. 43.

84. "L'Histoire médicale recueille les documents pathologiques relatifs à un particulier, à ses maladies, à ses médecins, les détails de ses obsèques, etc. C'est de la Pathologie historique à l'état fragmentaire, anecdotique et isolé" (Brachet, p. xii).

85. "Le médecin seul . . . peut entreprendre ce travail, faire en un mot: la philosophie de l'histoire, la médecine de l'histoire"; Maurice Beaujeu, *Psychologie des premiers Césars. Une étude de médecine légale dans l'histoire. Thèse pour obtenir le grade de docteur en médecine.* Faculté de médecine et de pharmacie de Lyon, no. 796 (Lyon: Imprimerie de A. Storck, 1893), p. 1.

86. "Tout un travail de commentation de l'histoire," "Les historiens nous rapportent les faits accomplis: rien de plus; ils ne cherchent point à les expliquer: ils ne le peuvent pas" (Beaujeu, p. 1).

87. "Modèle de la politique la plus perfide et la plus raffinée," "Une taciturnité sombre, une gravité dure et repoussante, les âpres inégalités d'un caractère plein de caprices et d'emportements, la recherche de la solitude, un regard oblique, le timide embarras d'une âme artificieuse, trahissent, dès la jeunesse, la disposition mélancolique de Louis XI"; Philippe Pinel, *Nosographie philosophique*, 6th ed., vol. 3 (Paris: J. A. Brosson, 1818 [1797]), p. 91; see esp. "Les Vésanies," pp. 51–159. Pinel also cited the examples of Pascal, victim of excessive work, and of Tasso and Rousseau, vicims of persecution anxiety.

88. "Le premier, il chercha l'origine des événements dans la constitution morbide des personnages, les causes des comédies politiques dans la santé de leurs acteurs"; Henry Céard, *La Chronique médicale*, 15 juillet 1898, qtd. in Dr. Cabanès, *Le cabinet secret de l'Histoire*, vol. 1 (Paris: Albin Michel, 1929 [1895–1898]), p. ix.

89. "mon problème historique posé comme *résurrection de la vie intégrale*, non pas dans ses surfaces, mais dans ses organismes intérieurs et profonds"; Jules Michelet, "Préface de 1869," *Œuvres complètes*, vol. 1 (Paris: Flammarion, 1893–1898), pp. i–xlv, at p. iv.

90. "Telles sont les phases bizarres du gouvernement personnel. Le règne de Louis XIV se partage en deux parts: *avant la fistule, après la fistule*. Avant, Colbert et les conquêtes;

après, madame Scarron et les défaites, la proscription de cinq cent mille Français. François Ier varie de même: *avant l'abcès, après l'abcès.* Avant, l'alliance des Turcs, etc.; après, l'élévation des Guises et le massacre des Vaudois, par lesquels finira son règne"; Jules Michelet, *Histoire de France,* in *Œuvres complètes,* vol. 8 (Paris: Flammarion, 1893–1898 [1855]), p. 409. See also Dr. Auguste Corlieu, *La fistule de Louis XIV* (Paris: Librairie Germer-Baillière, 1874).

91. "Francis I," *Encyclopædia Britannica Online,* 2011, http://www.britannica.com/EBchecked/topic/216656/Francis-I, accessed May 26, 2011.

92. Roland Barthes, *Michelet,* trans. Richard Howard (New York: Hill and Wang, 1987), p. 90 (*Michelet* [Paris: Seuil, 1995 (1954)], p. 73), translation modified. Roland Barthes inversed the problem and produced a physiology of the historian, who is "sick from history" (*malade de l'histoire*), p. 18 [p. 21]. See also Roland Barthes, "Michelet, l'Histoire et la Mort," *Esprit* 178 (1951): 497–511.

93. "le rôle du sexualisme dans l'Histoire est loin d'être négligeable" (Dr. Cabanès, *L'histoire éclairée,* p. 114). "Sexualisme" denotes a theory on sexuality and the whole of sexual practices. Here, however, Cabanès prefers this expression over "sexualité," which in the early twentieth century when he wrote was still used to describe sexual acts and behavior, rather than identity. Conversely, "sexuality," in our modern meaning of "identity," is a more adequate translation of Cabanès's meaning.

94. "a donné la place qu'elle méritait à la peinture des mœurs et des caractères, et il a montré combien les documents les plus secs peuvent devenir instructifs pour qui sait les interroger; il a insisté sur l'influence jusque-là négligée des causes physiologiques et pathologiques en histoire, et ouvert aux investigations une voie nouvelle, très dangereuse il est vrai, mais fertile en découvertes curieuses"; Gabriel Monod, *Les maîtres de l'histoire: Renan, Taine, Michelet,* 2nd ed. (Paris: Calman Lévy, 1894 [1875]), pp. 180–81.

95. "Il n'avait point de méthode qu'il pût enseigner et transmettre, car il ne procédait que par intuition et par divination" (Monod, p. 180).

96. "Sa méthode inquiète, en effet, les esprits positifs, qui aiment à ne mettre lentement le pied que sur un terrain solide; ils se sont demandé s'il fallait croire à ces inductions, à ces hypothèses, à ces analyses physiologiques si neuves"; Larousse, *Grand dictionnaire universel,* "Michelet (Jules)," vol. 11 (1874), pp. 223–24, at p. 224.

97. "Devinés d'après une phrase de chroniqueur, d'après un portrait ou une gravure" (Larousse, "Michelet," p. 224), "modèles à imiter" (Monod, p. 180).

98. "Médecin philologue"; Jacques Postel and Claude Quetel, eds., *Nouvelle histoire de la psychiatrie* (Paris: Dunod, 1994), p. 148.

99. "Le psychiatre viendra efficacement au secours de l'historien. Il ne s'agit plus, dès lors, de pénétrer un caractère à travers un tempérament, mais de montrer les conséquences que peut avoir sur les destinées d'une nation la santé cérébrale de celui qui la dirige" (Dr. Cabanès, *L'histoire éclairée,* pp. 125–26).

100. "La psychologie de l'Histoire" (Dr. Cabanès, *L'histoire éclairée,* p. 275).

101. Louis-Francisque Lélut, *Du démon de Socrate: Spécimen d'une application de la*

science psychologique à celle de l'histoire (Paris: Trinquart, 1836). His later work, *L'amulette de Pascal: Pour servir à l'histoire des hallucinations* (Paris: J.-B. Baillière, 1846), was just as famous.

102. "Subordonner les doctrines, les croyances, les convictions, à l'état maladif du corps, ce qui peut être vrai du caractère, de l'humeur, mais ce qui est de toute fausseté par rapport à l'âme"; Alexandre Brierre de Boismont, "Réponse à l'article d'Alfred Maury," *Annales médico-psychologiques. Journal de l'anatomie, de la physiologie et de la pathologie du systeme nerveux* 5 (mai 1845): 339–41, at p. 341.

103. Brachet is the author of several other grammars and dictionaries. His grammar was held in high regard, notably by Gaston Paris; see Ursula Bähler, *Gaston Paris et la philologie romane* (Genève: Droz, 2004), p. 91; Charles Ridoux, *Évolution des études médiévales en France de 1860 à 1914* (Paris: Honoré Champion, 2001), p. 446. See also the review of Brachet in *Revue des cours littéraires de la France et de l'Étranger*, 5e année, no. 9 (1er février, 1868): 152. Brachet was the translator of Friedrich Diez's *Grammaire des langues romanes* (1874–1876). In 1870, he published a very successful *Dictionnaire étymologique de la langue française* (Ridoux, pp. 432–33). He abruptly stopped his philological work after the 1874 publication of *Nouvelle grammaire française* (Ridoux, p. 446). He states that "this *Pathologie mentale des Rois de France* [was] undertaken in 1880 on the advice of my venerated master, Mr. Littré, the first four volumes of which I published in February 1896, after working slightly more than fifteen years" (cette *Pathologie mentale des Rois de France* [a été] entreprise en 1880 sur le conseil de mon vénéré maître M. Littré, et dont j'ai publié en février 1896, après un travail d'un peu plus de quinze années, les quatres premiers volumes) (Brachet, p. i).

104. "La *Pathologie historique* est proprement l'explication, par la science biologique, des données que nous fournissent les textes historiques, données réunies et contrôlées suivant les règles de la critique scientifique, dans le double but de servir, tantôt à la science médicale, tantôt à la science historique. Elle ne doit être confondue ni avec l'*Histoire de la médecine* qui relate la doctrine ou la vie d'un médecin . . . ni avec l'*Histoire médicale*, c'est-à-dire la biographie pathologique d'un individu. La Pathologie historique néglige la première et englobe la seconde qui lui sert d'étage" (Brachet, p. xii).

105. "des *loci desperati* de l'Histoire" (Brachet, p. cxx).

106. "La Pathologie historique proprement dite, en négligeant de ces faits ceux qui n'ont qu'un intérêt anecdotique, relie les *data* pathologiques que lui fournit à l'état fragmentaire l'Histoire médicale, pour en établir la succession, et en tirer la loi pathologique de l'individu ou, comme on dit en clinique, sa formule pathogénique, et de là psychique" (Brachet, p. xii).

107. "la reconstitution de la formule biologique (somatique et psychique) de chacun de ces *representative men* de l'humanité. Nous aurons à justifier l'utilité de cet ordre de recherches au triple point de vue de l'*histoire politique* (interprétation des actes des personnages), — de la *pathologie* (histoire et évolution des maladies), — de la *clinique* (constitution scientifique de l'hérédité)" (Brachet, p. vii).

108. "dossier pathologique" (Dr. Cabanès, *Les morts mystérieuses*, p. xvii).

109. "les archives cliniques du passé"; Dr. Cabanès, *Légendes et curiosités de l'histoire*, vol. 1 (Paris: Albin Michel, 1912), p. 1.

110. "pathologie mentale," "pathologie somatique" (Brachet, p. vii).

111. "la physiologie n'est pas *toute* la psychologie" (Dr. Cabanès, *L'histoire éclairée*, p. 53).

112. "La Pathologie historique doit être purement clinique et non philosophique" (Brachet, p. cxxxviii).

113. "une pathologie psychologique purement fantaisiste" (Brachet, p. lvi). See also "this slightly fanciful method" (*cette méthode un peu fantaisiste*), in Eugène de Mirecourt, *Michelet* (Paris: Gustave Havard, 1857), p. 63.

114. "pour les pies et les geais, oiseaux jaseurs, elle est inadmissible pour les grues et les chouettes, a fortiori pour les cerfs et les biches qui eussent difficilement crié 'Péronne' sur le passage du roi" (Brachet, p. cxi).

115. "au lieu de faire tordre le cou à ses insulteurs ailés, après avoir commencé par les faire plumer vifs, les entoure de soins, leur assigne, en échange de leurs quolibets, un traitement royal dans ses parcs d'Amboise" (Brachet, p. cxi).

116. "L'hypothèse (très improbable) du hasard," "La solution du moraliste qui pense que Louis aura par orgueil choisi ce moment pour cette mesure sans précedent, précisément pour faire sentir aux Parisiens ravis de son échec à Péronne, que, bien qu'ils le crussent vaincu, il était pour eux toujours le maître," "du neuropathologiste" (Brachet, p. cxviii). One of Louis XI's failed goals was for Charles the Bold to renounce his alliance with the English, a remnant of the Hundred Years' War, during which Paris was under the English rule from 1420 to 1436.

117. "l'impulsion irrésistible de la kleptomanie zoophilique" (Brachet, p. cxix).

118. "un dégénéré héréditaire, nul doute que les faits précités, loin d'être une mode aristocratique, sont une manifestation très nette de l'un des stigmates classiques des dégénérés: la zoophilie" (Brachet, p. cxv).

119. "fait de psychologie morbide," "organisation spéciale," "hommes supérieurs"; Jacques-Joseph Moreau de Tours, *La psychologie morbide dans ses rapports avec la philosophie de l'histoire ou De l'influence des névropathies sur le dynamisme intellectuel* (Paris: Victor Masson, 1859), p. 535. Moreau de Tours uses "representative men" in English.

120. "Beaucoup de grands hommes semblent n'avoir pas eu d'ancêtres et trop souvent ils ne font point souche" (Letourneau, p. 600).

121. Charles Féré, *La famille névropathique: Théorie tératologique de l'hérédité et de la prédisposition morbides et de la dégénérescence*, 2nd ed. (Paris: Félix Alcan, 1898 [1894], p. 41.

122. "toute l'instance contradictoire. Il aurait fallu relever, à côté des cas de talents transmis, tous les cas connus où les talents paternels n'ont pas été transmis" (Brachet, p. x). Sir Francis Galton, *Hereditary Genius: An Inquiry into its Laws and Consequences* (London: Macmillan, 1869); Théodule Ribot, *L'hérédité, étude psychologique sur ses phénomènes, ses lois, ses causes, ses conséquences* (Paris: Ladrange, 1873).

123. "le développement intellectuel amène les névropathies, et, comme leur conséquence, la dégénérescence et l'extinction de la race"; Dr. Paul Jacoby, *Études sur la sélection chez l'homme*, 2nd ed. (Paris: Felix Alcan, 1904 [1881]), p. xi.

124. "archives familiales" (Brachet, p. ix).

125. "Le but de mon livre est de reconstituer les faits et la symptomatologie des rois par les textes que nous offre l'histoire et cette symptomatologie assurée, reconstituer l'enchaînement pathologique de la famille capétienne" (Brachet, pp. xiii–xiv).

126. "Si l'histoire de la médecine a une grande utilité, la médecine de l'histoire a une importance non moins considérable. Elle est dans certains cas un profond enseignement pour la politique; elle nous montre, pour la famille des Valois, l'application irréfutable des grandes lois de la pathologie générale"; Dr. Auguste Corlieu, *La médecine de l'histoire. Étude médicale sur la dynastie des Valois* (Paris: Adrien Delahaye, 1872), p. 1.

CHAPTER 2

1. "De toute notre étude le rôle du médecin apparaît alors avec toute son importance. . . . Connaissant les antécédents héréditaires de ses clients, il sera à même de prévoir ce que seront leurs enfants; à ses yeux, l'adage: *Pater est quem morbi filiorum demonstrant* offrira bien plus de garanties que l'axiome fameux du droit romain: *Pater est quem nuptiae demonstrant*"; Dr. Fernand Debret, *La sélection naturelle dans l'espèce humaine. Contribution à l'étude de l'hérédité convergente* (Paris: G. Steinheil, 1901), p. 72.

2. Alexandre Lacassagne, "Consanguinité," in *Dictionnaire encyclopédique des sciences médicales*, ed. Amédée Dechambre (Paris: G. Masson, 1864–1889), 1ère série, t. 19 (Paris, 1876), pp. 652–716, at p. 663.

3. "Après avoir fait partie exclusivement du langage juridique, le mot [consanguinité] a été adopté par les médecins qui en ont fait le synonyme de parenté" (Lacassagne, p. 653).

4. "la parenté des individus nés d'un même tronc," "Provenant en descendance directe d'un auteur commun, les parents consanguins sont rattachés à la même famille par les liens de *sang*" (Lacassagne, p. 653).

5. "Le mariage est défendu par l'Église jusqu'au quatrième degré de consanguinité inclusivement, mais par la loi de nature la consanguinité n'est point un obstacle au mariage, excepté en ligne directe. La consanguinité finit au sixième ou au septième degré, excepté pour la Couronne, dans ce cas la consanguinité se perpétue à l'infini" (Furetière, qtd. in Lacassagne, p. 671).

6. "Ce mot [la consanguinité] s'emploie aussi d'une manière plus générale pour indiquer toute sorte de parenté" (Lacassagne, p. 653). See David Warren Sabean and Simon Teuscher, "Kinship in Europe: A New Approach to Long-Term Development," in *Kinship in Europe: Approaches to Long-Term Development (1300–1900)*, ed. David Warren Sabean, Simon Teuscher, and Jon Mathieu (New York: Berhahn Books, 2010 [2007]), pp. 1–32; David Warren Sabean, "Kinship and Class Dynamics in Nineteenth-Century Europe," in *Kinship in Europe*, ed. Sabean, Teuscher, and Mathieu, pp. 301–13.

7. Sara Eigen Figal, *Heredity, Race, and the Birth of the Modern* (New York: Routledge, 2008), esp. chap. 1; for the Middle Ages, see Zrinka Stahuljak, *Bloodless Genealogies*

of the French Middle Ages: Translatio, *Kinship, and Metaphor* (Gainesville: University Press of Florida, 2005).

8. Sabean and Teuscher, "Kinship in Europe."

9. See Michel Foucault, *"Il faut défendre la société," Cours au Collège de France, 1976* (Paris: Gallimard/Seuil, 1997); *Naissance de la biopolitique, Cours au Collège de France, 1978–1979* (Paris: Gallimard/Seuil, 2004).

10. See works on hematology, e.g., Maximien Parchappe, *Études sur le sang dans l'état physiologique et l'état pathologique* (Paris: J.-B. Baillière, 1856).

11. "Le sang est la pépinière des *races*, c'est le ciment de *l'organisme*"; Dr. Francis Devay, *Traité spécial d'hygiène des familles, particulièrement dans ses rapports avec le mariage, au physique et au moral et les maladies héréditaires*, 2nd rev. ed. (Paris, Labé, 1858), p. 32, emphasis added.

12. "le rôle considérable que joue le fluide nourricier sur la rénovation organique" (Devay, *Traité spécial d'hygiène*, p. 32).

13. "ce sol *des générations*, c'est le sang"; Dr. Francis Devay, *Du danger des mariages consanguins au point de vue sanitaire* (Paris: Labé; Lyon: M. Savy, 1857), p. iii, emphasis added.

14. "cette double mission du sang" (Devay, *Traité spécial d'hygiène*, p. 33), "c'est du sang que se tirent les matériaux qui réparent, qui cicatrisent; c'est en lui que se puisent également ceux qui édifient" (Devay, *Traité spécial d'hygiène*, p. 32).

15. "si le sang est infecté de quelque virus particulier, la matière séminale sera aussi viciée, et par conséquent le germe participera au virus dominant de son père"; Dr. Francis Devay, *Hygiène des familles ou perfectionnement physique et moral de l'homme considéré particulièrement dans ses rapports avec l'éducation et les besoins de la civilisation moderne*, 2 vols. (Lyon: Dorier, 1846), vol. 2, p. 119. This is the first edition of the *Traité spécial d'hygiène des familles.*

16. "Un chef de famille exalte la supériorité de sa race lorsqu'il parle de la beauté de son sang; il la voit dégénérer lorsque le sang s'y altère" (Devay, *Traité spécial d'hygiène*, p. 32).

17. "On accuse les alliances entre parents de même souche, d'amener, de créer, par le seul fait du non-renouvellement du sang, une cause spéciale de dégradation organique, fatal à la propagation de l'espèce" (Devay, *Traité spécial d'hygiène*, p. 246).

18. "outre qu'elles servent à distinguer les personnes historiques du même nom et de la même famille, elles montrent les liaisons de parenté, les successions, les droits, les prétentions"; Pierre Larousse, *Grand dictionnaire universel du XIXe siècle* (Paris: Administration du Grand dictionnaire universel, 1866–1877), vol. 8 (1872), "Généalogie," pp. 1131–32, at p. 1131.

19. "les familles qui s'unissent entre elles ne tardent pas à dégénérer et à s'abâtardir," "l'union de deux individus de mauvaise constitution . . . donne naissance à des enfants plus faibles, plus débiles"; Larousse, *Grand dictionnaire universel*, "Hérédité," vol. 9 (n.d.), pp. 217–18, at p. 218.

20. "être conseillées; il faut, au contraire, renouveler la constitution et le tempérament par un croisement bien entendu" (Larousse, "Hérédité," p. 218).

21. "la constatation des *généalogies* pour chaque race," "Cette institution . . . demeure stérile au point de vue du maintien et du perfectionnement des races parce que l'homme,

qui se dit intelligent, est trop fier, trop entiché de sa liberté individuelle pour appliquer sur lui-même les principes d'économie animale qui devraient présider à toute procréation" (Larousse, "Généalogie," p. 1132).

22. Alfred Bourgeois, *Quelle est l'influence des mariages consanguins sur les générations? Thèse pour le doctorat en médecine* (Paris: Rignoux, 1859).

23. "pour la famille des Valois, l'application irréfutable des grandes lois de la pathologie générale. On y voit en effet un chef de race, François Ier, dont la fécondité n'a pas été douteuse, n'avoir pu conserver la couronne dans sa famille que pendant 74 ans"; Dr. Auguste Corlieu, *La médecine de l'histoire. Étude médicale sur la dynastie des Valois* (Paris: Adrien Delahaye, 1872), p. 1.

24. Théodule Ribot, *L'hérédité, étude psychologique*, 5th corrected and rev. ed. (Paris: Felix Alcan, 1894 [1873]). It would be in its 10th edition in 1914.

25. "faisait de la généalogie une auxiliaire de la pathologie"; Dr. Cabanès, *L'histoire éclairée par la clinique* (Paris: Albin Michel, 1921), p. 288.

26. "les époques antérieures n'étant pas assez bien connues sous le rapport des généalogies"; Dr. Paul Jacoby, *Études sur la sélection dans ses rapports avec l'hérédité chez l'homme*, 2nd ed. (Paris: Felix Alcan, 1904 [1881]), p. 321.

27. "70 tables généalogiques"; Jules Dejerine, *L'hérédité dans les maladies du système nerveux* (Paris: Anselin et Huzeau, 1886).

28. "l'histoire de la folie est . . . l'histoire même de la civilisation"; Alexandre Cullerre, *Les frontières de la folie* (Paris: J.-B. Baillière, 1888), p. 318.

29. "l'influence d'une lourde hérédité et d'une mauvaise éducation sur l'état mental de ces trois souverains"; Émile Dusolier, *Psychologie des derniers Valois (François II, Charles IX, Henri III), Thèse présentée à la Faculté de médecine et de pharmacie de Lyon, et soutenue publiquement le mercredi 27 novembre 1895* (Lyon: Imprimerie de A.-H. Storck, 1895), p. 3.

30. "Bien plus dignes de notre pitié, j'entends de celle du médecin" (Dusolier, p. 3).

31. "on n'a pas cherché dans le lointain de leurs antécédents et dans la fatalité de leur naissance le secret de leur dégénérescence" (Dusolier, p. 3).

32. "La lecture des généalogies des familles souveraines a pu paraître aride, mais celles-ci constituent, avec les portraits, la base même de notre démonstration, en nous permettant de fixer la filiation exacte du personnage représenté et de montrer, alors même que nous n'avons pas le portrait de ses ancêtres, la conservation à travers les âges d'un type familial héréditaire, ou sa reviviscence"; Dr. Victor Galippe, *L'hérédité des stigmates de dégénérescence et les familles souveraines* (Paris: Masson, 1905), pp. 443–44.

33. Dr. Cabanès, *Le mal héréditaire I. Les descendants de Charles-Quint. Ouvrage orné de 63 illustrations* (Paris: Albin Michel, 1926); *Le mal héréditaire II. Les Bourbons d'Espagne. Ouvrage orné de 93 illustrations* (Paris: Albin Michel, 1927).

34. "Il suit de là que, dans bien des cas, les unions consanguines seront nuisibles, et d'autant plus dangereuses que les prédispositions morbides, communes aux deux conjoints, seront plus marquées" (Ribot, pp. 352–53).

35. "S'il en eût toujours été ainsi, la France n'aurait peut-être pas eu tant à souffrir sous

l'administration de races dégénérées, et l'hérédité pathologique est, à mon avis, l'un des plus puissants arguments contre l'hérédité dynastique" (Corlieu, p. 9).

36. Jacques-Joseph Moreau de Tours, *La psychologie morbide dans ses rapports avec la philosophie de l'histoire ou De l'influence des névropathies sur le dynamisme intellectuel* (Paris, Victor Masson, 1859), p. 535.

37. "Comment comparer . . . un aliéné de Charenton, simple plante annuelle, avec un fou tel que Charles VI dont nous pouvons étudier l'hérédité pendant six cents ans, dans dix-huit générations en ligne directe, non pour transformer en autant d'aliénés ces dix-huit générations de la maison de France, comme le feraient les historiens à la Michelet, mais pour voir par quelles graduelles métamorphoses morbides ces générations ont passé, et comment l'accumulation des tares morbides, soit seules, soit par les mariages consanguins . . . aboutit à un fou"; Auguste Brachet, *Pathologie mentale des rois de France. Louis XI et ses ascendants. Une vie humaine étudiée à travers six siècles d'hérédité, 852–1483* (Paris: Librairie Hachette, 1903 [1896]), p. cxlii.

38. "divers accidents" (Brachet, p. cl), "Les uns l'attribuent aux abus de tournois et de fêtes, d'autres à une enfance maladive, Michelet à une course équestre qui n'a jamais eu lieu" (Brachet, p. cxlix).

39. "Toute dégénérescence d'une race ne peut être que le résultat d'une hérédité, et l'un des facteurs importants de l'hérédité est précisément les consanguinités [*sic*] des géniteurs" (Brachet, p. cxxvi).

40. "Toutes les classes privilegiées, toutes les familles qui se trouvent dans des positions exclusivement élevées, partagent le sort des familles régnantes, quoiqu'à un degré moindre et qui est toujours en rapport direct avec la grandeur de leurs privilèges et la hauteur de leur position sociale" (Jacoby, p. 431).

41. "delusions of grandeur . . . 'Cesaritis,' as Professor Lacassagne calls it" (la monomanie des grandeurs . . . la 'césarite,' comme l'appelle le professeur Lacassagne); Maurice Beaujeu, *Psychologie des premiers Césars. Une étude de médecine légale dans l'histoire. Thèse pour obtenir le grade de docteur en médecine.* Faculté de médecine et de pharmacie de Lyon, no. 796 (Lyon: A. Storck, 1893), p. 51. See also "the exercise of absolute power, *Cesaritis*, as Professor Lacassagne so aptly named it" (l'exercice du pouvoir absolu, la *césarite*, comme l'a bien nommée le professeur Lacassagne); Dr. Cabanès, *Les morts mystérieuses de l'Histoire. Souverains et princes français de Charlemagne a Louis XVII* (Paris: A. Maloine, 1901), p. xviii. The first to connect power to insanity is Franz de Champagny, *Les Césars* (Paris: Renouard, 1841).

42. "le délire du pouvoir"; Gabriel Tarde, "Introduction," in Dr. Paul Jacoby, *Études sur la sélection chez l'homme*, 2nd ed. (Paris: Felix Alcan, 1904 [1881]), pp. i–iii, at p. ii.

43. "l'influence dégénérative du pouvoir" (Galippe, p. 444).

44. "plus il [le pouvoir] est grand, plus la décadence morale, intellectuelle et physique sera manifeste," "le pouvoir absolu . . . produit la dégénérescence la plus profonde et la plus rapide" (Jacoby, p. x).

45. "Comment aller jusqu'à affirmer que la dégénérescence est une *preuve* d'origine royale?" (Brachet, p. cxxii).

46. "ce sont les dynasties royales qui offrent au clinicien les sources les plus précieuses de documentation pathologique" (Brachet, p. ix).

47. "étudiant les familles *royales,* et n'en ayant *pas d'autres* à sa disposition pour une recherche scientifique, affirme qu'il y a des maladies mentales *de cause royale,*" "la *royalty* de la zoophilie de Louis XI," "*Toute* dégénérescence d'*une* race ne peut être que le résultat d'*une* hérédité" (Brachet, pp. cxxii, cxxvi, emphasis added).

48. "La Pathologie historique . . . relie les *data* pathologiques que lui fournit à l'état fragmentaire l'Histoire médicale, pour en établir la succession, et en tirer la loi pathologique de l'individu ou, comme on dit en clinique, sa formule pathogénique, et de là psychique" (Brachet, p. xii).

49. "pathologie individuelle . . . d'une collectivité d'individus" (Brachet, p. xii).

50. "tentative d'une constitution scientifique partielle de l'hérédité humaine," "une révolution en histoire" (Dr. Cabanès, *L'histoire éclairée*, p. 268).

51. "la croyance à la paternité, c'est-à-dire sur *un acte de foi,*" "les familles européennes" (Brachet, p. cxci).

52. "Mais il faut être en garde contre les absurdités de certains historiens, qui, par adulation, font remonter jusqu'aux temps héroïques l'origine des maisons ou des princes en faveur de qui ils écrivent" (Larousse, "Généalogie," p. 1131).

53. "On voit que les familles nobles, et surtout celles dont l'origine remonte à des temps éloignés, ne sont parvenues à continuer jusqu'à nous non pas leur lignée, mais leur nom, bien plus ancien chez la plupart d'entre elles que la descendance, qu'à l'aide de fictions de toute espèce et en se greffant sur des familles étrangères"; Louis-François Benoiston de Châteauneuf, *Mémoire sur la durée des familles nobles de France* (Extrait du compte-rendu des séances et travaux de l'Académie des sciences morales et politiques, 1845), *Annales d'hygiène publique et de médecine légale* 35 (janvier 1846): 27–58, at p. 30; see also Lacassagne, p. 696.

54. "des substitutions multipliées à l'infini," "des mariages contractés successivement deux, trois et quatre fois," "des demandes en dispense de vœux," "des transmissions du nom, par les femmes, dans des familles étrangères, etc.," "la légitimation d'un enfant naturel" (Châteauneuf, pp. 29–30).

55. "Pour l'histoire politique, qu'un roi soit ou non le fils de celui qu'on croit être son père, si personne que sa mère ne le sait, la politique suit son cours" (Brachet, p. cxcii).

56. "suffit à l'historien, elle est insuffisante pour le clinician" (Brachet, p. ccxiii): "En pathologie la maxime *Pater est quem nuptiae demonstrant* n'est pas de mise, puisque les conclusions à tirer de l'enchaînement et de la succession des faits reposent entièrement, avant toute étude, sur ce postulat essentiel que les souverains à étudier descendent bien l'un de l'autre" (Brachet, p. cxcii).

57. "La base, le postulat de cette étude est nécessairement la réalité de l'hérédité" (Brachet, p. cxcii).

58. "L'étude de l'hérédité reposant essentiellement sur le postulat de la certitude *biologique,* toute expérience sur l'hérédité est frappée de nullité dès que l'un des facteurs n'appartient pas sûrement à la chaîne généalogique" (Brachet, p. cxci).

59. "Par la connaisance de son hérédité *absolument négligée* jusqu'ici par les historiens et les médecins" (Brachet, p. ccxvi).

60. "tout fils d'un père fou doit être fou," "Charles VI, Louis [d'Orléans, l'amant d'Isabeau], ou un inconu" (Brachet, pp. cxcvii, ccxiii).

61. "La formule biologique d'Isabeau . . . nous permet d'affirmer que, quel que fût le père, cette formule était en soi *suffisante* pour expliquer les tares de Charles VII et de Louis XI" (Brachet, p. ccxvi).

62. "La *Pathologie historique* est proprement l'explication, par la science biologique, des données que nous fournissent les textes historiques" (Brachet, p. xii).

63. Cf. "It is always easy to prove that royal blood has been much spread amongst the people; it would be perhaps more challenging to establish that royal blood came in large part from the people. . . . Official paternity must therefore be accepted as dogma, legitimate paternity as an act of faith, which is not up for discussion" (Il est toujours facile de prouver que le sang royal s'est fort répandu dans le peuple; il serait peut-être plus malaisé d'établir que le sang royal est venu, en grande partie, du peuple. . . . Force est donc d'accepter la paternité officielle comme un dogme, la paternité légitime comme un acte de foi, qui ne se prête pas à la discussion) (Dr. Cabanès, *Le mal héréditaire*, 1:27).

64. "la biographie négative dans l'histoire d'un individu, c'est-à-dire n'admettre que l'existence de faits contrôlés" (Brachet, p. cxcii).

65. Georges Duby, *La société aux XIe et XIIe siècles dans la région mâconnaise* (Paris: A. Colin, 1953).

66. Georges Duby, "The Nobility in Medieval France," in *The Chivalrous Society*, trans. Cynthia Postan (Berkeley: University of California Press, 1980), pp. 94–111; "The Structure of Kinship and Nobility," in *The Chivalrous Society*, pp. 134–48; "Lineage, Nobility and Knighthood: The Mâconnais in the Twelfth Century: A Revision," in *The Chivalrous Society*, pp. 59–80.

67. Theodore Evergates, *The Aristocracy in the County of Champagne, 1100–1300* (Philadelphia: University of Pennsylvania Press, 2007), p. 85.

68. See also Amy Livingstone, *Out of Love for My Kin: Aristocratic Family Life in the Lands of the Loire, 1000–1200* (Ithaca, N.Y.: Cornell University Press, 2010).

69. Sabean and Teuscher, "Kinship in Europe," p. 6. See also David Warren Sabean, "From Clan to Kindred: Kinship and the Circulation of Property in Premodern and Modern Europe," in *Heredity Produced: At the Crossroads of Biology, Politics, and Culture, 1500–1870*, ed. Staffan Müller-Wille and Hans-Jörg Rheinberger (Cambridge, Mass.: MIT Press, 2007), pp. 37–59; *Le modèle familial européen: Normes, déviances, contrôle du pouvoir. Actes des séminaires organisés par l'École française de Rome et l'Università di Roma* (1984) (Rome: École française de Rome, 1986).

70. "Les vocables latins et français . . . se rapportent . . . à telle ou telle manière de considérer ou de fractionner l'ensemble aux contours mouvants que constitue la parentèle"; Anita Guerreau-Jalabert, "La désignation des relations et groupes de parenté en latin médiéval," *Bulletin du Cange* 46–47 (1988): 65–108, at p. 92; see also Anita

Guerreau-Jalabert, "La parenté dans l'Europe médiévale et moderne: À propos d'une synthèse récente," *L'homme* 110 (1989): 69–93, at p. 71.

71. Guerreau-Jalabert, "La désignation," p. 74. See also Anita Guerreau-Jalabert, "*Spiritus* et *caritas*. Le baptême dans la société médiévale," in *La parenté spirituelle*, ed. Françoise Héritier-Augé and Elisabeth Copet-Rougier (Paris: Éditions des archives contemporaines, 1995), pp. 133–203, especially pp. 171–83.

72. Guerreau-Jalabert, "La parenté," p. 73; Guerreau-Jalabert, "La désignation," p. 72.

73. Constance Brittain Bouchard, *Those of My Blood: Constructing Noble Families in Medieval Francia* (Philadelphia: University of Pennsylvania Press, 2001), p. 5.

74. R. Howard Bloch, *Etymologies and Genealogies: A Literary Anthropology of the French Middle Ages* (Chicago: University of Chicago Press, 1983); Gabrielle M. Spiegel, "Genealogy: Form and Function in Medieval Historiography," in *The Past as Text: The Theory and Practice of Medieval Historiography* (Baltimore: Johns Hopkins University Press, 1997 [1983]), pp. 99–110.

75. R. Howard Bloch, "Genealogy as a Medieval Mental Structure and Textual Form," in *La littérature historiographique des origines à 1500*, *Grundriss der romanischen Literaturen des Mittelalters*, ed. Hans Ulrich Gumbrecht, Ursula Link-Heer, and Peter-Michael Spangenberg, vol. 11, part 1 (Heidelberg: Carl Winter, 1986), pp. 135–56, at p. 140.

76. In her ground-breaking study of visual representations of medieval genealogy, Klapisch-Zuber showed that the image of the genealogical tree, despite all of the (anachronistic) symbolism inherent in it for us, did not exist before the fifteenth century; Christiane Klapisch-Zuber, *L'ombre des ancêtres: Essai sur l'imaginaire médiéval de la parenté* (Paris: Fayard, 2000). See also "La genèse de l'arbre généalogique," in *L'arbre. Histoire naturelle et symbolique de l'arbre, du bois et du fruit au Moyen Âge*, ed. Michel Pastoureau (Paris: Le Léopard d'Or, 1993), pp. 41–81. The case of the Trees of Jesse is just like that of the *arbor iuris* that, despite its name, does not correspond to a tree (Klapisch-Zuber, "Le corps," p. 46).

77. "Le sang *comme définition de la parenté* chez les nobles n'apparaît . . . que tardivement. Ce n'est qu'à partir du XIVe siècle que les descendants directs des rois seront appelés 'princes du sang.' Ce n'est que dans l'Espagne de la fin du XVe siècle qu'apparaît, à l'encontre des Juifs, la notion de 'pureté du sang'"; Jacques Le Goff (with Nicolas Truong), *Une histoire du corps au Moyen Âge* (Paris: Éditions Liana Levi, 2003), p. 40.

78. David Nirenberg, "Mass Conversion and Genealogical Mentalities: Jews and Christians in Fifteenth-Century Spain," *Past and Present* 174:1 (2002): 3–41.

79. "*sanc* peut recouvrir la parenté consanguine, mais cet usage semble se développer surtout à la fin du Moyen Âge"; Anita Guerreau-Jalabert, "Sang," in *Dictionnaire du Moyen Âge*, ed. Claude Gauvard, Alain de Libera, and Michel Zink (Paris: Presses universitaires de France, 2002), pp. 1280–81.

80. "Dans les traités sur la consanguinité et les prohibitions de mariage, les théologiens et canonistes des XIIe et XIVe siècles font pourtant moins allusion au sang commun qu'on ne l'attendrait"; Christiane Klapisch-Zuber, "Le corps de la parenté," *Micrologus* 1 (1993): 43–60, at p. 48.

81. Caroline Walker Bynum, *Wonderful Blood: Theology and Practice in Late Medieval*

Northern Germany and Beyond (Philadelphia: University of Pennsylvania Press, 2007), pp. 157–58.

82. Guerreau-Jalabert, "Sang"; Sabean and Teuscher, "Kinship in Europe," p. 5.

83. Stahuljak, *Bloodless Genealogies*; Christine Ferlampin-Acher and Denis Hüe, eds., *Lignes et lignages dans la littérature arthurienne. Actes du 3e colloque arthurien organisé à l'université de Haute-Bretagne, 13–14 octobre 2005* (Rennes: Presses Universitaires de Rennes, 2007).

84. Bynum, *Wonderful Blood*; Bettina Bildhauer, *Medieval Blood* (Cardiff: University of Wales Press, 2006); Peggy McCracken, *The Curse of Eve, the Wound of the Hero* (Philadelphia: University of Pennsylvania Press, 2003); Jacques Voisenet, "Le tabou du sang dans les pénitentiels du haut Moyen Âge," in *Le sang au Moyen Âge*, Montpellier, L'Université de Paul-Valéry Montepellier III: Les Cahiers du C.R.I.S.I.M.A., 4 (1999): 111–25; Michel Pastoureau, "Ceci est mon sang. Le christianisme médiéval et la couleur rouge," in *Le pressoir mystique*, ed. D. Alexandre-Bidon (Paris: Le Cerf, 1990), pp. 43–56; Jacques Le Goff, "Métiers licites et métiers illicites dans l'Occident médiéval," in *Pour un autre Moyen Âge* (Paris: Gallimard, 1977), pp. 91–107, at p. 93.

85. "Le bon sang, symbole de la vie humaine et du sacrifice divin, et le mauvais sang, image du péché et de la mort" (Voisenet, p. 111).

86. Duby, "Structure of Kinship," pp. 134–35, emphasis added. Duby repeats this statement in "French Genealogical Literature," in *The Chivalrous Society*, pp. 149–57, at p. 149, and in "The Nobility in Medieval France," p. 99.

87. Bouchard, p. 4; "the significance attached to one's varied biological or blood ties always had to be modified, weakened or strengthened" (Bouchard, p. 180).

88. Gabrielle M. Spiegel, "Foucault and the Problem of Genealogy," *Medieval History Journal* 4:1 (2001): 1–14, at p. 1.

89. "a genealogical analysis"; Michel Foucault, "Nietzsche, Genealogy, History," in *Language, Counter-Memory, Practice: Selected Essays and Interviews*, ed. Donald F. Bouchard, trans. Donald F. Bouchard and Sherry Simon (Ithaca, N.Y.: Cornell University Press, 1977), pp. 139–64, at p. 159 ("Nietzsche, la généalogie, l'histoire," *Dits et écrits I, 1954–1975*, ed. Daniel Defert and François Ewald [Paris: Quarto Gallimard, 2001 (1971)], pp. 1004–24, at p. 1020); translation occasionally modified. See Rudi Visker, *Michel Foucault: Genealogy as Critique*, trans. Chris Turner (London: Verso, 1995); Giorgio Agamben, *Signatura rerum. Sur la méthode*, trans. Joël Gayraud (Paris: Vrin, 2009).

90. Michel Foucault, "The Confession of the Flesh," in *Power/Knowledge: Selected Interviews and Other Writings, 1972–1977*, ed. Colin Gordon, trans. Colin Gordon, Leo Marshall, John Mepham, and Kate Soper (New York: Pantheon Books, 1980), pp. 194–228, at p. 197 ("Le jeu de Michel Foucault," *Dits et écrits II, 1976–1988*, ed. Daniel Defert and François Ewald [Paris: Quarto Gallimard, 2001 (1977)], pp. 298–329, at p. 301).

91. Paul Freedman and Gabrielle M. Spiegel, "Medievalisms Old and New: The Rediscovery of Alterity in North American Medieval Studies," *American Historical Review* 103:3 (1998): 677–704, at p. 701.

92. Michel Foucault, *History of Sexuality. Volume I: An Introduction*, trans. Robert

Hurley (New York: Vintage, 1990), p. 124 (*Histoire de la sexualité I: La Volonté de savoir* [Paris: Gallimard/Tel, 2005 (1976)], p. 164); see also Introduction.

93. This does not resolve Foucault's fantasy of origins about a free, sexually liberated Middle Ages that has been extensively analyzed and criticized; Karma Lochrie, "Desiring Foucault," *Journal of Medieval and Early Modern Studies* 27:1 (1997): 3–16; Bruce Holsinger, *The Premodern Condition: Medievalism and the Making of Theory* (Chicago: University of Chicago Press, 2005), esp. pp. 52–56. But because Foucault simultaneously worked on sexuality and genealogy, because genealogy allowed him to read the apparatus of sexuality, a future project could consist of reading medieval sexuality through genealogy. See Daniel Boyarin and Elizabeth A. Castelli, "Introduction: Foucault's *The History of Sexuality*: The Fourth Volume, or, a Field Left Fallow for Others to Till," *Journal of the History of Sexuality* 10:3–4 (2001): 357–74.

CHAPTER 3

1. For a classic history of the trial of the Templars, see Malcolm Barber, *The Trial of the Templars* (Cambridge: Cambridge University Press, 1978).

2. Charles Hippolyte Maillard de Chambure, *Règle et statuts secrets des Templiers, précédés de l'Histoire de l'établissement, de la destruction et de la continuation moderne de L'ordre du Temple* (Paris: Brockaus et Avenarius; Dijon: Lamarche, 1840), pp. 13–14.

3. "[Leur succès] fut immense, parce qu'il révélait (et cette révélation est prodigieuse après deux siècles de littérature classique) une littérature nationale"; Ch. Nodier, P. Lepeintre, L. Thiessé, and A. Lesourd, eds., *Bibliothèque dramatique, ou Répertoire universel du théâtre français* (Paris: Dabo-Butschert, 1824), p. 4; also p. 157. See also Pierre Larousse, *Grand dictionnaire universel du XIXe siècle* (Paris: Administration du Grand dictionnaire universel, 1866–1877), "Templier," vol. 14 (1875), pp. 1586–88, at p. 1588.

4. François Just Marie Raynouard, *Les Templiers, tragédie. Précédée d'un précis historique sur les Templiers* (Paris: Giguet et Michaud, 1805).

5. "à l'espoir de résoudre un problème historique, il ne s'était joint un motif plus noble et plus impérieux: j'ai pensé que j'ajouterais peut-être aux titres de gloire de ma patrie, et au juste respect qu'inspire le caractère national, si je recueillais, avec zèle et impartialité, les preuves de l'innocence de l'Ordre et des Chevaliers"; François Juste Marie Raynouard, *Monumens historiques relatifs à la condamnation des Chevaliers du Temple, et l'abolition de leur ordre* (Paris: A. Égron, 1813), n.p.

6. "vénérables victimes" (Raynouard, *Monumens historiques*, p. 164).

7. "suivant les uns, une statue, suivant d'autres une tête, suivant d'autres encore, un ensemble de plusieurs têtes ou même une peinture sur bois"; Salomon Reinach, "La tête magique des Templiers," in *Cultes, mythes et religions*, ed. Hervé Duchêne (Paris: Robert Laffont, 2000 [1911]), pp. 1015–25, at p. 1017.

8. "n'était pas le nom de l'idole, mais bien celui d'un hyéroglyphe [*sic*] empreint sur elle. La tête . . . avait été un symbole, l'image du Père éternel en état de repos, tel que les

anciens Gnostiques le représentaient"; Jules Loiseleur, *La doctrine secrète des Templiers* (Cazilhac: Belisane, 1999 [1873]), p. 107. See also Elizé de Montagnac, *Histoire des Chevaliers Templiers et de leurs prétendus successeurs, suivie de l'histoire des ordres du Christ et de Montesa* (Paris: Auguste Aubry, 1864), p. 119.

9. "L'Orient, et surtout la Syrie, était le refuge de tous les sectaires et hérésiarques proscrits par l'empire d'Occident. Quel foyer pour le *gnosticisme*! C'est là, sans aucun doute, que les templiers virent se corrompre leurs traditions religieuses, en se laissant aller au leurre d'un prétendu *christianisme rectifié*"; Prosper Mignard, *Éclaircissements sur les pratiques occultes de Templiers* (Dijon: Douillier, 1851), p. 25.

10. "Ainsi l'ordre qui avait représenté au plus haut degré le génie symbolique du moyen âge mourut d'un symbole non compris"; Jules Michelet, *Histoire du Moyen Âge*, in *Œuvres complètes*, vols. 1–6 (Paris: Flammarion, 1893–1898 [1833–1844]), vol. 3, p. 161.

11. "Des sectes mystiques, hétérodoxes, comme les francs-maçons et les rose-croix, [qui] ont glorifié les chevaliers du Temple d'une partie des crimes . . . afin de se rattacher à une tige ancienne: elles ont voulu voir de la profondeur dans le symbolisme inepte que les accusateurs du Temple ont décrit"; Charles-Victor Langlois, *Saint Louis, Philippe le Bel, les derniers Capétiens directs (1226–1328)*, *Histoire de France depuis les origines jusqu'à la Révolution*, ed. Ernest Lavisse, vol. 3 (Paris: Hachette, 1901), p. 180.

12. "à la formation d'un fantôme, qui, après avoir été exploité contre l'orthodoxie des Templiers, a tourné la tête de plus d'un archéologue" (Reinach, p. 1016).

13. Joseph von Hammer-Purgstall, introduction and trans., *Ancient Alphabets and Hieroglyphic Characters Explained: With an Account of the Egyptian Priests, their Classes, Initiation, and Sacrifices* (London, 1806), p. i.

14. "comment la connoissance de l'écriture hiéroglyphique, perdue, à ce qu'il paroît, longtemps avant que les Arabes fussent maîtres de l'Egypte, avoit pu sortir des ténèbres qui la couvroient pour passer entre leurs mains"; Silvestre de Sacy, "Livres divers" (Review of Hammer-Purgstall), *Magasin encyclopédique* 6 (1810): 145–75, at p. 155. Silvestre de Sacy was the founder of modern Orientalism but also an active public policymaker in the matters of the Orient; see Henri Dehérain, *Silvestre de Sacy, ses contemporains et ses disciples*, *Bibliothèque archéologique et historique*, vol. 27 (Paris: Librairie orientaliste Paul Geuthner, 1938).

15. "et que ce nom, dans l'esprit des Templiers et de leurs accusateurs, a dû être celui de *Mahomet*" (Sacy, p. 159).

16. "Dans la déposition des deux témoins entendus à Carcassonne, qui parlent de *Figura Baffometi*, il est évident que c'est par une faute d'orthographe ou de prononciation que ce mot est ainsi écrit au lieu de *Mahometi*, soit que dans les provinces du midi, on prononçat ainsi le nom de Mahomet, soit que le copiste ait écrit par erreur *Baffometi*, comme il a écrit en même temps *asorare* pour *adorare*; et ce qui doit ne laisser aucun doute à cet égard, c'est que le témoin prétend qu'on lui fit prononcer *Yalla*, mot des Sarrazins, dit-il qui signifie Dieu" (Raynouard, *Monumens historiques*, p. 301).

17. Qtd. in Montagnac, p. 120, emphasis added. There is further evidence of substitution of "Bafometz" for "Mahomet" in a poem "Senhors, per los nostres peccatz" by troubadour Gavaudan (1195): "ab Luy venseretz totz los cas / Cuy *Bafometz* a escarnitz / e·ls

renegatz outrasalhitz" (with his [Jesus'] help you will defeat all the dogs whom *Mahomet* has led astray and the impudent renegades; vv. 34–36); qtd. in Michael Routledge, "The Later Troubadours," in *The Troubadours: An Introduction*, ed. Simon Gaunt and Sarah Kay (Cambridge: Cambridge University Press, 1999), pp. 99–112, at p. 112, n. 8, emphasis added. "De Bafomet" is also the title of one of four surviving chapters of an Occitan translation of Ramon Llull's earliest known work, the *Libre de la doctrina pueril*; see Diego Zorzi, "Un frammento provenzale della *Doctrina Pueril* di Raimondo Lull," *Aevum* 28:4 (1954): 345–49.

18. "Comme le symbole que les Templiers sont censés rejeter et même souiller est le crucifix, le *Christ*, on qualifie du nom de *Mahomet* le symbole qu'ils préfèrent et opposent à celui-là, sans songer que les musulmans eux-mêmes n'avaient pas d'images" (Reinach, p. 1025).

19. Joseph von Hammer-Purgstall, "Mysterium Baphometis revelatum," in *Fundgruben des Orients, bearbeitet durch eine Gesellschaft von Liebhabern* [Treasures of the Orient, exploited by an Amateur Society], vol. 6 (Wien: Anton Schmid, 1818), pp. 3–120. This volume also contains a response to Raynouard, "Gegenrede wider die Einrede der Vertheidiger der Templer" [A Reply to the Plea of the Defenders of the Templars], pp. 445–92.

20. "le baptême de l'esprit, le baptême gnostique, qui ne se faisoit point par l'eau de la rédemption, mais qui étoit une lustration spirituelle par le feu: *Bafomet* signifie donc l'illumination de l'esprit"; Just Marie François Raynouard, "*Fundgruben des Orients*" (Review), *Journal des sçavans* (mars 1819): 151–61, at p. 153.

21. "baptême de l'intelligence" (Loiseleur, p. 110).

22. "des vases gnostiques, des calices; . . . il avance que le roman du *Saint Graal*, ou sainte Coupe, est un roman symbolique qui cache et prouve à la fois l'apostasie, la doctrine gnostique des Templiers" (Raynouard, "*Fundgruben des Orients*" [Review], p. 153).

23. "simples et . . . vaines conjectures"; Just Marie François Raynouard, "*Fundgruben des Orients*" (Review), *Journal des sçavans* (avril 1819): 221–29, at p. 229; see also Chambure, p. 52. The esoteric interpretation by Hammer-Purgstall of Templar Gnosticism will also fuel the speculations that the Templars survived their trial and persecution, and that they reemerged in the eighteenth and nineteenth centuries, either under the guise of Freemasons or in the New Order of the Templars. This myth is also indirectly maintained throughout the nineteenth century (and beyond, to this day, with books such as *The Da Vinci Code*) in the association of Templars with Freemasonry: "the caliphs practicing masonic mysteries and communicating them to the knights of the Temple" (les califes pratiquant les mystères maçonniques et les communiquant aux chevaliers du Temple); "*Masonry* came out of the Crusades. . . . It was the noble princes and barons of Christianity who invented it in the Holy Land for the betterment of religion" (La *maçonnerie* est issue des croisades. . . . Ce sont les nobles princes et barons de la chrétienté qui l'ont inventée en terre sainte pour le plus grand bien de la religion); Larousse, *Grand dictionnaire universel*, "Franc-Maçonnerie," vol. 8 (1872), pp. 764–69, at pp. 764, 765. This Oriental hypothesis is among many on the origins of the free-masonic cult. Though a tenacious one, it is upheld by some loggias themselves: "The supposedly Templar origin of *Masonry*, a tradition just as apocryphal as that of the Freemason crusaders. The Templars were never Freemasons. . . . But what is true is that

the French and German Freemasons from the middle of the last [eighteenth] century, on the contrary, made the order of the Temple enter into *Masonry*, by natural extension of its supposedly chivalrous origins" (la prétendue origine templière de la *maçonnerie*, tradition tout aussi apocryphe que celle des croisés francs-maçons. Les templiers n'ont jamais été francs-maçons. . . . Mais ce qui est vrai, c'est que les francs-maçons français et allemands du milieu du siècle dernier ont, au contraire, fait entrer l'ordre du Temple dans la *maçonnerie*, par l'extension naturelle des prétendues origines chevaleresques) (Larousse, "Franc-Maçonnerie," p. 765); see also Montagnac, pp. 139–43, on the claims of Freemasons to the succession of the Templars; and Chambure, pp. 15–19, on the New Order of the Templars.

24. Joseph von Hammer-Purgstall, *Mémoire des deux coffrets gnostiques du Moyen Âge, du cabinet de M. le Duc de Blacas* (Paris: Librairie orientale de Dondey-Dupré père et fils, 1832). The duke himself was a member of the ultra-royalist faction. For a history of publications on Baphomet, see Salomon Reinach, "Bas-reliefs énigmatiques," *Revue africaine* 52:269 (1908): 155–70, and "La tête magique des Templiers"; see also Peter Partner, *The Murdered Magicians: The Templars and Their Myth* (Oxford: Oxford University Press, 1982).

25. "La solution de cette grave question de *métaphysique anti-chrétienne* . . . est cachée dans les mille replis des sectes nombreuses du *gnosticisme*"; Prosper Mignard, *Monographie du coffret de M. le Duc de Blacas* (Paris: Dumoulin, 1852), p. 47; also Mignard, *Éclaircissements*. For a refutation, see Montagnac, p. 117: "Doubtless in the goal of finding an explanation for the absurd and incomprehensible crimes by which the Order had been condemned, they imagine to associate it with all sorts of impious sects, such as the Gnostics, the Ophians, the Manicheans, the Assassins, etc., etc." (Dans le but, sans doute, de trouver une explication aux crimes absurdes et incompréhensibles pour lesquels l'Ordre avait été condamné, ils imaginent de le rattacher à toutes espèces de sectes impies, telles que les Gnostiques, les Ophites, les Manichéens, les Assassins, etc., etc.). Cf. "Such questions, despite the ingenious conjectures of the Moderns, will remain forever obscure in the insufficiency of monuments" (De telles questions, malgré les ingénieuses conjectures des modernes, resteront toujours obscures dans l'insuffisance des monuments) (Michelet, 3:104).

26. "la statue androgyne baphométique" (Hammer-Purgstall, *Mémoire*, p. 13). See also "the Baphometic statue" (*la statue baphométique*), "the androgynous statue" (*la statue androgyne*), "the androgynous genie" (*le génie* [*androgyne*]) (Hammer-Purgstall, *Mémoire*, pp. 22, 4, 11, 5).

27. "l'amour androgyne" (Hammer-Purgstall, *Mémoire*, p. 32).

28. "pratiques impures" (Mignard, *Éclaircissements*, p. 6), "ces graves désordres avaient été importés d'Orient et introduits dans l'ordre par quelques-uns des chefs les plus pervertis; et moralement même il est impossible de se refuser à le croire!" (Mignard, *Monographie*, p. 36). As other authors explain, the accusation of sodomy and heresy is one of the oldest in political history and it has little grounding in reality; see J. M. Plane, *Apologie des Templiers et des Francs-Maçons* (Meudon, 1797), pp. 7–8; Nodier, pp. 155–56. Since Voltaire, such collective crimes were deemed impossible, though authors acknowledged willingly that individual disorders were always possible: "I have no doubt that several young Templars abandoned themselves to excesses that have in all ages been shared by youth; and it is

these passing vices that are far better ignored than punished" (Je ne doute nullement que plusieurs jeunes templiers ne s'abandonnassent à des excès qui de tout temps ont été le partage de la jeunesse; et ce sont de ces vices passagers qu'il vaut beaucoup mieux ignorer que punir); Voltaire, *Essai sur les mœurs et l'esprit des nations, et sur les principaux faits de l'histoire, depuis Charlemagne jusqu'à Louis XIII*, in *Œuvres*, vols. 15–18, ed. M. Beuchot (Paris: Werdet et Lequien, 1829 [1756–1763]), vol. 16, pp. 289–90. Authors who wished to clear the Templars of the crime of sodomy dismissed the confessions because of torture or resorted to the strategy employed by Voltaire, which was to deny the possibility of a collective and ritual practice of sodomy: "While admitting that certain knights allowed themselves to give unto *Oriental mores*, it is impossible to suppose that they [these mores] were publicly accepted in the Order, as if sanctioned by the Order's rules, and that the permission to indulge in a vice that is shameful and against nature could have been given to the knights" (En admettant que certains chevaliers se soient laissé gagner par *les mœurs orientales*, il est impossible de supposer qu'elles aient été acceptées publiquement dans l'Ordre, en quelque sorte à l'état de règlement, et que la permission de se livrer à un vice honteux et contre nature ait été donnée aux chevaliers) (Montagnac, p. 70, emphasis added). Cf. "Such things should not be written. Corruption enters into an Order by mutual and tacit complicity" (De telles choses ne s'écrivent pas. La corruption entre dans un ordre par connivence mutuelle et tacite) (Michelet, 3:108). In the tradition of Voltaire, H. C. Lea in his seminal *History of the Inquisition of the Middle Ages* (1888) and Charles-Victor Langlois in his *Saint Louis, Philippe le Bel, les derniers capétiens directs, 1226–1328* (1901) cleared the Templars of their crimes.

29. Éliphas Lévi (Alphonse Constant), *Dogme et rituel de la haute magie*, 2 vols. (Paris: G. Baillière, 1856).

30. "essayèrent des superstitions orientales, de la magie sarrasine" (Michelet, 3:107).

31. "Comme ils se passaient de femmes, ils se passaient aussi de prêtres," "Leur vrai dieu, ce semble, devint l'ordre même," "L'ordre . . . tomba dans une farouche religion de soi-même, dans un satanique égoïsme" (Michelet, 3:107).

32. "le dieu ou le génie androgyne et barbu"; Henri Martin, *Histoire de France*, 17 vols. (Paris: Furne, 1855–1860 [1838–1853]), vol. 4, p. 477. Martin wrote the first volume of the first edition of his *Histoire* with Paul Lacroix in 1833. He published the second edition alone from 1834 to 1836. He revised and expanded the *Histoire* in a third edition between 1838 and 1853. The *Histoire de France* won several prizes from the Académie des Inscriptions et Belles-Lettres, the Académie française, and the Institut de France.

33. "un amalgame d'idées et de rites empruntés moitié aux récentes hérésies musulmanes, moitié à de vieilles hérésies chrétiennes qui ne s'étaient jamais éteintes complètement en Orient, amalgame où dominait un sentiment de réaction contre la *spiritualité* chrétienne au profit d'une religion de la force et de la matière" (Martin, 4:478).

34. "Si les templiers renièrent le Christ, ce fut pour se venger de ce qu'il les avait reniés, de ce qu'il avait abandonné la cause de ses serviteurs" (Martin, 4:476). Martin's interpretation is directly opposed to Michelet's: where Michelet considers that the order was destroyed because of their great pride, considering themselves untouchable because "too rich and too powerful" (*Trop riche et trop puissant*) (Michelet, 3:108), Martin finds that they

were humbled and humiliated and hence sold their faith "to the victorious Saracens" (*aux Sarrasins victorieux*) (Martin, 4:476, n. 1).

35. "tolère chez ses adorateurs des voluptés immondes," "Les mœurs des templiers n'étaient pas moins exécrables que leur impiété: ils foulaient aux pieds la loi naturelle aussi bien que la loi divine; ils étaient initiés par une cérémonie immonde et dégoûtante; ils prétaient serment de s'abstenir du commerce des femmes, mais on leur permettait entre eux le péché contre nature" (Martin, 4:477, 473).

36. Edward W. Said, *Orientalism* (New York: Vintage Books, 2003 [1978]), p. 137. Even though the English could likewise claim its medieval link to the Near East—English French-speaking kings were also among the main participants in the Crusades (e.g., Richard Cœur de Lion)—it is possible that the contemporary British imperialism in India, compared to the French imperialism in North Africa, displaced this past for the English that often focused on antiquity. See John M. Ganim, *Medievalism and Orientalism* (New York: Palgrave Macmillan, 2005).

37. "l'Orient, soit comme image, soit comme pensée, est devenu pour les intelligences autant que pour les imaginations une sorte de préoccupation générale à laquelle l'auteur de ce livre a obéi peut-être à son insu," "cette autre mer de poésie"; Victor Hugo, "Préface," *Les Orientales* (Paris: Librairie de L. Hachette, 1858 [1829]), pp. 5, 6.

38. "jusqu'ici on a beaucoup trop vu l'époque moderne dans le siècle de Louis XIV et l'antiquité dans Rome et la Grèce: ne verrait-on pas de plus haut et de plus loin en étudiant l'ère moderne dans le moyen âge et l'antiquité dans l'Orient?" (Hugo, p. 6).

39. For Cauzons, Baphomet is in all likelihood a corruption of Mahomet, although "his legend was perhaps born of the unavoidable relationships between the Templars and Saracens" (sa légende était peut–être née des relations obligatoires entre Templiers et Sarrasins); Thomas de Cauzons, *La magie et la sorcellerie en France*, 3 vols. (Paris: Dorbon ainé, 1910–1912), vol. 2, pp. 246–47. For V.-É. Michelet, Baphomet is the origin of all orders of chivalry; Victor-Émile Michelet, *Le secret de la Chevalerie* (Paris: Ch. Bosse, 1928).

40. "mahométans cachés" (Plane, p. 41; Roy, p. 220).

41. "des rapports nombreux . . . existaient entre cette secte secrète [les Ismaëliens] et l'ordre des Templiers"; Gustave Brunet, "Tribunaux secrets," in *Le Moyen Âge et la Renaissance: Histoire et description des mœurs et usages, du commerce et l'industrie, des sciences, des arts, des littératures et des beaux-arts en Europe*, ed. Paul Lacroix and Ferdinand Seré (Paris, 1850), vol. 3, n.p. See a late example of association of Gnosticism and sodomy in Alexandre Lacassagne, "Pédérastie," in *Dictionnaire encyclopédique des sciences médicales*, ed. Amédée Dechambre (Paris: G. Masson, 1864–1889), 3e série, t. 22 (1886), pp. 239–59, at p. 243. See also Dr. Jaf (Dr. Fauconney), *Physiologie du vice, son histoire à travers les âges* (Paris: C. Offenstadt, 1903), p. 54: "It was heresies that brought with them this prodigious laxism in Christianity" (Ce furent les hérésies qui amenèrent avec elles ce prodigieux relâchement dans la chrétienté).

42. "Tous les croisés qui revenaient de Syrie ne parlaient que des trahisons des Templiers, de leurs liaisons avec les infidèles. Ils étaient notoirement en rapport avec les Assassins de Syrie; le peuple remarquait avec effroi l'analogie de leur costume avec celui des sectateurs

du Vieux de la Montagne. Ils avaient accueilli le Soudan dans leurs maisons, permis le culte mahométan" (Michelet, 3:109–10).

43. "le contact des voluptés syriennes corrompit leurs mœurs; l'orgueil, l'avidité, la turbulence, l'égoïsme, remplacèrent le dévouement austère des premiers temps" (Martin, 4:468).

44. "la vie molle et delicieuse"; Just-Jean-Etienne Roy, *Les Templiers. Histoire et procès* (Paris: Jean de Bonnot, 1995 [1848]), p. 217. See also "Soon pride and love of luxury, sensual pleasures, and good fare replaced the old austerity of this order" (Bientôt l'orgueil, le luxe, la volupté, et les plaisirs de la table firent place à l'ancienne austérité de cet ordre) (Plane, p. 27).

45. "un Ordre célèbre, mais dégénéré" (Nodier, p. 165). Sodomy was not the only "Oriental custom" (*habitude orientale*) that the Templars were said to adopt. The other was the custom of public baths; Pierre Dufour (Paul Lacroix), *Histoire de la prostitution, chez tous les peuples du monde depuis l'antiquité la plus reculée jusqu'à nos jours* [History of prostitution among all of the peoples of the world from the earliest antiquity to the current day], 6 vols. (Paris: Seré; Martinon, 1851–1853), vol. 6, p. 22.

46. "mille bruits infâmes circulaient sur leurs actes secrets, leurs initiations, leur vie intérieure, l'altération de leur foi, mélangée de superstition orientale et de magie sarrasine, sur les vices dégradants qu'ils avaient rapportés de l'Orient, sur les idoles qu'ils adoraient" (Larousse, "Templier," p. 1587).

47. "En ce qui concerne les mœurs, il est croyable que l'infamie qu'on leur reprochait n'était pas générale. Quant à l'accusation d'hérésie, ces chevaliers, hommes d'action avant tout, s'occupaient sans doute peu du dogme, et l'adoration d'une idole appelée Baphomet ne semble guère vraisemblable" (Larousse, "Templier," p. 1587).

48. David Arnold, ed., *Imperial Medicine and Indigenous Societies* (Manchester: Manchester University Press, 1988); Roy MacLeod and Milton Lewis, eds., *Disease, Medicine, and Empire: Perspectives on Western Medicine and the Experience of European Expansion* (New York: Routledge, 1988); Ann Laura Stoler, *Race and the Education of Desire: Foucault's History of Sexuality and the Colonial Order of Things* (Durham, N.C.: Duke University Press, 1995); Linda Bryder, "Sex, Race, and Colonialism: An Historiographical Review," *International History Review* 20:4 (1998): 806–22; Mary P. Sutphen and Bridie Andrews, eds., *Medicine and Colonial Identity* (New York: Routledge, 2003).

49. "On sait que, dès la plus haute antiquité, ces crimes contre nature existèrent en Orient chez toutes les nations idolâtres"; Larousse, *Grand dictionnaire universel*, "Pédérastie," vol. 12 (1874), pp. 491–92, at p. 491. See also Rudi Bleys, *The Geography of Perversion: Male-to-Male Sexual Behaviour Outside the West and the Ethnographic Imagination, 1750–1918* (London: Cassell, 1996), p. 163: "the claim . . . that a close affinity existed between sodomitical practices and Islam"; and Robert Aldrich, *Colonialism and Homosexuality* (New York: Routledge, 2003).

50. It is conceivable that the introduction in the nineteenth century of the element of race into the study of medieval religion can be dated to this moment as well; witness the current debate on race in contemporary medieval studies. Biological notions of race are a

phenomenon of modernity and emphasis on continuity between medieval and modern genealogies introduces a transhistorical notion of race: "Where we might expect racial designations, medieval texts tend to take religion as a primary marker of difference"; Sharon Kinoshita, "Deprovincializing the Middle Ages," in *The Worlding Project: Doing Cultural Studies in the Era of Globalization*, ed. Rob Wilson and Christopher Leigh Connery (Santa Cruz, Calif.: New Pacific Press, 2007), pp. 75–89, at p. 82. Difference is expressed in terms of environmental influence, and cultural and social components—customs, language, and law—to ethnic identities; see Robert Bartlett, "Medieval and Modern Concepts of Race and Ethnicity," *Journal of Medieval and Early Modern Studies* 31:1 (2001): 39–56. See also Robert Bartlett, *The Making of Europe: Conquest, Colonization, and Cultural Change, 950–1350* (Princeton, N.J.: Princeton University Press, 1993); and the entire special issue "Race and Ethnicity in the Middle Ages," ed. Thomas Hahn, *Journal of Medieval and Early Modern Studies* 31:1 (2001).

51. "Transmission des caractères de races, de nation"; Larousse, *Grand dictionnaire universel*, "Hérédité," vol. 9 (n.d.), pp. 217–18, at p. 217.

52. "On donne le nom de *race* à toute collection d'individus de même espèce qui présentent un ensemble de différences de même ordre que dans la variété, différences prononcées et qui une fois produites se reproduisent dans un certain nombre de circonstances qui ne sont pas complètement identiques"; Larousse, *Grand dictionnaire universel*, "Race," vol. 13 (1875), pp. 595–600, at p. 596. Larousse copied the definition of race from Émile Littré and Charles Robin, *Dictionnaire de médecine, de chirurgie, de pharmacie, de l'art vétérinaire et des sciences qui s'y rapportent*, 14th rev. ed. (Paris: J.-B. Baillière, 1878), "Race," p. 1326. On polygenism, see George W. Stocking, *Race, Culture, and Evolution: Essays in the History of Anthropology* (New York: Free Press, 1968).

53. "À toutes les époques, on trouve notre pays à la tête des grands mouvements qui ont entraîné l'humanité vers les chimères généreuses, qu'il s'agisse de délivrer le tombeau du Christ ou de prêcher à l'Europe la Déclaration des Droits de l'homme"; G. Beaumont, *Discours de distribution des prix au Collège de Saint-Dié, le 12 juilliet 1919* (Saint-Dié: Imprimerie G. Freisz, 1919), p. 7. I am grateful to Virginie Greene for this reference.

54. "Quelques écrivains ont soutenu qu'une race supérieure physiquement et moralement, lorsqu'elle s'unit à une race inférieure, s'abaisse sans l'élever, de sorte que chacune de ces alliances serait une perte pour la civilisation"; Théodule Ribot, *L'hérédité, étude psychologique sur ses phénomènes, ses lois, ses causes, ses conséquences* (Paris: Ladrange, 1873), p. 411. In a revised edition, Ribot structures the argument of cross-breeding differently; Théodule Ribot, *L'hérédité, étude psychologique*, 5th corrected and rev. ed. (Paris: Felix Alcan, 1894), esp. pp. 343–84.

55. "Je pense donc que le mot *dégénéré*, s'appliquant à un peuple, doit signifier et signifie que ce peuple n'a plus la valeur intrinsèque qu'autrefois il possédait, parce qu'il n'a plus dans ses veines le même sang, dont des alliages successifs ont graduellement modifié la valeur; autrement dit, qu'avec le même nom, il n'a pas conservé la même race que ses fondateurs; enfin, que l'homme de la décadence, celui qu'on appelle l'homme *dégénéré*, est un produit différent, au point de vue ethnique, du héros des grandes époques. . . . Il

mourra définitivement, et sa civilisation avec lui, le jour où l'élément ethnique primordial se trouvera tellement subdivisé et noyé dans des apports de races étrangères, que la virtualité de cet élément n'exercera plus désormais d'action suffisante"; Arthur de Gobineau, *Essai sur l'inégalité des races humaines*, 2 vols., 2nd ed. (Paris: Firmin-Didot, 1884 [1853–1855]), vol. 1, p. 24.

56. Jean-Alexandre-Napoléon Périer, *Essai sur les croisements ethniques* (Paris: Martinet, 1865); Paul Broca, *Recherches sur l'hybridité animale en général et sur l'hybridité humaine en particulier* (Paris: J. Claye, 1860).

57. Owen White, *Children of the French Empire: Miscegenation and Colonial Society in French West Africa* (Oxford: Oxford University Press, 1999), pp. 98–99. On miscegenation and polygenism, see also George W. Stocking, *Race, Culture, and Evolution.*

58. "Nous allons voir, en effet, en étudiant l'hérédité dans ce mode de reproduction, qu'un faible degré de métissage relève ordinairement la vigueur et presque toujours la fécondité des couples"; Yves Delage, *L'hérédité et les grands problèmes de la biologie générale*, 2nd ed. (Paris: Librairie C. Reinwald, 1903 [1895]), pp. 269–70.

59. "Les croisements extraordinaires . . . *ceux de peuples à peuples, de races,* de familles trop opposées sous le rapport des mœurs, des institutions, du génie individuel, des caractères physiques mêmes, sont dangereux, et ne peuvent amener cet heureux équilibre, cette pondération dans les facultés et dans les énergies humaines, qui constituent *la civilisation*"; Eugène Dally, *Des dangers attribués aux mariages entre consanguins.* Extrait de la *Gazette hebdomadaire de médecine et de chirurgie* 9 (1862): 1–45, at p. 9, emphasis added.

60. "En tout cas, c'est pure supposition d'admettre qu'un sang étranger puisse régénérer les familles en état de déchéance organique, et c'est dépasser les droits de la médecine scientifique, que d'affirmer en son nom que les croisements sont le remède suprême. À une race se substitue une autre race, à un type un type nouveau; mais le croisement, le mélange, sont en général illusoires. La race pure et le type individuel doivent subsister inaltérés, ou ils doivent disparaître" (Dally, p. 21).

61. "l'étude des croisements de races humaines, loin d'être favorable à cette théorie, montre que les métis sont d'autant moins féconds que les différences sont plus profondes entre leurs parents" (Dally, p. 25).

62. Ann Laura Stoler, *Carnal Knowledge and Imperial Power: Race and the Intimate in Colonial Rule* (Berkeley: University of California Press, 2002), p. 144. Elsa Dorlin has offered a poignant corrective on the imbrication of sexuality, colony, and gender in *La matrice de la race. Généalogie sexuelle et coloniale de la nation française* (Paris: Éditions la Découverte, 2006). Gender is neither a preexisting fixed notion nor a result of nineteenth-century transformations in family, nation, and colony, but ideas of sexual and racial difference mutually shape each other.

63. Paul Lacroix (see Chapter 5), a popularizer and prolific writer (*polygraphe*), is the only author to express the isolated opinion that the origin of sodomy and syphilis is linked: "We have said how abominable aberrations of the senses had been able to multiply the powers of the virus by tenfold, in exceptional cases, by moving it to the parts of the organism the least appropriate to receive it" (On a dit aussi comment d'abominables aberrations

des sens avaient pu, en cas exceptionnel, centupler les forces du virus, en le portant dans les parties de l'organisme les moins propres à le recevoir) (Lacroix, *Histoire*, 4:337).

64. "l'idéal de la croisade semblait réalisé dans l'ordre du Temple" (Michelet, 3:101).

65. "On vit . . . l'occident s'arracher de ses fondemens pour se précipiter sur l'orient"; Alexandre Brierre de Boismont, "De l'influence de la civilisation sur le développement de la folie," *Annales de l'hygiène publique et de médecine légale* 1:21 (1839): 242–97, at pp. 248–49.

66. "Des souverains qui veulent maintenir leurs sujets dans leurs habitudes et leurs croyances originelles doivent bien se garder de permettre à un grand nombre d'entre eux de séjourner longtemps en pays étranger. Ces déplacements sont toujours funestes aux vieilles habitudes"; J.-A. Dulaure, *Histoire physique, civile et morale de Paris*, 7th ed. (Paris: Dutertre, 1867 [1821–1825]), p. 314.

67. "Il était naturel que le relâchement s'introduisit parmi des moines guerriers, des cadets de la noblesse, qui couraient les aventures loin de la chrétienté, souvent loin des yeux de leurs chefs, entre les périls d'une guerre à mort et les tentations d'un climat brûlant, d'un pays d'esclaves, de la luxurieuse Syrie" (Michelet, 3:106).

68. "le vice contre nature, que le séjour des Français en Palestine avait acclimaté en France" (Lacroix, *Histoire*, 4:58). In his *Physiologie du vice*, Dr. Fauconney plagarizes entire sections from Lacroix; see Dr. Fauconney, p. 61.

69. "on les trouve dans le long séjour des Templiers en Orient, où le vice contre nature est presque endémique, et où la crainte de la lèpre, du mal des ardents et de diverses affections cutanées ou organiques, est toujours attachée au commerce des femmes" (Lacroix, *Histoire*, 4:64).

70. "L'Orient, à la même époque, était largement livré aux pratiques antinaturelles, et le livre de Moll donne quelques précieuses indications sur les mœurs des Turcs et des Persans au moyen âge"; Léon-Henri Thoinot, *Attentats aux mœurs et perversions du sens génital* [Offense against public decency and perversions of genital orientation] (Paris: Octave Doin, 1898), p. 291.

71. "En Orient, vers cette époque, la pédérastie florissait. À Constantinople, des maisons publiques existaient, où l'on pouvait se procurer des hommes aussi facilement que des femmes"; Dr. Riolan, *Pédérastie et homosexualité, Nouvelle collection exclusive d'hygiène et de médecine*, vol. 10 (Paris: P. Fort, 1909), p. 24.

72. "La condamnation des Templiers par Philippe-le-Bel et Boniface VIII ne fut qu'une terrible mesure de justice envers le vice contre nature que les chevaliers avaient rapporté de l'Asie Mineure"; Julien Chevalier, *Une maladie de la personnalité: L'inversion sexuelle* (Paris: G. Masson, 1893), p. 110.

73. "L'existence de la société elle-même pouvait paraître menacée, lorsque Philippe-le-Bel . . . se proposa d'arrêter les progrès de la sodomie. . . . Telle fut la principale cause du procès des Templiers" (Lacroix, *Histoire*, 4:59).

74. "Mais c'est surtout après le retour de Terre-Sainte des Croisés que le phénomène se répandit. Philippe le Bel s'éleva contre ces pratiques, et c'est sous son règne qu'eut lieu le fameux procès contre les templiers que l'on accusa, au milieu de bien d'autre griefs, de

pédérastie"; Émile Dusolier, *Psychologie des derniers Valois (François II, Charles IX, Henri III). Thèse présentée à la Faculté de médecine et de pharmacie de Lyon* (Lyon: A.-H. Storck, 1895), p. 63; see Dr. Fauconney, p. 202.

75. "Les croisades furent le plus beau moment de la chevalerie," "que ce prodigieux rassemblement d'hommes de tous les rangs et de tous les pays [a] réchauffé dans son sein le germe corrupteur de la luxure" (Dr. Fauconney, p. 62). Cf. Georges Cuvier, *Histoire des sciences naturelles* (Paris: Fortin, Masson et Cie, 1841), p. 398: "The Crusades procured a great many more and much more instructive ties with the Orient, and we will see that these enterprises, . . . considering only the goal in which they were attempted, contributed diversely to the progress of civilization and to the perfecting of the sciences" (Les croisades procurèrent des rapports bien plus nombreux et bien plus instructifs encore avec l'Orient, et nous verrons que ces entreprises, . . . à ne considérer que le but dans lequel elles furent tentées, contribuèrent diversement au progrès de la civilisation et au perfectionnement des sciences).

76. "depuis la plus haute antiquité," "l'épouvantable débauche romaine" (Lacroix, *Histoire*, 4:333, 334).

77. "s'était successivement combinée avec d'autres maladies, et surtout avec la lèpre, qui lui avait donné une physionomie toute nouvelle" (Lacroix, *Histoire*, 4:364).

78. "qu'on regardait alors comme un mal nouveau," "qu'une dernière transformation de la lèpre"; P.-L. Jacob, bibliophile (Paul Lacroix), *Recherches historiques sur les Maladies de Vénus, dans l'antiquité et le Moyen Âge* (Bruxelles: Brancart, 1883), at p. vii. This is an 1883 reprinting of one of the earliest summaries of the long debate on the origin of venereal disease.

79. "La débauche la plus audacieuse gangrena ces hordes indisciplinées . . . la plupart rapportant avec eux les fruits amers de la vie orientale: on peut assurer que la maladie vénérienne n'était alors qu'une des formes de la lèpre." This quote does not appear in the original edition of Lacroix's *Histoire de la prostitution*. It appears instead in a reprint of the original edition, P. L. Jacob, bibliophile, "Origines du Mal de Naples," in *Curiosités de l'histoire des croyances populaires au Moyen Âge* (Paris: Adolphe Delahays, 1859), pp. 239–323, at pp. 253–54.

80. "une erreur manifeste"; Larousse, *Grand dictionnaire universel*, "Syphilis," vol. 14 (1874), pp. 1337–41, at p. 1337. Modern scientists and historians of medicine reject this opinion of the link between leprosy and syphilis. There is even debate on "the role played by the Crusades in the dissemination of leprosy. It is certain, on the one hand, that the disease was present in the West before these military expeditions toward the Holy Land and, on the other hand, that it was between the eleventh and fourteenth centuries that endemic leprosy reached its peak" (le rôle joué par les Croisades dans la dissémination de la lèpre. Il est certain, d'une part, que la maladie était présente en Occident avant ces expéditions militaires vers la Terre Sainte et, d'autre part, que c'est entre le XIe et le XIVe siècles que l'endémie lépreuse atteignit son apogée); Mirko D. Grmek, *Les maladies à l'aube de la civilisation occidentale* (Paris: Payot, 1983), p. 254.

81. "Longtemps on a cru que la syphilis provenait de l'Amérique et qu'elle avait été

inconnue à l'antiquité"; Émile Littré and Charles Robin, *Dictionnaire de médecine, de chirurgie, de pharmacie, des sciences accessoires et de l'art vétérinaire de P.-H. Nysten*, 10th ed. (Paris: J.-B. Baillière, 1855), "Syphilis," pp. 1221–27, at p. 1221.

82. Joseph Rollet, "Syphilis (Synonymie; Historique)," in *Dictionnaire encyclopédique des sciences médicales*, ed. Dechambre, 3e série, t. 14 (1884), pp. 256–90, at p. 287. See also Larousse, "Syphilis," p. 1337.

83. "quand on considère la date de la découverte du nouveau monde et la date de l'explosion des accidents syphilitiques, il est impossible de ne pas reconnaître que . . . la communication par le coït n'aurait pas suffi pour propager cette maladie qui, peu de temps après sa première apparition, se montra d'une manière formidable en Italie, en France, en Espagne, en Allemagne, en Angleterre" (Littré and Robin, "Syphilis," p. 1222).

84. "L'origine américaine est une fable inventée dans un intérêt de honteuse politique, fable contredite par l'autorité des dates, et impuissante à expliquer l'explosion presque simultanée de l'épidemie dans dix pays à la fois"; Dr. Michel Villemont, *Histoire des maladies vénériennes* (Paris: Librairie des publications nouvelles, n.d. [1882]), p. 276.

85. "nos maîtres de la Grèce, de Rome et du moyen âge, n'ont pas décrit la syphilis" (Rollet, p. 271).

86. "des foyers anciens et en quelque sorte primitifs de la maladie" (Rollet, p. 256).

87. "Cette partie du monde ne l'a pas vue naître dans son sein" (Rollet, p. 256).

88. "Elle nous est arrivée alors par importation," "par contagion, et c'est ainsi qu'elle a gagné l'Europe" (Rollet, p. 256). Rollet concedes for some colonies that "it is difficult to decide if the disease is located there in the form of an old endemic, or if it was introduced by Europeans" (il est difficile de décider si la maladie s'y trouve à l'état d'endémie ancienne, ou si elle y a été introduite par les Européens) (Rollet, p. 258).

89. "C'est d'ailleurs une doctrine conforme à ce que nous savons aujourd'hui sur le développement des maladies contagieuses, dont on n'admet plus la spontanéité, et sur l'importation des affections exotiques et la facilité de leur dissémination hors de leur foyer originel" (Rollet, p. 283).

90. "la dissémination finale de la syphilis dans le monde entier," "la raison de son éclosion et de sa concentration sur des points limités du globe" (Rollet, p. 256). The origin of syphilis to this day remains unresolved; see Grmek, *Les maladies à l'aube*, p. 200. See also F. E. Rabello, "Les origines de la syphilis," *Nouvelle presse médicale* 2:20 (1973): 1376–80; Francisco Guerra, "The Dispute over Syphilis: Europe versus America," *Clio Medica* 13:1 (1978): 39–61; Francisco Guerra, "The Description of Syphilis in Avicenna," *27th International Congress of the History of Medicine, 1980* (Barcelona: Academia de Ciencies Mediques de Catalunya i Balears, 1981), pp. 731–33; Mirko D. Grmek, ed., *Histoire de la pensée médicale en Occident*, vol. 3, trans. Louise Lambrichs (Paris: Seuil, 1999), pp. 271–93.

91. "[Les maladies vénériennes] n'ont pas de nationalité et . . . leur origine est aussi ancienne que le monde, ou du moins que le débordement des mœurs"; Dr. Bouglé, *Les vices du peuple, suivis de l'histoire des maladies vénériennes* (Paris: Georges Carré, 1888), p. 138.

92. "Ce fut la Prostitution qui . . . servit d'auxiliaire énergique à ce fléau. . . . Voilà comment la lèpre vénérienne éclata en même temps, avec la même fureur, en France, en

Italie, en Espagne, en Allemagne et en Angleterre, au moment où Christophe Colomb était à peine de retour du premier voyage qu'il fit à l'île Espagnole" (Lacroix, *Histoire*, 4:364).

93. "la perversité, c'est-à-dire le vice proprement dit, où aucune impulsion n'impose au sujet des actes de débauche raffinée, car celui-là a conscience de son immoralité et peut ne pas s'en rendre coupable, s'il le veut bien" (Dr. Fauconney, pp. 6–7).

94. Albert Moll, *Die conträre Sexualempfindung* (Berlin: H. Kornfeld, 1891). It was translated into French as *Perversions de l'instinct génital* and published in 1893.

95. "Détraqués, dégénérés, invertis sont des malades"; Dr. Laupts, *Perversion et perversité sexuelles* (Paris: Georges Carré, 1896), p. 5.

96. See Laure Murat, *La loi du genre: Une histoire culturelle du "troisième sexe"* (Paris: Fayard, 2006).

97. "dégénérescences de l'espèce humaine"; Benedict-Augustin Morel, *Traité des dégénérescences physiques, intellectuelles et morales de l'espèce humaine* (Paris: J. B. Baillière, 1857).

98. "une déviation maladive du type normal de l'humanité," "une déviation maladive d'un type primitif"; Valentin Magnan and Paul-Maurice Legrain, *Les dégénérés, état mental et syndromes épisodiques* (Paris: Rueff, 1895), p. 74.

99. "il va de soi que les mêmes causes qui déterminent l'extinction des individus, doivent nécessairement déterminer l'extinction des groupes dont ces derniers font partie"; Dr. René Larger, *Théorie de la contre-évolution ou dégénérescence par l'hérédité pathologique* (Paris: Félix Alcan, 1917), p. 7. On the impact of degeneration in all areas of culture, thought, and art, see J. Edward Chamberlin and Sander L. Gilman, eds., *Degeneration: The Dark Side of Progress* (New York: Columbia University Press, 1985); Daniel Pick, *Faces of Degeneration: A European Disorder, c.1848–c.1918* (Cambridge: Cambridge University Press, 1989).

100. "les dégénérés ne constituent donc pas de races viables"; Dr. Fernand Debret, *La sélection naturelle dans l'espèce humaine (Contribution à l'étude de l'hérédité convergente)* (Paris: G. Steinheil, 1901), p. 17.

101. "L'histoire du vice contre nature dans ses diverses manifestations se perd dans la nuit des temps" (Dr. Fauconney, p. 128).

102. "Force est de reconnaître que la déviation a été de tous les temps, de tous les pays," "et l'on peut ajouter qu'elle a frappé toutes les races et s'est manifestée dans tous les pays" (Laupts, pp. 31, 37).

103. "à partir de la Grèce antique, les poètes se sont chargés de nous éclairer sur les hautes turpitudes de leurs concitoyens. C'est surtout à Rome qu'on était passé maître dans l'art du raffinement de la lubricité" (Dr. Fauconney, p. 128).

104. "la chaîne ininterrompue, reliant la Rome antique à la Renaissance, où toutes les traditions grecques et romaines refleurirent et s'épanouirent" (Thoinot, p. 291).

105. "C'est à l'établissement des races normandes dans les provinces gallo-franques que l'on doit en attribuer l'apparition en France; il fut ensuite ravivé par les croisades, et le séjour des Français en Palestine" (Chevalier, p. 109), "l'influence de l'an 1000, des invasions normandes, des Flagellants, des Croisades, de l'ordre des Templiers" (Lacassagne, p. 243).

106. "Each nation denied having engendered this illness, by attributing to it the name of the neighboring nation to which popular opinion ascribed the principle of illness. Thus,

the Italians, Germans, and British, who accused France of being the birthplace of *syphilis*, called it the *mal français* [the French disease]: *mal francese, frantzosen*, ou *frantzosichen poeken*, French pox; the French later saw fit to take revenge, calling it the *mal napolitan* [the Neapolitan disease]; the Flemish and Dutch, the Africans and the Moors, the Portuguese and the Navarrese cursed the *mal espagnol* or *castillan* [the Spanish or Castilian disease]; . . . the Orientals called it the *mal des chrétiens* [the disease of Christians]; Asians, the *mal des Portugais* [the disease of the Portuguese]; Persians the *mal des Turcs* [the disease of the Turks]; the Poles, the *mal des Allemands* [the disease of the Germans] and Muscovites [Russians], the *mal des Polonais* [the disease of the Poles]" (Chaque nation se défendait d'avoir engendré cette maladie, en lui attribuant le nom de la nation voisine à laquelle l'opinion populaire attribuait le principe du mal. Ainsi, les Italiens, les Allemands, et les Anglais, qui accusaient la France d'avoir été le berceau de la *grosse vérole*, l'appelaient *mal français*: *mal francese, frantzosen* ou *frantzosichen poeken, french pox*; les Français s'avisèrent plus tard de se revancher, en l'appelant *mal napolitain*; les Flamands et les Hollandais, les Africains et les Maures, les Portugais et les Navarrais maudissaient le *mal espagnol* ou *castillan*; . . . les Orientaux le nommaient *mal des chrétiens*; les Asiatiques, *mal des Portugais*; les Persans, *mal des Turcs*; les Polonais, *mal des Allemands,* et les Moscovites, *mal des Polonais*) (Lacroix, *Histoire*, 4:366–67). In that sense, John Boswell's controversial *Christianity, Social Tolerance, and Homosexuality: Gay People in Western Europe from the Beginning of the Christian Era to the Fourteenth Century* (Chicago: University of Chicago Press, 1980) can be read as an attempt to put a stop to the myth of foreignness of homosexuality by describing an indigenous "gay" culture.

107. For the same point, see Doctors Dusolier; Laupts; Thoinot; Fauconney; Riolan; and Dr. Alibert, *Pédérastie et aberrations sexuelles, Bibliothèque médicale populaire et d'hygiène*, vol. 11 (Paris: Librairie de la nouvelle France, n.d. [1908]). The Middle Ages also provided a preview of Renaissance debauchery, in the person of Gilles de Rais, "un monstrueux exemple de délire génésique" (Lacassagne, p. 243; see also Chevalier, pp. 111–15), followed by "the arrival of the Italians in France after Catherine de Medici. . . . Under Henri III, with the *mignons*, it was the *beau vice*, the *Italian vice*, the *offense of the back spine*" (l'arrivée des Italiens en France à la suite de Catherine de Medicis. . . . Sous Henri III, avec les *mignons,* ce fut le *beau vice*, le *vice italien,* le *délit de l'épine du dos*) (Lacassagne, p. 243).

108. Jacques Postel and Claude Quetel, eds., *Nouvelle histoire de la psychiatrie* (Paris: Dunod, 1994), pp. 133–51.

109. "Ainsi font les peuples. Leur origine est obscure; ils grandissent, donnent la mesure de leur force, puis, fatalement, en viennent à cette période où ils n'appartiennent plus qu'à l'histore" (Ribot, 5th rev. ed., p. 271).

110. "La Rome des Césars devint ainsi un des plus beaux foyers d'inversion sexuelle dont l'histoire nous ait conservé l'exemple" (Thoinot, p. 288).

111. "De cette revue, il ressort que les principales causes de la dépravation des mœurs dans l'antiquité païenne furent les opinions en philosophie et en morale et l'autorité absolue du maître sur l'esclave" (Chevalier, p. 106).

112. "phrénopathies, névropathies, singularités et aberrations tant intellectuelles que

morales, vices de conformation, stérilité, dégénérescence, mort prématurée, et enfin extinction de la race"; Dr. Paul Jacoby, *Études sur la sélection chez l'homme*, 2nd ed. (Paris: Felix Alcan, 1904 [1881]), p. 321.

113. "Rome, en sa décadence, avait tellement abusé des sens, qu'il fallait de longues années d'enseignement spirituel pour rétablir le culte de la pudeur"; Émile de la Bédollière, *Histoire des mœurs et de la vie privée des Français. Usages, coutumes, institutions, physionomie de chaque époque, depuis l'origine de la monarchie jusqu'à nos jours*, 3 vols. (Paris: V. Lecou, 1847–1849), vol. 2, p. 81.

114. "Tel avait été le sort des grands peuples civilisés de l'antiquité, tel sera celui des nations civilisées de l'Europe, et bien des symptômes indiquent déjà que nous sommes entrés dans cette période d'épuisement et de décadence" (Jacoby, p. 607).

115. "L'Europe n'est peut-être pas encore bien vieille, mais elle est tout au moins dans son âge mûr, et le jour viendra où comme tout ce qui subit l'évolution de la vie, il lui faudra *faire place aux jeunes*"; Alexandre Cullerre, *Les frontières de la folie* (Paris: J.-B. Baillière, 1888), pp. 330–31.

116. "La dégénérescence n'est, en somme, que la conséquence de notre évolution" (Dusolier, p. 46).

117. "Par le fait de la sélection et de la loi fatale de l'extinction des races privilégiées les peuples se civilisent d'abord, montent au faîte de la grandeur, puis déclinent rapidement et disparaissent épuisés, surmenés, anéantis, retombent dans la barbarie et sont remplacés par des peuples plus jeunes" (Jacoby, p. 607).

118. "les peuples privilégiés . . . le groupe des nations qui marchent à la tête de la civilisation . . . : *la vieille Europe*" (Cullerre, p. 330).

119. "Nos États modernes doivent-ils périr comme les grandes nations primitives?"; Charles Letourneau, "Hérédité," in *Dictionnaire encyclopédique des sciences médicales*, ed. Dechambre, 4e série, t. 13 (1888), pp. 588–605, at p. 604.

120. "La dégénérescence est donc un état *pathologique* et non un état *régressif, une anomalie réversive,* ainsi que la comprennent certains auteurs" (Magnan and Legrain, p. 74).

121. "un type morbide . . . ne possède plus en lui-même les moyens de se régénérer," "la perfectibilité est une qualité de tout être qui évolue normalement" (Magnan and Legrain, pp. 76, 75).

122. "médecine préventive et perfective"; Dr. Francis Devay, *Hygiène des familles ou perfectionnement physique et moral de l'homme considéré particulièrement dans ses rapports avec l'éducation et les besoins de la civilisation moderne*, 2 vols. (Lyon: Dorier, 1846), vol. 1, p. 152.

123. "leur degré de moralité" (Letourneau, p. 604).

124. "Le progrès incessant est la condition de la durée, et une nation peut vivre indéfiniment, à la seule condition d'en être digne" (Letourneau, p. 605).

125. "un nouveau développement d'attitudes plus puissantes, de goûts plus délicats et de penchants mieux réglés dans les actes qu'ils suscitent"; Littré and Robin, *Dictionnaire de médecine*, 14th rev. ed., "Hérédité," pp. 729–30, at p. 730.

126. "l'exaltation morbide de l'instinct sexuel qui est une des marques de notre époque," "la corruption des mœurs" (Chevalier, p. 125, 127).

127. "instabilité de nos institutions politiques" (Chevalier, p. 126).

128. "sans vouloir affirmer, comme je l'ai entendu faire souvent, que ce vice soit de plus en plus répandu" (Tardieu, p. 120).

129. "encore trop souvent dans la société moderne, il est certain qu'il était bien plus commun dans l'antiquité" (Larousse, "Pédérastie," p. 491).

130. "si, de tout temps, la perversion sexuelle a existé, elle a été plus fréquente autrefois, et qu'en résumé, la moralité publique est meilleure de nos jours"; Alexandre Lacassagne, "Préface," in Chevalier, *Une maladie de la personnalité*, pp. i-vi, at p. iii. See Chevalier, p. 134: "Nothing proves on the other hand that vice is today more widespread than before. Despite the freedom of mores in the great centers of population, our modern society . . . would need to change quite a bit to arrive at the degree of depravity in Greek and Roman societies" (Rien ne prouve d'autre part, que le vice soit de nos jours plus répandu qu'autrefois. Malgré la licence des mœurs dans les grands centres de population, notre société moderne . . . aurait beaucoup à faire pour en arriver au degré de dépravation des sociétés grecques et romaines).

131. "que de nos jours elle est relativement rare, et que jamais, peut-être, elle n'a été considérée avec plus de mépris" (Laupts, p. 31).

132. "une passion très-ordinaire," "ils ont soin de s'envelopper dans l'ombre et le mystère, ils rougissent même d'être soupçonnés" (Larousse, "Pédérastie," p. 491).

133. "les Européens . . . refusent souvent d'ajouter foi aux relations des voyageurs les plus véridiques. . . . On sait aussi combien ces passions épouvantables sont en faveur auprès des Arabes et des peuples leurs voisins. Kæmpfer dit qu'au Japon il existe, pour faciliter ces rapprochements monstrueux, des maisons publiques qui sont tolérées par le gouvernement; il en est à peu près de même en Chine" (Larousse, "Pédérastie," p. 491).

134. "Les croisades et l'émancipation des communes qui marquèrent l'apogée du gouvernement féodal, sont les deux événements qui ont le plus contribué à le détruire," "vertueuse indépendance, sa simplicité et ses mœurs"; Charles Nodier, "Introduction," in Jean-Baptiste de La Curne de Sainte-Palaye, *Mémoires sur l'ancienne chevalerie*, ed. Charles Nodier, 2 vols. (Paris: Girard, 1826 [1759–1781]), vol. 1, pp. i–xxv, at pp. xi–xii.

135. "en parcourant l'histoire, j'ai pu voir que chaque race royale, après avoir présenté de grands politiques, de sages administrateurs, de brillants capitaines, ou de profonds penseurs, finissait par des princes obscurs, sans vertus mais non pas sans vices, le plus souvent" (Dusolier, p. 41).

136. "Puis, avec le cours des siècles, les idées, les institutions, les habitudes se modifient, et les mœurs changent comme elles. Elles étaient toutes guerrières, toutes empreintes de l'âpreté, de la rudesse des camps. Elles s'en dépouillent peu-à-peu; elles deviennent plus douces, plus polies, et finissent par être molles, efféminées, licencieuses jusqu'au scandale, prodigues jusqu'à la ruine"; Louis-François Benoiston de Châteauneuf, *Mémoire sur la durée des familles nobles de France* (Extrait du compte-rendu des séances et travaux de l'Académie des sciences morales et politiques, 1845), *Annales d'hygiène publique et de médecine légale* 35 (janvier 1846): 27–58, at p. 38.

137. "La morale et la médecine . . . ont depuis long-temps placé le dépérissement des

hautes classes dans le dépérissement des mœurs" (Châteauneuf, p. 41). See also J. J. Virey, *De la femme sous ses rapports physiologique, moral et littéraire* (Paris: Crochard, 1823), p. 405: "Nonetheless, nobility chiefly vanished due to moral depravity, the inevitable result of a powerful fortune and of a talent for pleasures" (Toutefois, la noblesse s'est principalement évanouie par la dépravation morale, suite inévitable d'une puissante fortune et de la facilité des jouissances).

CHAPTER 4

1. "On divisa habilement le royaume de Satan. Contre sa fille, son épouse, la Sorcière, on arma son fils, le Médecin"; Jules Michelet, *La sorcière.*

2. See the objections to this thesis on the role of Gilles de Rais in Jacques Heers, *Gilles de Rais* (Paris: Perrin, 1994), p. 14: "Not a great war leader and less so a loyal partisan of Joan of Arc, nor a very rich lord" (Ni grand chef de guerre et pas davantage fidèle partisan de Jeanne d'Arc, ni très riche seigneur).

3. "Ces hommes de sang, qui, peu à peu, rentraient dans leurs manoirs après la guerre, la continuaient, et plus atroce encore, contre les pauvres gens sans défense"; Jules Michelet, *Histoire du Moyen Âge,* in *Œuvres complètes,* vols. 1–6 (Paris: Flammarion, 1893–1898 [1833–1844]), vol. 5, p. 192. See also Jules Michelet, *La sorcière* (Paris: GF Flammarion, 1966 [1862]), p. 70.

4. "degré de perversité auquel étaient parvenus quelques nobles, quelques grands seigneurs, qui s'abandonnaient à toutes les aberrations de la débauche"; Pierre Dufour (Paul Lacroix), *Histoire de la prostitution, chez tous les peuples du monde depuis l'antiquité la plus reculée jusqu'à nos jours* [History of prostitution among all of the peoples of the world from the earliest antiquity to the current day], 6 vols. (Paris: Seré; Martinon, 1851–1853), vol. 4, p. 324.

5. "ancien compagnon de Jeanne d'Arc," "en quels bas fonds de barbarie rampaient les hautes classes elles-mêmes"; Drs. Caufeynon and Jaf (Dr. Fauconney), *Les messes noires. Le culte de Satan-Dieu* (Paris: Pancier, 1905), p. 113.

6. "L'imagination la plus monstreusement dépravée n'avait sans doute jamais rêvé ce que révélèrent les débats de son procès"; Henri Martin, *Histoire de France depuis 1789 jusqu'à nos jours,* 17 vols. (Paris: Furne, 1855–1860 [1838–1853]), vol. 6, p. 397.

7. "la dépravation des mœurs la plus criminelle," "infâmes voluptés," "plaisirs abominables," "horribles débauches"; H. Audiffret, "Retz (Gilles de Laval, seigneur de)," in *Biographie universelle, ancienne et moderne,* vol. 37 (Paris: L. G. Michaud, 1824), pp. 398–401, at p. 400.

8. "crimes atroces," "licences effrénées," "sublimes horreurs," "les plus vils instincts," "le crime contre nature"; Armand Guéraud, *Notice sur Gilles de Rais* (Rennes: Imprimerie de A. Marteville et Oberthur, 1855), pp. 3–74; reprinted, with minor revisions, in A. Guéraud, "Rais (Gilles, baron de)," in *Biographie bretonne,* ed. P. Levot, vol. 2 (Vannes: Cauderan; Paris: Dumoulin, 1857), pp. 677–91, at pp. 678, 690, 686.

9. "goût du sang, de la souffrance et de la mort"; Eugène Bossard, *Gilles de Rais, Maréchal de France, dit Barbe-Bleue, 1404–1440, d'après des documents inédits* (Grenoble: Jérôme Millon, 1992 [1885]), p. 150.

10. "voluptés atroces et insensées," "le délire de la férocité"; Auguste Vallet de Viriville, *Histoire de Charles VII, roi de France, et de son époque, 1403–1461*, 3 vols. (Paris: Renouard, 1862–1865), vol. 2, pp. 413, 415.

11. "D'autres peut-être étaient aussi coupables," "ces *écorcheurs* des campagnes, ces terribles petits rois de châteaux" (Michelet, *Histoire du Moyen Âge*, 5:192, 195).

12. "Le cas de Gilles de Rais ne mérite l'étude que parce qu'il n'est point exceptionnel"; Charles Petit-Dutaillis, *Charles VII, Louis XI et les premières années de Charles VIII (1422–1492)*, in *Histoire de France depuis les origines jusqu'à la Révolution*, ed. Ernest Lavisse, vol. 4, part 2 (Paris: Hachette, 1902), p. 184.

13. "La barbarie reparaissait, moins ce qu'elle eut de bon, la simplicité et la foi. La féodalité revenait, mais non ses dévouements, ses fidélités, sa chevalerie. Ces revenants féodaux apparaissaient comme des damnés qui rapportaient de là-bas des crimes inconnus" (Michelet, *Histoire du Moyen Âge*, 5:193).

14. "La féodalité, qui avait été grande, n'avait plus ni force ni dignité, et les crimes se commettaient impunément au grand jour, ou se dissimulaient dans l'ombre des châteaux" (Guéraud, "Rais," pp. 690–91).

15. "cette fille du peuple" (Michelet, *Histoire du Moyen Âge*, 5:38). The English translation of this portion of Michelet's *Histoire de France* is available in Jules Michelet, *Joan of Arc*, trans. Albert Guérard (Ann Arbor: University of Michigan Press, 1967), except for this introductory paragraph.

16. Alphonse de Lamartine, "Joan of Arc," in *Memoirs of Celebrated Characters*, vol. 2 (New York: Harper, 1854), pp. 49–118, at pp. 49–50 (Jeanne d'Arc," in *Vies de quelques hommes illustres. Œuvres complètes de Lamartine publiées et inédites*, vol. 35 [Paris: L'auteur, 1863 (1855–1856)], pp. 121–95, at p. 121).

17. "excitée par l'amour de la patrie"; Alexandre Bertrand, "De l'état d'extase. Jeanne d'Arc," *Le Globe, Journal littéraire* (April 23, 1825): 489–91, at pp. 489.

18. "comme l'effet d'une exaltation de sentiments patriotiques et religieux"; Alexandre Brierre de Boismont, *Des hallucinations ou Histoire raisonnée des apparitions, des visions, des songes, de l'extase, du magnétisme et du somnambulisme* (Paris: Germer Baillière, 1845), p. 427.

19. "dégénérescences de l'espèce humaine," "déviations morbides du type normal de l'humanité"; Benedict-Augustin Morel, *Traité des maladies mentales* (Paris: V. Masson, 1860), p. ii. See also Benedict-Augustin Morel, *Traité des dégénérescences physiques, intellectuelles et morales de l'espèce humaine* (Paris: J.-B. Baillière, 1857).

20. "divers états morbides du système nerveux," "à l'origine des facultés mentales exceptionnelles," "Le *génie*, c'est-à-dire la plus haute expression, le *nec plus ultrà* [*sic*] de l'activité intellectuelle, qu'une *névrose*? Pourquoi non?"; Jacques-Joseph Moreau de Tours, *La psychologie morbide dans ses rapports avec la philosophie de l'histoire ou De l'influence des névropathies sur le dynamisme intellectuel* (Paris: Victor Masson, 1859), p. 464.

21. Louis-Francisque Lélut, *Du démon de Socrate: Spécimen d'une application de la science psychologique à celle de l'histoire* (Paris: Trinquart, 1836).

22. "il avait un génie ou un démon dont il entendait la voix, et qui le dirigeait dans ses actions"; Émile Littré, "Socrate et Pascal. Pathologie mentale," in *Médecine et médecins*, 2nd ed. (Paris: Didier, 1872 [1836–1848]), pp. 82–110, at p. 82; also p. 83.

23. "tous ceux qui ont de fausses perceptions, fussent-ils Socrate, Jeanne d'Arc, Pascal, etc., sont des visionnaires, des hallucinés, des aliénés, des fous"; Alexandre Brierre de Boismont, *De l'hallucination historique ou Étude médico-psychologique sur les voix et les révélations de Jeanne d'Arc* (Paris: Germer Baillière, 1861), p. 7.

24. "délire de l'intelligence"; François Leuret, *Fragmens psychologiques sur la folie* (Paris: Crochard, 1834).

25. "ces idées qui sortent à l'improviste de notre être et qui viennent assiéger notre esprit malgré nous-mêmes," "se complique d'un trouble de la raison" (Littré, "Socrate et Pascal," p. 104).

26. "le délire et le génie ont de communes racines," "surexcitation des centres nerveux" (Moreau de Tours, pp. 385, 466). Cesare Lombroso later posited that artistic genius is a form of hereditary madness; see Cesare Lombroso, *L'uomo di genio in rapporto alla psichiatria, alla storia ed all'estetica* [Man of genius] (Turin: Fratelli Bocca, 1888).

27. "les rangs élevés de la société, dans les familles qui comptent parmi leurs membres des hommes exceptionnels," "les désordres névropathiques de toute espèce, la folie, l'hystérie, etc." (Moreau de Tours, p. 307).

28. "Plus d'énergie dans les fonctions de cet organe, mais aussi plus de chances d'aberration et de déviation de ces mêmes fonctions" (Moreau de Tours, p. 386).

29. "les phénomènes de la psychologie morbide" (Moreau de Tours, p. 536).

30. "le génie, le talent hors ligne, sont des membres de la grande famille névropathique"; Dr. Paul Jacoby, *Études sur la sélection chez l'homme*, 2nd rev. ed. (Paris: Felix Alcan, 1904 [1881]), p. 450.

31. "lorsque cette maladie est le résultat de phénomènes pathologiques transmis par hérédité, elle constitue un véritable état dégénératif" (Morel, *Traité des maladies mentales*, p. iv).

32. "déséquilibrés de l'intelligence"; Valentin Magnan and Paul-Maurice Legrain, *Les dégénérés, état mental et syndromes épisodiques* (Paris: Rueff, 1895), pp. 106–7, 109.

33. "parmi les maladies intellectuelles nous comprenons la criminalité, qu'il est impossible de distinguer de l'aliénation: les criminels sont des dégénérés qui, à quelque âge qu'on les considère, ne diffèrent des autres ni par leurs caractères biologiques ni par leur caractères anatomiques"; Charles Féré, *La famille névropathique: Théorie tératologique de l'hérédité et de la prédisposition morbides et de la dégénérescence*, 2nd ed. (Paris: Félix Alcan, 1898 [1894]), p. 236. See also Charles Féré, *Dégénérescence et criminalité: Essai physiologique* (Paris: F. Alcan, 1888).

34. "Par le fait même de son état mental, le *dégénéré devient souvent un délinquant*" (Magnan and Legrain, p. 181).

35. Cesare Lombroso, *L'uomo delinquente, studiato in rapporto alla antropologia, alla medicina legale ed alle discipline carcerarie* (Milan: U. Hoepli, 1876).

36. "naissent non pas criminels, mais avec les aptitudes maximum à le devenir" (Magnan and Legrain, p. 191).

37. "altérations du sens moral" (Magnan and Legrain, p. 113).

38. Georges Bataille, *The Trial of Gilles de Rais*, trans. Richard Robinson (Los Angeles: Amok Books, 2004), p. 128 (*Le procès de Gilles de Rais* [Paris: Éditions Pauvert, 1959], p. 155). A note to the French and English editions specifies that the texts of the two trials are "based on the minutes and annotated by Georges Bataille." The Latin text of the eccesiastical trial was translated by Pierre Klossowski.

39. "prétendus devins et magiciens"; also "Cruelty is not sufficient to condemn men to such executions, also required is fanaticism composed of superstition and ignorance . . . ; and, in Brittany, the Marshal of Retz was put to death, accused of magic and of having slit the throats of children supposedly to make charms from their blood" (Ce n'est pas assez de la cruauté pour porter les hommes à de telles exécutions, il faut encore ce fanatisme composé de superstition et d'ignorance . . . ; et en Bretagne, on fit mourir le maréchal de Retz accusé de magie et d'avoir égorgé des enfants pour faire avec leur sang de prétendus enchantements); Voltaire, *Essai sur les mœurs et l'esprit des nations, et sur les principaux faits de l'histoire, depuis Charlemagne jusqu'à Louis XIII*, in *Œuvres*, ed. M. Beuchot (Paris: Werdet et Lequien, 1829 [1756–1763]), vol. 16, p. 410. The irregularity and unfairness of medieval trials of inquisition will periodically be invoked as reasons for Gilles de Rais's innocence; see Henry Charles Lea, "Gilles de Rais," in *History of the Inquisition of the Middle Ages*, vol. 3 (New York: Harper, 1888), pp. 468–88; Salomon Reinach, "Gilles de Rais," in *Cultes, mythes, et religions*, ed. Hervé Duchêne (Paris: Robert Laffont, 1996 [1904]), pp. 1026–49; Gabriel Monod, "Opinion de G. Monod," in Salomon Reinach, "Gilles de Rais," pp. 1045–46; Fernand Fleuret, *De Gilles de Rais à Guillaume Apollinaire* (Paris: Mercure de France, 1933); John de Saint-Jorre, *Fernand Fleuret et ses amis* (Coutances: Imprimerie P. Bellée, n.d. [1945?]), on the retrial staged by lawyer Maurice Garçon; Gilbert Prouteau, *Gilles de Rais ou La gueule du loup* (Monaco: Ed. du Rocher, 1992).

40. "Nous pourrions continuer les citations en puisant dans les ouvrages de Daru, Mellinet, Leboyer, Massé Isidore, Verger, Loudun, Chevas, etc.; mais tout cela ne nous apprendrait rien, si ce n'est que ces écrivains modernes se sont copiés sans critique les uns les autres" (Guéraud, *Notice*, p. 55).

41. "les nombreuses erreurs commises au sujet de Gilles de Rais," "complètement dénuée de fondement" (Guéraud, "Rais," p. 688).

42. See Annie Le Brun, *Les châteaux de la subversion* (Paris: J.-J. Pauvert, 1982). For a complete history of Gilles de Rais in the literary and historical imagination (but not medical literature), see Michel Meurger, *Gilles de Rais et la littérature* (Rennes: Terre de Brume, 2003).

43. "combien est fausse l'histoire telle que l'a faite la fantaisie des romanciers" (Bossard, p. 163).

44. "Romans! Dira-t-on, purs romans faits tout exprès pour effrayer l'imagination. –Eh bien! Il faut l'affirmer: il n'y a rien là qui ne soit vrai, appuyé sur les preuves les moins contestables. . . . L'homme s'est fait démon" (Bossard, pp. 168–69).

45. J.-K. Huysmans, *Là-Bas: A Journey into the Self*, trans. Brendan King (Cambs, UK: Dedalus, 2001), pp. 32, 33, translation modified and emphasis added (*Là-bas* [Paris: Gallimard/Folio, 1985 (1891)], pp. 44, 45).

46. "Ce livre est de l'Histoire découpée en roman"; Aimé Giron and Albert Tozza, *La bête de luxure: Gilles de Rais*, 4th ed. (Paris: Ambert, 1907), p. 8.

47. Émile Gabory, *Alias Bluebeard: The Life and Death of Gilles de Raiz*, trans. Alvah C. Bessie (New York: Brewer & Warren, 1930), p. 296 (*La vie et la mort de Gilles de Raiz, dit—à tort—Barbebleue* [Paris: Perrin, 1926], p. 230).

48. In 1824, "the number of them would seem to be incalculable" (Le nombre en paraîtra incalculable) (Audiffret, p. 400). In 1842, it is still "incalculable"; Charles Mourain de Sourdeval, "Les Sires de Retz et le château de Machecoul," *Mémoires de la Société archéologique de Touraine* 1 (1842): pp. 27–70, at p. 48. In 1885, Bossard states: "Several have brought up to seven and eight hundred the count of these victims: enormous numbers, which are perhaps exaggerated, and yet one would dare to say that it seems not unrealistic, if one listens to Gilles's avowals and those of his accomplices" (Plusieurs ont porté jusqu'à sept et huit cents le nombre de ces victimes: chiffre énorme, qui est peut-être exagéré, et cependant l'on oserait dire qu'il n'a rien d'invraisemblable, si l'on fait attention aux aveux de Gilles et de ses complices) (Bossard, p. 165). In 1959, Bataille declared the question "unsolvable" (p. 139 [p. 171]).

49. "Sans exemple comme sans imitation, ces crimes sont tels . . . que . . . Gilles de Rais tient une place à part, presque sans analogie, ni avec ceux qui l'ont précédé, ni avec ceux qui l'ont suivi" (Bossard, p. 132).

50. "horreurs impossibles à dire"; Prosper Mérimée, *Notes d'un voyage dans l'Ouest de la France* (Paris: Librairie de Fournier, 1836), p. 302.

51. "des choses impossibles à raconter"; Pître-Chevalier, *La Bretagne ancienne et moderne* (Paris: W. Coquebert, 1844), p. 480.

52. "défient toute expression"; Auguste Vallet de Viriville, "Rais ou Retz (Gilles de)," in *Nouvelle biographie générale*, ed. Dr. Hoefer, vol. 41 (Paris: Firmin Didot Frères, 1862), pp. 498–501, at p. 500.

53. "Je me dispenserai de répéter les détails atroces ou obscènes de ce procès"; Stendhal, *Mémoires d'un touriste*, 2 vols. (Paris: A. Dupont, 1838]), vol. 2, p. 63.

54. "Nous avons donc laissé dans l'ombre la plupart des actes épouvantables"; P. L. Jacob, bibliophile (Paul Lacroix), "Le Maréchal de Rais," in *Curiosités de l'histoire de France, procès célèbres* (Paris: Adolphe Delahays, 1858), pp. 1–119, at p. 4. The preface is dated to 1846.

55. "Pour la vérité complète de l'histoire, on doit avertir le lecteur que l'on passe forcément sous silence certains détails d'une immoralité telle, que des juges seuls pouvaient les entendre. On n'oserait les répéter, même en citant les graves documents du procès" (Bossard, p. 169).

56. "de ressembler à un roman à la fois atroce et froid" (Stendhal, p. 64).

57. "la vie de ce premier des Don Juans" (Stendhal, p. 62).

58. "Ce personnage, qui se rendit tristement célèbre par ses cruautés horribles, a inspiré

en partie à Perrault son conte de *Barbe-Bleue*. À l'article Barbe-Bleue, nous avons raconté la vie de ce féroce maniaque, resté en Bretagne comme un type légendaire"; Pierre Larousse, *Grand dictionnaire universel du XIXe siècle* (Paris: Administration du Grand dictionnaire universel, 1866–1877), "Rais (Gilles de), maréchal de France," vol. 13 (1875), p. 1087.

59. "ils ne se ressemblent que sur un point, la cruauté" (Guéraud, "Rais," p. 689).

60. See Meurger, pp. 92, 122–24. On Charles Perrault's Bluebeard, see Maria Tatar, *Secrets beyond the Door: The Story of Bluebeard and His Wives* (Princeton, N.J.: Princeton University Press, 2004).

61. "Dans le pays, en effet, dont il fut la terreur, toutes les autorités, l'histoire, la poésie, c'est-à-dire les traditions écrites—et surtout les traditions orales qui, seules, se doivent de faire définitivement foi dans ces matières—nous affirment, par un témoignage unanime, universel et constant, que le véritable Barbe-Bleue est le héros de ce livre, Gilles Rais, égorgeur de femmes et d'enfants, jugé pour ces crimes et brûlé à Nantes. Voilà ce qui a survécu même après l'apparition du conte du Perrault et ce que le conte de Perrault a laissé subsister intact. Voila ce qui remonte bien au-delà du XVIIe siècle" (Bossard, p. 302).

62. "l'étroite liaison, qui existe entre Barbe-Bleue et Gilles de Rais, vient du peuple et se trouve en lui," "mais le roman a envahi l'histoire" (Bossard, pp. 307, 317).

63. "la perversité humaine" (Guéraud, "Rais," p. 678), "la perversité bestiale" (Reinach, p. 1045).

64. "peu à peu l'homme étant parvenu à détruire en soi tout ce qu'il avait de l'homme" (Michelet, *Histoire du Moyen Âge*, 5:189).

65. "on ne peut guère prétendre l'adapter aux classifications des hommes, puisque peut-être il n'est pas un homme"; Robert Soueix, *Gilles de Raiz devant les médecins, essai médico-psychologique* (Bordeaux: Imprimerie de l'Académie et des Facultés, 1931), no. 83, p. 62.

66. "mal infini" (Michelet, *Histoire du Moyen Âge*, 5:190), "une lubricité sans nom" (Pître-Chevalier, p. 480), "dépravation infinie" (Viriville, "Rais," p. 499).

67. "Il chercha au-delà du réel, au-delà de la nature et du possible, un monde de voluptés atroces et insensées" (Viriville, *Charles VII*, p. 413).

68. "impudente mystification" (Reinach, p. 1047), "fausse érudition" (Fleuret, *Gilles de Rais*, p. 14).

69. Gabory, p. 183 [p. 147]. This portion of the 1926 book, on Gilles's criminal responsibility, was originally published in Émile Gabory, *La psychologie de Gilles de Rais, extrait de la Revue du Bas-Poitou* (Fontenay-le-Comte: H. Lussaud, 1925).

70. Ludovico Hernandez (Fernand Fleuret), *Le procès inquisitorial de Gilles de Rais, maréchal de France avec un essai de réhabilitation* (Paris: Bibliothèque des curieux, 1921); revised and reprinted in Fleuret, *Gilles de Rais*. For a short version of this argument, see Zrinka Stahuljak, "'Archéologie pathologique': Le Moyen Âge et l'histoire de la médecine," in *Fantasmagories du Moyen Âge, Senefiance* 56, ed. Élodie Burle-Errecade and Valérie Naudet (Aix-en-Provence: Publications de l'université de Provence, CUER MA, 2010), pp. 209–16.

71. "était . . . une fantaisie comme une autre, une fantaisie innocente," "lui a épargné les injures" (Fleuret, *Gilles de Rais*, p. 7).

72. "unanimité compétente désigne Gilles de Raiz comme le plus grand criminel de l'histoire" (Soueix, p. 10).

73. Michelet, *Histoire du Moyen Âge*, 5:188. This expression, though Michelet only used it once, had incredible longevity; see, for instance, Bossard, pp. 139, 313, and elsewhere; Gabory, p. 247 [p. 197].

74. "la bête de luxure" (Giron and Tozza); Bossard, pp. 150, 172; Michelet, *Histoire du Moyen Âge*, 5:189.

75. Richard von Krafft-Ebing, *Psychopathia sexualis: A Clinical-Forensic Study*, trans. F. J. Rebman from the twelfth and final edition, complete and unabridged (Burbank, Calif.: Bloat, 1999 [1886]), pp. 81–82.

76. Albert Moll, *Die conträre Sexualempfindung* (Berlin: H. Kornfeld, 1891) (*The Perversions of the Sex Instinct: A Study of Sexual Inversion*, trans. Maurice Popkin [Newark, N.J.: Julian Press, 1931]).

77. "je dois néanmoins confesser que je fus induit à me retirer en mes terres de Rays, par certaine furieuse passion et convoitise que je sentois envers votre propre dauphin, tellement que je faillis un jour l'occire comme j'ai depuis occis nombre de petits enfants par secrète tentation du diable"; Julien Chevalier, *Une maladie de la personnalité: L'inversion sexuelle* (Paris: G. Masson, 1893), p. 114.

78. "ce goût particulier des images licencieuses"; Léon-Henri Thoinot, *Attentats aux mœurs et perversions du sens génital* (Paris: Octave Doin, 1898), p. 131.

79. "perversion sanguinaire," "Tibère, Néron et d'autres Césars furent des sadiques avérés, comme ils étaient aussi des types achevés d'autre perversions génitales (inversion, etc.)" (Thoinot, p. 442).

80. "'Je ne sais . . . mais j'ai de moi-même et de ma propre tête, sans conseil d'autrui, pris ces imaginations d'agir ainsi seulement par plaisance et délectation de luxure; de fait j'y trouvais *incomparable jouissance*, sans doute par l'instigation du diable. Il y a huit ans que cette idée diabolique me vint. . . . Or, étant d'aventure en la librairie dudit château, je trouvai un livre latin de la vie et des mœurs des Césars de Rome par un savant historien qui a nom Suétonius; ledit livre était orné d'images fort bien peintes, auxquelles se voyaient les déportements de ces empereurs païens, et je lus en cette belle histoire comment Tibérius, Caracalla et autres Césars s'ébattaient avec les enfants et *prenaient plaisir singulier à les martyriser.* Sur quoi je voulus imiter lesdits Césars, et le même soir je commencai à ce faire en suivant les images de la leçon et du livre'" (Thoinot, p. 442).

81. Dr. Hayès, *La pédérastie, historique, conséquences funestes de ce vice honteux*, Bibliothèque d'hygiène des deux sexes, vol. 7 (Paris: Librairie des publications modernes, 1891), p. 78–80; Dr. Cabanès, *La flagellation dans l'histoire et la littérature* (Clermont: Imprimerie Daix Frères, 1899), p. 16. For a longer development, see Dr. Cabanès, "Le légendaire Barbe-Bleue," in *Légendes et curiosités de l'Histoire* (Paris: Albin Michel, 1922 [1912]), vol. 1, pp. 45–147, at p. 72.

82. "La lecture de la *Vie des douze empereurs romains*, par Suétone, avait excité ce puissant seigneur à imiter leurs monstrueuses débauches"; Dr. Jaf (Dr. Fauconney), *Physiologie du vice, son histoire à travers les âges* (Paris: C. Offenstadt, 1903), p. 65.

83. Dr. B.-A. Morel, "Folie héréditaire. Condamnation pour outrage public aux mœurs, d'un individu qui comptait des aliénés dans son ascendance paternelle et maternelle. Considérations préliminaires sur les actes immoraux et pervers, commis par des individus non aliénés" [Hereditary insanity. Condemnation for public indecency by an individual who included the mentally ill in his paternal and maternal lineage. Preliminary considerations on the immoral and perverted acts, committed by individuals not insane], *Gazette hebdomadaire de médecine et de chirurgie* 7:46 (15 novembre 1861): 734–37, and 8:48 (29 novembre 1861): 768–70, at p. 736.

84. "Ainçois, je désire et vouldroye par vous en savoir la pure verité pour les causes que je vous ay ja souvent dictes," "Vrayement il n'y avoit autre cause, fin ni intencion que ce que je vous ay dit" (Bataille, p. 193 [p. 239]).

85. "Le procès du maréchal de Rays n'a jamais été publié; il est raconté très-sommairement dans les recueils de causes célèbres; il est indiqué seulement dans les histoires de France: M. Michelet est le seul écrivain qui ait osé donner une idée des crimes qui firent condamner au bûcher un parent du duc de Bretagne, un maréchal de France" (Lacroix, *Curiosités*, p. 4).

86. "de représenter scrupuleusement la physionomie d'un procès" (Lacroix, *Curiosités*, p. 4).

87. "la vie privée des empereurs romains," "et c'était sans doute à cette source impure que Gilles de Rais avait emprunté ses monstrueux forfaits" (Lacroix, *Curiosités*, p. 74).

88. "'Je ne sais, mais j'ai de moi-même et de ma tête, sans conseil d'autrui, pris ces imaginations d'agir ainsi, seulement par plaisance et délectation de luxure; de fait, j'y trouvais incomparable jouissance, sans doute par l'instigation du diable. Il y a huit ans environ que cette idée diabolique me vint; ce fut l'année même où feu mon aïeul le sire de la Suze alla de vie à trépas. Je quittai lors le camp du roi de France, pour me rendre à Chantocé, afin de mettre la main sur les biens et héritages du défunt: or, étant d'aventure en la librairie dudit château, je trouvai un livre latin de la vie et mœurs de césars de Rome, par un savant historien qui a nom Suétonius; ledit livre était orné d'images fort bien peintes, auxquelles se voyaient les déportements de ces empereurs païens, et je lus en cette belle histoire comment Tibérius, Caracalla et autres césars s'ébattaient avec des enfants et prenaient singulier plaisir à les martyriser. Sur quoi, je voulus imiter lesdits césars, et le même soir, je commençai à ce faire, en suivant les images de la leçon et du livre'" (Lacroix, *Curiosités*, pp. 93–94).

89. "They say that he read Suetonius: this is perhaps just a tradition; but it is certain that he had a *Valerius-Maximus*, a book *On Properties of Things* [*De proprietatibus rerum/Le livre des propriétés des choses* by Bartholomeus Anglicus], the *Metamorphoses* by Ovid, and the *City of God* by Saint Augustine, in Latin and in French. These are the only books in his library that are designated in the trial minutes found in Orleans" (On dit qu'il lisait Suétone: ce n'est peut-être qu'une tradition; mais il est certain qu'il avait un *Valère-Maxime*, un livre *des Propriétés*, les *Métamorphoses d'Ovide*, et la *Cité de Dieu* de saint Augustin, en latin et en français. Ce sont les seuls livres de sa bibliothèque qui soient désignés dans les *minutes* trouvées à Orléans) (Bossard, p. 17; also p. 169).

90. "jamais un manuscrit de Suétone n'a été illustré de miniatures indécentes"

(Reinach, p. 1048). See also Fleuret, *Gilles de Rais*, p. 22, n. 1: "Or at least there aren't any except in the *Curiosités de l'Histoire de France* of the bibliophile Jacob, where the forger confuses Caracalla with Caligula. . . . It is easy to see that Paul Lacroix was thinking of d'Hancarville's engravings, a famous eighteenth-century hoax, where, moreover, no scenes of cruelty are depicted" (Ou du moins il n'en est question que dans les *Curiosités de l'Histoire de France* du bibliophile Jacob, où le faussaire confond Caracalla et Caligula. . . . Il est aisé de se rendre compte que Paul Lacroix songeait aux camées de d'Hancarville, fameuse supercherie du XVIIIe siècle, où ne figure d'ailleurs aucune scène de cruauté). Fleuret is referring here to Pierre-François Hugues, a French pseudo-aristocrat who renamed himself Baron d'Hancarville and published *Monumens de la vie privée des douze Césars, d'après une suite de pierres gravées sous leur règne* (Rome, 1780). Lacroix may also have been inspired by Michelet: "Neither the Nerons of the Empire, nor the tyrants of Lombardy, would have had anything to offer in comparison" (Ni les Nérons de l'Empire, ni les tyrans de Lombardie, n'auraient eu rien à mettre en comparaison) (Michelet, *Histoire du Moyen Âge*, 5:188). Also, the analogy with Caesars was made in 1776 by Baculard d'Arnaud (Meurger, p. 54).

91. Huysmans, p. 58 [p. 74]. See also J.-K. Huysmans, *La magie en Poitou. Gilles de Rais* (Ligugé, 1899), p. 6; Bossard, p. 15.

92. "Les historiens ont parlé de sa grande intelligence, de son érudition, de son raffinement, de son sens esthétique infaillible" (Soueix, p. 34).

93. Dr. M. Hamblin-Smith, "Introduction," in A. L. Vincent and Clare Binns, *Gilles de Rais: The Original Bluebeard* (London: A. M. Philpot, 1926), pp. 7–35, at pp. 8, 23.

94. Lacroix also compared Gilles de Rais to Sade: "There is but one man in the world who equaled, if not surpassed, the Marshal of Rays in the domain of moral depravity, it is the Marquis de Sade!" (Il n'y a qu'un homme au monde qui ait égalé, sinon surpassé le maréchal de Rays en fait de dépravation morale, c'est le marquis de Sade!) (Lacroix, *Curiosités,* p. 4).

95. "les crimes inexplicables" (Guéraud, "Rais," p. 685), "le plus grand criminel de l'histoire" (Soueix, p. 10).

96. "Gilles de Rays est un assassin de volupté, un 'sadique' sanguinaire" (Dr. Cabanès, "Le légendaire Barbe-Bleue," p. 102).

97. "la vie sédentaire, la nourriture animale, les boissons fortes, les épices [qui] ont un contre-coup sur la vie sexuelle" (Dr. Cabanès, "Le légendaire Barbe-Bleue," p. 105), "échauffé sans cesse par les mets succulents, les liqueurs alcooliques, les veilles et les émotions" (Guéraud, "Rais," p. 684).

98. "L'effet n'attend qu'une provocation, une impression venue du dehors, pour se produire. Pour Moreau de Tours, l'organisme, *le terrain physique et moral,* est tout, quant aux causes qui déterminent l'éclosion du vice, elles ne sont qu'accessoires. . . . En un mot, l'hérédité joue un grand rôle dans la question, puisque les causes objectives, les causes provocatrices déterminantes, auxquelles il faut reconnaître un certain rôle, n'agissent pas indistinctement sur tous les individus; il leur faut un terrain spécial, un terrain préparé" (Dr. Hayès, pp. 109–10).

99. "la fatale et pernicieuse lecture," "C'est le mécanisme bien simple d'une

obsession s'abattant sur un terrain propice mal défendu par une volonté débile ou pervertie, s'enracinant peu à peu"; Frédéric-Henri Bernelle, *La psychose de Gilles de Rais, Sire de Laval, Maréchal de France, 1404–1440. Thèse pour le doctorat en médecine*, no. 218 (Paris, 1910), p. 70.

100. "Les pernicieuses lectures dont il a été question, ont réveillé la bête en rut qui sommeillait, ont entraîné la volonté chancelante sur la pente où elle s'est laissée glisser. N'est-ce pas un choc le plus souvent insignifiant qui déclanche la névrose latente?" (Dr. Cabanès, "Le légendaire Barbe-Bleue," p. 105).

101. Bernelle, p. 80; Dr. Ch. Coubard, "Gilles de Rais devant la médecine et devant la mystique," *Aesculape, Revue mensuelle illustrée des lettres et des arts dans leurs rapports avec les sciences et la médecine* 1 (janvier 1931): 9–16, 40–47, 148–52, at p. 12; Soueix, p. 60; Mila Simeonova Strezova, *Gilles de Rais. Étude médico-légale et psychiatrique. Thèse présentée pour le doctorat en médecine* (Faculté de médecine de Strasbourg, 1934), p. 29.

102. "Sa vive intelligence, ses autres qualités éliminent d'emblée la vaste sous-classe des débiles, des pauvres d'esprit, en même temps que ses tendances sentimentales vers la puissance, la richesse, l'Au-delà, le mystérieux, nous ramènent vers le déséquilibre, le dégéneré supérieur nettement défini par Magnan" (Bernelle, pp. 79–80); also Soueix, p. 39. Only a few authors in the twentieth century will challenge this idea, notably Gabory: "Nor did he display unusual intelligence" but rather "boundless credulity" (p. 154 [p. 125]) and "simplicity" (p. 163 [p. 132]). Following Gabory, Bataille insisted on Gilles's lack of intelligence, his naïveté and credulity, in short, his mediocrity that *did not* set him apart from his contemporaries. Gabory said as much of Gilles: "He appears to us to have been distinctly of his period" (p. 153 [p. 124]).

103. "Or, Gilles, par sa grande intelligence, par son sens esthétique, par sa culture et son érudition, s'écarte d'autant des dégénérés moyens et inférieurs qu'il s'en rapproche par la nature de ses meurtres" (Soueix, p. 61). See also: "Let us still note however that the impulses brought on by an inferior psyche, based on sensations and various aberrations, bring him closer to average degenerates and his tendency toward necrophilia could bring him down even further" (Notons toutefois que les impulsions relevant du psychisme inférieur, basées sur des sensations, aberrations diverses, le rapprochent des dégénerés moyens et que ses tendances à la nécrophilie pourraient le faire descendre plus bas encore) (Bernelle, p. 80); and "The truly monstrous impulsions that are mentioned in the trial: sadism, necrophilia, vampirism, are ordinarily the acts of degenerates of the most inferior category, meaning idiots, morons, innocents. . . . And yet . . . Giles de Rais, with his brilliant intellectual qualities and his refined temperament of an artist had likely nothing in common with such people" (Les impulsions vraiment monstrueuses dont il est fait mention au procès: sadisme, nécrophilie, vampirisme, sont d'ordinaire le fait des dégénérés de la plus basse catégorie, c'est-à-dire des idiots, des crétins, des innocents. . . . Or, . . . Gilles de Rais, avec ses brillantes qualités intellectuelles et son tempérament affiné d'artiste n'avait sans doute rien de ces gens-là) (Coubard, p. 13).

104. "Ces malades sont fous, mais ne paraissent pas fous parce qu'ils s'expriment avec lucidité. Ils sont fous dans leurs actes plutôt que dans leurs paroles. . . . Ils sont lucides

jusque dans leurs conceptions délirantes. Leur folie est lucide"; Ulysse Trélat, *La folie lucide étudiée et considérée au point de vue de la famille et de la société* (Paris: A. Delahaye, 1861), pp. xii–xiii. See Bossard for an early use of this medical term, p. 272; also Bernelle, p. 82; Soueix, p. 63; Dr. Cabanès, "Le légendaire Barbe-Bleue," p. 106.

105. For instance, Ambroise Tardieu, *Étude médico-légale sur la folie* (Paris: J.-B. Baillière, 1872); Dr. Richard von Krafft-Ebing, *La responsabilité criminelle et la capacité civile dans les états de trouble intellectuel*, trans. Dr. Chatelain (Paris: G. Masson, 1875); Paul Dubuisson and A. Vigouroux, *Responsabilité pénale et folie, étude médico-légale* (Paris: Librairie Felix Alcan, 1911). See Alexandre Cullerre, *Les frontières de la folie* (Paris: J.-B. Baillière, 1888), p. 307: "To admit a certain legal responsibility in the matter of mental alienation is to commit a scientific heresy and desire to relinquish to the arbitrary and random the life and the honor of individuals and families" (Admettre une certaine responsabilité légale en matière d'aliénation mentale, c'est commettre une hérésie scientifique et vouloir livrer à l'arbitraire et au hasard la vie et l'honneur des individus et des familles).

106. "Comment la *pédérastie*, . . . avec toutes ses horreurs dégoûtantes et ignominieuses, est-elle possible dans une civilisation avancée comme la nôtre? La débauche effrénée, la sensualité blasée… ; mais il est difficile, dans bien des cas, de ne pas admettre une véritable aliénation des facultés morales"; Larousse, *Grand dictionnaire universel*, "Pédérastie," vol. 12 (1874), pp. 491–92, at p. 492.

107. "une renaissance du mysticisme"; Dr. Jules Regnault, *La sorcellerie (ses rapports avec les sciences biologiques)* (Bordeaux: Imprimerie du Midi, P. Cassignol, 1896), p. 5.

108. Qtd. in Nancy Erber, "Queer Follies: Effeminacy and Aestheticism in Fin-De-Siècle France, the Case of Baron d'Adelsward Fersen and Count de Warren," in *Disorder in the Court: Trials and Sexual Conflict at the Turn of the Century*, ed. George Robb and Nancy Erber (New York: New York University Press, 1999), pp. 186–208, at p. 188.

109. "où Roland Brévannes veut montrer l'office infernal en sa période de suprématie, au Moyen Âge . . . et son dernier avatar contemporain, dans les salons d'Axel Wartz" (Meurger, p. 185).

110. "Le déséquilibré mental s'accompagne fréquemment d'une tendance toute particulière à l'imitation," "La littérature et l'histoire y participent pour une large part, car elles fournissent des leçons fameuses que les imaginations déréglées savent merveilleusement exploiter. . . . En nous reportant du XVe au XXe, et des terres de Machecoul aux boudoirs parisiens, nous retrouvons cette même influence de la littérature et de l'histoire"; Dr. Gabriel Dromard, *Les fausses 'messes noires.' Causeries psychologiques et médico-légales sur quelques scandales modernes* (Paris: Éditions de la Revue libre, n.d. [1904]), pp. 75, 76, 78. Jean de Cherveix, writing at the time of the affair Fersen, also links Gilles de Rais's readings to murders in *Deux cents viols par un Maréchal de France* (Paris: Offenstadt, 1903).

111. "C'est le triste processus psychique que notre vie moderne calque et reproduit sans mesure des jeunes tarés, futurs escarpés, s'hypnotisant sur une scène, violemment coloriée, de meurtre, à la devanture d'un kiosque à journaux et recevant du fait divers bêtement illustré le choc pernicieux qui tôt ou tard leur mettra—irrésistiblement—l'eustache au poing" (Bernelle, p. 70).

112. "Aux quatorzième, quinzième et seizième siècles, époque d'explosion de mysticisme religieux et de folie génésique, on voit aussi apparaître les passions contre nature"; Alexandre Lacassagne, "Pédérastie," in *Dictionnaire encyclopédique des sciences médicales*, ed. Amédée Dechambre (Paris: G. Masson, 1864–1889), 3e série, t. 22 (1886), pp. 239–59, at p. 243. See also Chevalier, p. 124: "What were therefore . . . during the Middle Ages, the causes of this widespread priapism, which is its sallient trait? We must . . . incriminate above all religious asceticism, monastic life, mysticism that we never encounter throughout history without eroticism" (Quelles furent donc . . . pendant le Moyen-Âge, les causes de ce priapisme universel, qui en est le trait saillant? Il faut . . . incriminer surtout l'ascétisme religieux, la vie monastique, le mysticisme qu'on ne rencontre jamais dans l'histoire sans érotisme); also Dr. Laupts, *Perversion et perversité sexuelles* (Paris: Georges Carré, 1896), p. 37.

113. "De même que l'émotion religieuse, l'émotion esthétique n'est pas si loin qu'on pourrait le croire de l'émotion charnelle. Disons mieux: cette émotion esthétique n'est vraisemblablement qu'un dérivé de l'instinct sexuel, au point de vue général de l'évolution" (Dromard, p. 36).

114. "une recherche prolongée de mysticisme, d'ascétisme," "la cause de toutes ces folies génésiques" (Laupts, p. 36).

115. Étienne Esquirol, *Mental Maladies: A Treatise on Insanity*, trans. E. K. Hunt (New York: Hafner, 1965 [facsimile of 1845 English ed.]), p. 109 (*Des maladies mentales, considérées sous les rapports médical, hygiénique et médico-légal* [Paris: J.-B. Baillière, 1838 (1817)], p. 198). See also Leuret, p. 333.

116. "La démonomanie, la sorcellerie, la croyance aux incubes et aux succubes, aux maléfices et vilainies [*sic*] du diable remplissent cette époque, s'accompagnant d'un priapisme universel. La loi qui prétend que l'ascétisme et le mysticisme religieux ne vont pas sans toutes espèces de désordres sexuels trouva une confirmation éclatante" (Chevalier, p. 111).

117. "Au quinzième siècle la folie porta surtout l'empreinte des idées superstitieuses, des doctrines théologiques alors en renom"; Louis-Florentin Calmeil, *De la folie considérée sous le point de vue pathologique, philosophique, historique et judiciaire: Depuis la renaissance des sciences en Europe jusqu'au dix-neuvième siècle*, 2 vols. (Paris: J.-B. Baillière, 1845), vol. 1, p. 124.

118. Émile Littré, "Des tables parlantes et des esprits frappeurs," in *Médecine et médecins* (1856), pp. 41–81, at p. 72.

119. Boismont, *Des hallucinations*, p. 427; Calmeil, *De la folie*.

120. "Satan ne serait qu'une maladie!" (Michelet, *La sorcière*, p. 222).

121. See for instance G. Prouteau; Matei Cazacu, *Gilles de Rais* (Paris: Tallandier, 2005); Élisabeth Roudinesco, *La part obscure de nous-mêmes: Une histoire des pervers* (Paris: Albin Michel, 2007).

122. "The Tragedy of Gilles de Rais," in Bataille, *The Trial of Gilles de Rais*, pp. 13–66 [pp. 11–78]. For the relationship and indebtedness of Bataille's theory to the Middle Ages, see Bruce Holsinger, *The Premodern Condition: Medievalism and the Making of Theory* (Chicago: University of Chicago Press, 2005).

123. "Le caractère de la noble fille dont nous allons étudier les hallucinations, a été

très bien tracé par les éminents historiens de notre époque" (Boismont, *De l'hallucination historique*, p. 10).

124. "l'absurde, l'irrationnel, le contraire de ce que tout le monde peut constater dans le domaine de l'observation médicale et pathologique"; Dr. Louis Boucher, *Jeanne d'Arc au point de vue médical* (n.p.: Union typographique, n.d.), p. 12.

125. "la santé de tel personnage avait exercé une action sérieuse sur les idées, les doctrines d'une époque"; Alfred Maury, "De l'hallucination envisagée au point de vue philosophique et historique, ou Examen critique de l'opinion émise par M. Brierre de Boismont, touchant les caractères auxquels on doit reconnaître l'hallucination chez certains personnages célèbres de l'histoire" [On hallucination considered from a philosophical and historical standpoint, or Critical examination of the opinion stated by M. Brierre de Boismont, concerning the characters which must be considered hallucinations in certain celebrities throughout history], *Annales médico-psychologiques. Journal de l'anatomie, de la physiologie et de la pathologie du système nerveux* 5 (mai 1845): 317–38, at p. 318.

126. "le rôle souvent prépondérant que l'hallucination a joué dans les événements politiques et religieux," "il n'a échappé à aucun observateur . . . que les hallucinations que nous observons chez nos malades ont absolument le même caractère, se présentent absolument de la même façon que celles que l'on a constatées chez les personnages historiques. Dès lors on se trouve en présence de ce dilemme: ou tous ces personnages étaient fous, fous au même titre que ceux que nous renfermons dans nos asiles d'aliénés, ou au contraire ils n'étaient pas fous, et dans ce cas il faut admettre qu'il existe deux espèces d'hallucinations, l'une compatible avec la raison la plus parfaite, l'autre caractérisant l'aliénation mentale"; J. Christian, "Hallucination," in *Dictionnaire encyclopédique des sciences médicales*, ed. Dechambre, 4e série, t. 12 [1886]), pp. 77–121, at p. 103.

127. "De tous les personnages qui peuvent faire comprendre le phénomène si controversé et pourtant si réel de l'hallucination physiologique, il n'en est pas de comparable à l'illustre Jeanne d'Arc" (Boismont, *De l'hallucination historique*, p. 3).

128. "hallucinations spontanées, accidentelles, sans aliénation mentale et indépendamment de toute autre affection capable de déterminer du délire," "à la suite d'une méditation profonde, d'une concentration de la pensée, d'une tension de l'esprit sur une idée exclusive"; Claude-François Michéa, *Des hallucinations. Thèse à la Faculté de médecine de Paris*, no. 159 (Paris: Righoux, 1837), pp. 10–11.

129. "L'hallucination peut se manifester avec l'intégrité parfaite de la raison. . . . L'hallucination est un symptôme pathognomonique, mais non point *constant et nécessaire* de l'aliénation mentale"; Claude-François Michéa, "Des hallucinations, de leurs causes, et des maladies qu'elles caractérisent," *Mémoires de l'Académie royale de médecine*, vol. 12 (Paris: J.-B. Baillière, 1846), pp. 241–71, at p. 267.

130. "on peut, moyennant un grand caractère, de l'héroïsme et du génie, accomplir de grandes choses tout en cédant à l'élan, à l'inspiration d'un véritable délire" (Calmeil, p. 128).

131. "Le succès de ses armes ne prouvera jamais qu'elle fit preuve d'une bonne logique en ajoutant foi à ses propres visions. Un halluciné peut à la rigueur avoir les plus grandes vues dans l'esprit; mais ce qui fait qu'on est obligé de le considérer comme malade, c'est

qu'il aperçoit ce qui n'existe pas et qu'il croit fermement que ses propres idées lui sont soufflées dans les oreilles par des êtres qu'il estime autres que lui. Tel est le cas pathologique de Jeanne d'Arc" (Calmeil, p. 132).

132. "il arriva qu'en prenant des erreurs de l'imagination et du jugement pour des faveurs célestes, on sauva un royaume" (Calmeil, p. 131).

133. "ce caractère vague, indécis, nuancé entre le catholique et le philosophique qui a si bien chassé de son œuvre l'unité de doctrine et la puissance de déduction qui en eussent fait le mérite" (Maury, "De l'hallucination," p. 323), "les rationalistes du temps" (Boismont, *De l'hallucination historique*, 4).

134. "les hallucinations peuvent atteindre un grand nombre de personnes, sans que celles-ci puissent être soupçonnées de folie," "hommes célèbres" (Boismont, *Des hallucinations*, p. 419).

135. "celles de son siècle," "dans le corps social, et non dans les individus," "folies générales," "les croyances du temps," "des aliénations propres à chaque personne" (Boismont, *Des hallucinations*, pp. 422, 426, 427).

136. "croyances erronées," "les hallucinations de la raison" (Boismont, *Des hallucinations*, p. 444).

137. "les hallucinations compatibles avec la raison, et celles qu'on observe chez les aliénés ne diffèrent pas dans leur essence; mais de ce qu'elles ont les mêmes manifestations extérieures, est-on en droit de conclure qu'elles forment une même espèce pathologique?" (Boismont, *De l'hallucination historique*, p. 14).

138. "Si l'erreur générale explique l'hallucination de ces grands hommes, elle ne la justifie pas et ne la rend pas pour cela raisonnable" (Maury, "De l'hallucination," p. 332).

139. "Les hallucinations que nous avons constatées chez de grands esprits ont pu se concilier avec une raison saine et forte sur d'autres points, avec des actes parfaitement raisonnables toutes les fois que les hallucinations n'en étaient point le principe: c'est ce que la médecine psychologique a constaté et constate sans cesse" (Maury, "De l'hallucination," p. 332).

140. "Rien de plus affligeant, en effet, que d'attribuer à la folie des doctrines sublimes, des actions héroïques, des entreprises glorieuses"; Alexandre Brierre de Boismont, "Réponse," *Annales médico-psychologiques. Journal de l'anatomie, de la physiologie et de la pathologie du système nerveux* 5 (mai 1845): 339–341, at p. 340.

141. "Oui, Jeanne a accompli de grandes choses, et elle a fait preuve, dans leur accomplissement, de génie, d'une volonté intrépide; c'est qu'elle avait la foi de ses visions: c'est qu'elle se croyait l'instrument de Dieu et que sa croyance reposait sur l'inébranlable fondement d'hallucinations nombreuses, sur un fait psychique morbide qui récèle en lui une force d'impulsion, d'entraînement extrême, irrésistible, un stimulant moral et intellectuel, capable de transporter les montagnes, et qui, en réalité, plaçait Jeanne, par rapport à ceux qui l'entouraient, dans une sphère morale supérieure" (Moreau de Tours, p. 536). See earlier arguments in J. Baillarger, "Des hallucinations, des causes qui les produisent, et des maladies qui les caractérisent," *Mémoires de l'Académie royale de médecine*, vol. 12 (Paris: J.-B. Baillière, 1846), pp. 273–475, at p. 474: "The point of origin for hallucinations is and

can only be in the intellect" (Le point de départ des hallucinations est et ne peut être que dans l'intelligence).

142. "Folie et génie sont congénères" (Moreau de Tours, p. 493).

143. "la véritable origine des hautes facultés intellectuelles," "en aucune manière, le désordre général, absolu de l'intelligence" (Moreau de Tours, pp. 490, 491).

144. "au sein des plus sublimes facultés de l'âme" (Moreau de Tours, p. 490).

145. "intelligences ordinaires, de ces hallucinés vulgaires qui peuplent nos asiles" (Moreau de Tours, p. 491).

146. "parce que l'hallucination a été intimement mêlée à la vie politique et religieuse des peuples" (Christian, p. 78).

147. "aussi est-ce du peuple que sortait la libératrice de la France, celle qui affranchira son sol de l'étranger et relèvera la première l'étendard des nationalités et de la régénération des races" (Boismont, *De l'hallucination historique*, p. 3).

148. "Les partisans de cette opinion [qui déclare la Pucelle aliénée] ont prétendu qu'elle avait des extases secrètes que favorisait sans doute l'absence fréquente de tout écoulement menstruel" (Boismont, *De l'hallucination historique*, p. 34).

149. One of the best works debunking the myths connected to Joan of Arc is by historian Colette Beaune, *Jeanne d'Arc, vérités et légendes* (Paris: Perrin, 2008), who also wrote a recent biography, *Jeanne d'Arc* (Paris: Perrin, 2004). Karen Sullivan's *The Interrogation of Joan of Arc* (Minneapolis: University of Minnesota Press, 1999) examines the reliability and accuracy of the trial record, and Françoise Meltzer's *For Fear of the Fire: Joan of Arc and the Limits of Subjectivity* (Chicago: University of Chicago Press, 2001) probes the fascination of twentieth-century secular theorists for Joan's case.

150. "Nous ne faisons ce rapprochement entre Gilles de Rays et Jeanne d'Arc, ni pour établir un contraste entre ces deux extraordinaires figures de notre histoire, ni pour augmenter l'intérêt de notre récit, mais parce que le fait s'impose à notre conviction"; A. Bourdeault, "Chantoncé: Gilles de Rais et les ducs de Bretagne," *Mémoires de la Société d'histoire et d'archéologie de Bretagne* 5 (1924): 41–150, at p. 66.

151. "cette créature d'exception" (Dr. Cabanès, "Le légendaire Barbe-Bleue," p. 102).

152. "le souvenir du grand capitaine," "le souvenir du brigand"; P. L. d'Arc, "Gilles de Rais, maréchal de France, dit *Barbe-Bleue* (1404–1440), par l'abbé Eugène Bossard," *Revue catholique des institutions et du droit* 30 (1888): 463–67, at p. 465. See also Michelet, *Histoire du Moyen Âge*; Sourdeval, p. 41; Petit-Dutaillis, p. 184; J. Hebert, "Une cause célèbre au XVe siècle: Gilles de Rais," *Bulletin de la Société académique de Brest* 2:18 (1892–1893): 183–303, at pp. 198, 206.

153. "le *compaing*, le pur chevalier de Jeanne, et le bandit, l'assassin monstrueux"; Charles Derennes, *Amours et crimes* (Paris: Éditions de France, 1927), p. 10.

154. "un si beau rôle à côté de la jeune guerrière," "de si grands crimes" (Bossard, p. 33).

155. "le digne émule" (Guéraud, "Rais," p. 690).

156. "Ici paraît finir la carrière militaire et honorable du maréchal de Retz. Il ne nous reste plus que la tâche pénible d'offrir le tableau des extravagances, des vices et des crimes monstrueux qui ont plus contribué que ses exploits à sa malheureuse célébrité" (Audiffret, p. 399).

157. "Ainsi donc, Gilles ne s'adonna à l'étude soutenue de la magie qu'après la mort de Jeanne d'Arc, peut-être même seulement en 1432" (Bossard, p. 109).

158. Michel Tournier, *Gilles & Jeanne*, trans. Alan Sheridan (London: Minerva, 1987), p. 35 (*Gilles & Jeanne* [Paris: Gallimard/Folio, 1983], p. 45).

159. "l'exercice du pouvoir absolu, la *césarite*"; Dr. Cabanès, *Les morts mystérieuses de l'Histoire. Souverains et princes français de Charlemagne à Louis XVII* (Paris: A. Maloine, 1901), p. xviii. See also Maurice Beaujeu, *Psychologie des premiers Césars. Une étude de médecine légale dans l'histoire. Thèse pour obtenir le grade de docteur en médecine.* Faculté de médecine et de pharmacie de Lyon, no. 796 (Lyon: A. Storck, 1893), p. 51.

160. "le délire du pouvoir"; Gabriel Tarde, "Introduction," in Jacoby, *Études sur la sélection chez l'homme*, pp. i–iii, at p. ii.

161. "plus il [le pouvoir] est grand, plus la décadence morale, intellectuelle et physique sera manifeste" (Jacoby, p. x).

162. "La possession des privilèges et du pouvoir semble avoir eu à toutes les époques les plus désastreux effets sur la santé intellectuelle et morale de ceux qui en étaient investis. . . . N'avoir plus à se contenir, à vaincre ses désirs, à résister à ses passions, à réfréner ses mauvais instincts, fait subir, à la longue, aux facultés morales, une désorganisation profonde. La conscience s'endort, la notion du bien et du mal s'efface, les penchants bas se satisfont à l'aise, la personnalité s'exalte, un égoïsme d'autant plus puissant qu'il s'ignore, étouffe tout sentiment généreux. . . . Aussi voit-on le vice névropathique sous toutes ses formes: les névroses, la déséquilibration mentale, les anomalies psychiques, le crime, la débauche, les affections cérébrales, les maladies chroniques, les infirmités, les vices de conformation et la stérilité s'abattre successivement sur les races privilégiées et les éteindre avec une rapidité parfois foudroyante" (Cullerre, p. 319–20).

163. "par l'annoblissement des roturiers"; Émile Dusolier, *Psychologie des derniers Valois (François II, Charles IX, Henri III). Thèse présentée à la Faculté de médecine et de pharmacie de Lyon, et soutenue publiquement le mercredi 27 novembre 1895* (Lyon: Imprimerie de A.-H. Storck, 1895), p. 44; also Cullerre, p. 320.

164. "S'il en eût toujours été ainsi, la France n'aurait peut-être pas eu tant à souffrir sous l'administration de races dégénérées, et l'hérédité pathologique est, à mon avis, l'un des plus puissants arguments contre l'hérédité dynastique"; Dr. Auguste Corlieu, *Étude médicale sur la dynastie des Valois* (Paris: Adrien Delahaye, 1872), p. 9.

165. "Nous venons de voir les difficultés que la critique, en se fondant sur l'expérience, peut élever contre la noblesse considerée comme *fait naturel.* Ce qu'elle a valu comme *institution*, il n'y a pas lieu de le rechercher ici"; Théodule Ribot, *L'hérédité, étude psychologique sur ses phénomènes, ses lois, ses causes, ses conséquences* (Paris: Ladrange, 1873), p. 520. In the revised edition, Ribot structures the argument differently: "Of course, from an *entirely ideal point of view*, the institution of nobility can have an excellent reputation . . . : it would have been a procedure of rigorous selection, additionally with education. But this is beautiful only in dreams" (Certes, à un *point de vue tout idéal*, l'institution de la noblesse peut être réputée excellente . . . : c'eût été un procédé de sélection rigoureuse, avec l'éducation en

plus. Mais cela n'est beau qu'en rêve); Théodule Ribot, *L'hérédité, étude psychologique*, 5th corrected and rev. ed. (Paris: Felix Alcan, 1894), p. 367.

CHAPTER 5

1. This text first appeared in chapter form in Pierre Dufour (Paul Lacroix), *Histoire de la prostitution, chez tous les peuples du monde depuis l'antiquité la plus reculée jusqu'à nos jours*, [History of prostitution among all of the peoples of the world from the earliest antiquity to the current day], 6 vols. (Paris: Seré; Martinon, 1851–1853), vol. 4 (1852), p. 116. It was republished, with minor modifications, as "Les rues honteuses au Moyen-Âge," in P.-L. Jacob, bibliophile (Paul Lacroix), *Curiosités de l'histoire du vieux Paris* (Paris: Adolphe Delahays, 1858), pp. 115–85. Although the BNF catalogue specifies that *Histoire de la prostitution* is "a work generally attributed to Paul Lacroix, who has always disowned it" (*Ouvrage généralement attribué à Paul Lacroix, qui l'a toujours désavoué*), this reprint under his name proves otherwise, as does the entry in Pierre Larousse, *Grand dictionnaire universel du XIXe siècle* (Paris: Administration du Grand dictionnaire universel, 1866–1877), "Lacroix (Paul)," vol. 10 (1873), p. 40: "Aside from the pseudonym of the Bibliophile Jacob, he took on several others, notably those of Pierre Dufour and Antony Dubourg" (Outre le pseudonyme de Bibliophile Jacob, il en a pris plusieurs autres, notamment ceux de Pierre Dufour et d'Antony Dubourg).

2. "c'est surtout aux livres relatifs à la prostitution que convient le titre d'ouvrage pornographique"; Larousse, *Grand dictionnaire universel*, "Pornographie," vol. 12 (1874), pp. 1438–39, at p. 1439.

3. "en cet an de grâce des expropriations, démolitions, reconstructions et transformations municipales," "Le vieux Paris n'existe déjà plus: tous les jours il disparaît sous le nouveau. . . . Les débris du vaisseau, ce sont les noms des rues; les mâts, ce sont les tours de Notre-Dame; et nous, pauvres archéologues, attachons-nous aux reliques de ce grand naufrage"; Bibliophile Jacob (Paul Lacroix), *Physiologie des rues de Paris. Suivie de notes et renseignements et d'un plan de Paris et de ses faubourgs*, by Ch. Piquet (Paris: Martinon; Bureau du Musée des familles, 1842), pp. 21–22. Reprinted in *Curiosités de l'histoire du vieux Paris* (Paris: Adolphe Delahays, 1858), pp. 5–34.

4. "l'aspect extérieur de ces séjours de débauche, . . . leurs noms et leurs enseignes (du moins pour le plus grand nombre), . . . le système ordinaire de leur organisation impudique, . . . leur physionomie intérieure" (Lacroix, *Histoire*, 4:66).

5. "inventaire archéologique" (Lacroix, *Histoire*, 4:67), "on doit étudier leurs noms modifiés par la routine, réformés par arrêté municipal, changés par les événements, comme une langue morte qui se corrompt, qui se perd de jour en jour, et qui n'aura bientôt plus un seul interprète" (Lacroix, *Physiologie des rues*, pp. 53–54).

6. "voyage pornographique," "carte routière des rues honteuses de la capitale," "la topographie de la prostitution parisienne au moyen âge" (Lacroix, *Histoire*, 4:91, 66).

7. "on pourrait résumer les annales des arts, des lettres, des sciences, du crime et de la

vertu, avec des noms d'hommes inscrits à la tête des rues, aussi noblement que sur les tables de bronze du Panthéon"; also: "The seventeenth century had named abundant royal streets, where the great king showed the tip of his ear; the eighteenth made literary and philosophical streets; the nineteenth began the baptism of streets according to victories; but, presently, money alone baptizes our streets, our squares and our boulevards; and, money is named Véro or Dodat" (Le dix-septième siècle avait nommé force rues royales, où le grand roi montrait le bout de l'oreille; le dix-huitième fit des rues littéraires et philosophes; le dix-neuvième a commencé le baptême des rues par des victoires; mais, à présent, c'est l'argent seul qui baptise nos rues, nos places et nos boulevards; or l'argent se nomme Véro ou Dodat) (Lacroix, *Physiologie des rues*, p. 6).

8. "qui formait l'entrée de la rue actuelle du Grand-Hurleur . . . exclusivement reservée à la Prostitution depuis son origine jusqu'à nos jours, . . . ne devait pas son nom . . . à un chevalier, nommé Hugo Lupus (en vieux français, *Heu-leu*), lequel vivait au douzième siècle et fit plusieurs donations à l'église de Saint-Magloire; mais bien aux huées qui accompagnaient alors les gens simples ou crédules que le hasard amenait dans ce lieu infâme" (Lacroix, *Histoire*, 4:140).

9. "Il est certain pourtant que Pierre de Bordelles avait été qualifié ainsi dans les actes, parce qu'il possédait une maison dans cette rue, qui fut nommée *Bordelles*, *Bourdelle* et *Bordel*, en raison de son usage primitif et des nombreuses *bordes* que l'enceinte de Paris avait comprises dans ses murs" (Lacroix, *Histoire*, 4:104–5).

10. "leur surnom devait être le signe infaillible de leur naissance, de leur naturel, de leur état, en un mot, tout leur portrait physiologique," "Ces rues-là ne trompaient personne . . . un nom de rue tenait lieu de police et de réverbères" (Lacroix, *Physiologie des rues*, pp. 17, 51).

11. "Cependant ces noms de rues, que donne ou consacre tous les jours la préfecture, n'ont la plupart aucun retentissement, aucune sympathie dans le peuple, qui les adopte avec indifférence et qui les respecte par habitude" (Lacroix, *Physiologie des rues*, p. 3).

12. "Allez donc à present, sur la foi des noms, vous loger rue *Gracieuse*. . ., cueillir des cerises rue de la *Cerisaie*, voir l'heure rue du *Cadran*" (Lacroix, *Physiologie des rues*, p. 19).

13. "L'histoire morale et physique de Paris est liée à celle de ses rues" (Lacroix, *Physiologie des rues*, p. 53). For a contemporary history of the streets of Paris, see Daniel Milo, "Le nom des rues," in *Les lieux de mémoire*, ed. Pierre Nora, 3 vols. (Paris: Gallimard/Quarto, 1997 [1986]), vol. 2, pp. 1887–918; see also Michel de Certeau, "Walking in the City," *The Cultural Studies Reader*, ed. Simon During (New York: Routledge, 1993), pp. 126–33 ("Marcher dans la ville," in *Arts de faire. L'invention du quotidien*, vol. 1 [Paris: Gallimard, 1990 (1980)], pp. 139–64).

14. "ce mot *Prostitution* doit être ramené à son étymologie (*Prostitum*) et s'entendre alors de toute espèce de trafic obscène du corps humain" (Lacroix, *Histoire*, 1:6). In particular, *Histoire de la prostitution* inscribes itself in the trends of public health policy and social analysis in the wake of Alexandre-Jean-Baptiste Parent-Duchâtelet's *De la prostitution dans la ville de Paris* (1836).

15. BNF catalogue accessed on May 5, 2010. Lacroix states in his manuscript "Discours

pour sa reception [*sic*] supposée à l'Académie Francaise [*sic*]" (Speech for his presumed reception to the French Academy) that "I had written and printed more than five hundred volumes of all genres, which makes around ten volumes per year" (j'avais écrit et imprimé plus de cinq cents volumes en tous genres, ce qui fait environ une dizaine de volumes par an); Bibliothèque de l'Arsenal, MS 13427, 1942, no. 1.253, fol. 16. This speech is not dated, but since Lacroix mentions the seat of Henri Martin, who died in 1883, this could be the date of the writing of the "speech," toward the very end of his career and life. The entry in the *Grand dictionnaire universel* indicates that "Mr. Paul Lacroix wrote, translated, annotated, compiled, edited so much, a bit everywhere, that the complete list of his various publications is almost impossible to establish" (M. Paul Lacroix a tant écrit, traduit, annoté, compilé, arrangé, édité un peu partout, que la liste complète de ses productions diverses est presque impossible à dresser) (Larousse, "Lacroix [Paul]," p. 40).

16. Eugène de Mirecourt, *Bibliophile Jacob (Paul Lacroix)*, coll. Histoire contemporaine, no. 10 (Paris: Achille Faure, 1867), p. 31. Jean Richer published two letters from Victor Hugo to Paul Lacroix, which may contain proof of his influence on Hugo's *Le Roi s'amuse* and Vigny's *La Maréchale d'Ancre*. Richer also asserts that it is well known that Nerval's *Nicolas Flamel* is taken from Lacroix's *Soirées de Walter Scott à Paris* (1829); Jean Richer, "Deux lettres de Victor Hugo à Paul Lacroix: P. Lacroix, Vigny, Hugo. Une source probable du *Roi s'amuse*," *Revue des sciences humaines* 98–99 (1960): 199–201. On the historical-literary context, see Maurice Samuels, *The Spectacular Past: Popular History and the Novel in Nineteenth-Century France* (Ithaca, N.Y.: Cornell University Press, 2004).

17. Pierre Larousse, *Grand dictionnaire universel du XIXe siècle, Supplément* II (Paris: Administration du Grand dictionnaire universel, 1890), "Lacroix (Paul)," p. 1492: "The bibliophile Jacob leaves behind curious unpublished memoirs, the set of which will offer no fewer than twelve volumes, of which a very interesting one that discusses his relationship with H. de Balzac appeared in the journal 'le Livre'" (Le bibliophile Jacob laisse de curieux mémoires inédits, dont l'ensemble ne formera pas moins de douze volumes et dont une partie très intéressante, celle qui se rapporte à ses relations avec H. de Balzac, a paru dans la revue "le Livre"). This article, "Simple histoire de mes relations littéraires avec Honoré de Balzac," published in *Le Livre* in 1883, one year before his death, together with the information given to Dutacq and Michel Lévy in the preparation for the first edition of Balzac's *Œuvres diverses*, is "the starting point of many errors that wronged the image of Balzac and Balzac studies" (à l'origine de beaucoup d'erreurs qui firent du tort à l'image de Balzac et aux études balzaciennes); Honoré de Balzac, "Portrait de P.-L. Jacob," in *Œuvres diverses 2*, ed. Pierre-Georges Castex, Roland Chollet, and René Guise, Bibliothèque de la Pléiade (Paris: Gallimard, 1996), pp. 654–57, 1462–64, at p. 1463.

18. Larousse, *Grand dictionnaire universel*, "Martin (Bon-Louis-Henri)," vol. 10 (1873), pp. 1278–79, at p. 1278: "Through the course of his relations with the bibliophile Jacob, Mr. Henri Martin conceived the project of a *History of France by its Principal Historians*, and our two authors immediately set to work . . . but, after the publication of the first volume, Mr. Paul Lacroix abandoned the enterprise. Mr. Henri Martin then resolved to pursue it alone" (Dans le cours de ses relations avec le bibliophile Jacob, M. Henri Martin conçut le plan

d'une *Histoire de France par les principaux historiens*, et nos deux auteurs se mirent aussitôt à l'œuvre . . . mais, après la publication du premier volume, M. Paul Lacroix abandonna l'entreprise. M. Henri Martin résolut alors de la poursuivre seul). Martin wrote the second edition from 1834 to 1836, which he revised and expanded in a third edition between 1838 and 1853. In his "Discours pour sa reception [*sic*] supposée à l'Académie Francaise [*sic*]," Lacroix gives this version of events: "The blame for it falls on a bookstore keeper, a company of bookstores that felt the need to publish a new History of France and who asked me to provide its outline. . . . Henry [*sic*] Martin, gentlemen, consented to undertake alone, under my guidance and with my advice, this immense task which would have given pause to a Benedictine from the olden days" (La faute en est à un libraire, à une compagnie de libraires qui éprouvèrent le besoin de publier une nouvelle Histoire de France et qui me demandèrent d'en faire le plan. . . . Henry [*sic*] Martin, messieurs, consentit à entreprendre seul, sous mes yeux et avec mes conseils, cette immense tâche qui eut [*sic*] fait reculer un bénédictin d'autrefois); Arsenal MS 13427, 1942, no. 1.253, fol. 16.

19. Henry Martin, *Catalogue des manuscrits de la Bibliothèque de l'Arsenal, Histoire de la Bibliothèque de l'Arsenal*, vol. 8 (Paris: Librairie Plon, 1899), p. 564 (not to be confused with historian Henri Martin). Lacroix knew Janin well and catalogued his library; Paul Lacroix, *La bibliothèque de Jules Janin* (Paris: Librairie des bibliophiles, 1877).

20. Charles Ridoux, *Évolution des études médiévales en France de 1860 à 1914* (Paris: Honoré Champion, 2001), pp. 118, 861. The general tone reserved for Romantic medievalists is one of contempt.

21. Lacroix is well respected by historians of prostitution. The most up-to-date history of medieval prostitution available, Jacques Rossiaud's *La prostitution médiévale* (Paris: Flammarion, 1988), quotes from Lacroix. His *Histoire* is also abundantly cited in the chapter on medieval times in a classical work on prostitution in Paris, Jill Harsin, *Policing Prostitution in Nineteenth-Century Paris* (Princeton, N.J.: Princeton University Press, 1985).

22. "amour mystique: on ne parle que de cet amour-là, c'est le seul qu'on mette en avant. . . . Il faudrait étudier ce problème, sans tenir compte de la littérature, dans les chroniques secrètes, dans les procès criminels, dans les documents relatifs aux cloîtres"; André Gide, Letter to Georges Hérelle, dated July 14, 1934, transcribed by G. Hérelle, Bibliothèque municipale de Troyes, MS 3188, fol. 359. I thank Monique Nemer for bringing this document to my attention.

23. "histoire des mœurs" (Lacroix, *Histoire*, 6:90), "l'histoire de la Prostitution . . . ce sujet, délicat et suspect à la fois, se rattache de tous côtés à l'histoire des religions, des lois et des mœurs" (Lacroix, *Histoire*, 1:9).

24. "des belles études de mœurs qu'il avait depuis longtemps entreprises sur la société française" (Larousse, *Supplément*, "Lacroix," p. 1492).

25. *Encyclopaedia Britannica*, 11th ed. (London: Encyclopaedia Britannica, 1910–1911), p. 54.

26. Pascal Ory, *Histoire culturelle* (Paris: Presses universitaires de France, 2004), p. 30. In fact, there are several authors who precede Voltaire in this idea of universal history of customs and morals, e.g., abbé Claude Fleury's *Les mœurs des Israélites* (1681) and *Les mœurs*

des Chrétiens (1682), Louis Le Gendre's *Mœurs et coutumes des François, dans les différens tems de la monarchie* (1712), and Jean Frédéric Bernard and Bernard Picart's *Cérémonies et coutumes religieuses de tous les peuples du monde* (1723–1737), which offered a culturally relativizing model for a comparative history of religions because it turned "belief in one unique, absolute, and God-given truth into 'religion,' that is, into individual ceremonies and customs that reflected the truths relative to each people and culture"; Lynn Hunt, Margaret C. Jacob, and Wijnand Mijnhardt, *The Book That Changed Europe: Picart & Bernard's Religious Ceremonies of the World* (Cambridge, Mass.: Belknap Press of Harvard University Press, 2010), pp. 1–2.

27. "le dégoût que vous cause l'Histoire moderne," "en quelle année un prince indigne d'être connu succéda à un prince barbare chez une nation grossière," "ce qui mérite d'être connu de vous; l'esprit, les mœurs, les usages des nations principales"; Voltaire, *Essai sur les mœurs et l'esprit des nations, et sur les principaux faits de l'histoire, depuis Charlemagne jusqu'à Louis XIII*, in *Œuvres*, vols. 15–18, ed. M. Beuchot (Paris: Werdet et Lequien, 1829 [1756–1763]), vol. 15, p. 245.

28. "l'Orient, berceau de tous les arts, et qui a tout donné à l'Occident" (Voltaire, 15:247).

29. "avant que nous en sussions assez pour connaître que nous étions barbares" (Voltaire, 15:248).

30. "la moitié s'appelait la langue d'*Oui* ou d'*Oil*, et l'autre la langue d'*Oc*," "un pays" (Voltaire, 16:437).

31. "Un malheureux pays dépourvu de lois fixes, agité par des guerres civiles, sans commerce, sans police, sans coutumes écrites, et gouverné par mille coutumes différentes" (Voltaire, 16:437), "Les usages les plus ridicules et les plus barbares étaient alors établis" (Voltaire, 16:138).

32. "un extérieur plus brillant que les autres nations" (Voltaire, 16:437).

33. This bibliography is by no means exhaustive and a historiography of the *histoire des mœurs* from its eighteenth-century beginnings through the mid-nineteenth century remains to be written. Pierre Poivre, *Voyages d'un philosophe, ou Observations sur les mœurs et les arts des peuples de l'Afrique, de l'Asie et de l'Amérique* (1768); Jean François de la Harpe, *Abrégé de l'Histoire générale des voyages, contenant ce qu'il y a de plus remarquable, de plus utile et de mieux avéré dans les pays où les voyageurs ont pénétré: Les mœurs des habitans, la religion, les usages, arts et sciences, commerce, manufactures*, 23 vols. (1780–1786); Joseph Malliot, *Recherches sur les costumes, les mœurs, les usages religieux, civils et militaires des anciens peuples*, 3 vols. (1804); H. M. G. Grellmann, *Histoire des Bohémiens, ou Tableau des mœurs, usages et coutumes de ce peuple nomade, suivie de recherches historiques sur leur origine, leur langage et leur première apparition en Europe*, trans. M. Jansen (1810); Antoine-Laurent Castellan, *Mœurs, usages, costumes des Othomans, et abrégé de leur histoire* (1812); René-Claude Geoffroy de Villeneuve, *L'Afrique, ou Histoire, mœurs, usages et coutumes des Africains* (1814); Alphonse Cerfberr de Medelsheim, *Les Juifs, leur histoire, leurs mœurs* (1847); Louis Viardot, *Histoire des Arabes et des Mores d'Espagne, traitant de la constitution du peuple arabe-espagnol, de sa civilisation, de ses mœurs et de son influence sur la civilisation moderne*, 2 vols. (1851); Eugène

Daumas, *Mœurs et coutumes de l'Algérie: Tell, Kabylie, Sahara* (1853); Xavier Eyma, *Les Peaux Rouges, scènes de la vie des Indiens* (1854); Antoine Brun-Rollet, *Le Nil Blanc et le Soudan: Études sur l'Afrique centrale, mœurs et coutumes des sauvages* (1855); Auguste Wahlen, *Mœurs, usages et costumes de tous les peuples du monde, d'après des documents authentiques et les voyages les plus récents*, 4 vols. (1843–1844); Maxime de Montrond, *Missions d'Amérique, d'Océanie et d'Afrique* (1846); Maxime de Montrond, *Missions d'Amérique, d'Océanie et d'Afrique, lettres, récits et fragments divers*, 2nd ed. (1853).

34. Jean-Baptiste de La Curne de Sainte-Palaye, *Mémoires sur l'ancienne chevalerie* (1759–1781); Pierre Jean-Baptiste Legrand D'Aussy, *Histoire de la vie privée des Français depuis l'origine de la nation jusqu'à nos jours* (1782); Louis-Sébastien Mercier, *Tableau de Paris* (1782); Jacques-Antoine Dulaure, *Singularités historiques, ou Tableau critique des mœurs, des usages et des événements de différens siècles, contenant ce que l'histoire de la capitale et des autres lieux de l'Isle-de-France offre de plus piquant et de plus singulier, pour servir de suite aux descriptions de Paris et de ses environs* (1788); *Histoire de la vie privée des Français ou Tableau des mœurs, caractères, coutumes et usages de nos ancêtres aux différentes époques de la monarchie* (1817); J.-A. Dulaure, *Histoire physique, civile et morale de Paris depuis les premiers temps historiques jusqu'à nos jours* (1821); François Guizot, *Histoire de la civilisation en Europe, depuis la chute de l'Empire romain jusqu'à la Révolution française* (1828); F. Guizot, *Histoire de la civilisation en France, depuis la chute de l'Empire romain* (1832); Émile de la Bédollière, *Histoire des mœurs et de la vie privée des Français*, 3 vols. (1847–1849); Paul Lacroix (le Bibliophile Jacob), *Costumes historiques de la France d'après les monuments les plus authentiques, précédé de l'Histoire de la vie privée des Français, depuis les temps les plus reculés jusqu'à nos jours, comprenant l'histoire des mœurs, usages et costumes de la nation française pour servir d'introduction aux costumes historiques de la France, suivi d'un Recueil curieux de pièces originales, rares ou inédites, en prose et en vers, sur le costume et les révolutions de la mode en France*, 10 vols. (n.d. [1852]); *Adolphe* Chéruel, *Dictionnaire historique des institutions, mœurs et coutumes de la France*, 2 vols. (1855). There were a number of histories on the eighteenth century, especially works by Rétif de la Bretonne, e.g., *Tableaux de la vie, ou les Mœurs du dix-huitième siècle* (1767), and *Les Parisiennes, ou XL caractères généraux pris dans les mœurs actuelles, propres à servir à l'instruction des personnes du sexe* (1787). Antoine Caillot's work illustrates well the purpose of such works: *Mémoires pour servir à l'histoire des mœurs et usages des Français, depuis les plus hautes conditions jusqu'aux classes inférieures de la société, pendant le règne de Louis XVI, sous le Directoire exécutif, sous Napoléon Bonaparte, et jusqu'à nos jours*, 2 vols. (1827).

35. Paul Lacroix and Ferdinand Seré, eds., *Le Moyen Âge et la Renaissance: Histoire et description des mœurs et usages, du commerce et l'industrie, des sciences, des arts, des littératures et des beaux-arts en Europe*, 5 vols. (Paris, 1848–1851). According to the *Grand dictionnaire universel du XIX siècle*, his chief collaborators were Philarète Chasles, P. Mérimée, Mary-Lafon, A. Le Roux de Lincy, Alfred Michiels, Champollion-Figeac, and Lassus.

36. Achille Jubinal, *Jongleurs et trouvères, ou Choix de saluts, épîtres, rêveries et autres pièces légères des XIIIe et XIVe siècles* (Paris: J. A. Merklein, 1835); A. Jubinal, *Recherches sur l'usage et l'origine des tapisseries à personnages dites historiées, depuis l'antiquité jusqu'au XVIe*

siècle inclusivement (Paris: Challamel, 1840); A. Jubinal, *Nouveau recueil de contes, dits, fabliaux et autres pièces inédites des XIIIe, XIVe et XVe siècles, pour faire suite aux collections de Legrand d'Aussy, Barbazan et Méon*, 2 vols. (Paris: E. Pannier, 1839–1842); Francisque Michel and Édouard Fournier, *Histoire des hôtelleries, cabarets, hôtels garnis, restaurants et cafés, et des anciennes communautés et confreries d'hôteliers, de marchands de vins, de restaurateurs, de limonadiers, etc.* (Paris: Seré, 1851); Francisque Michel, *Recherches sur le commerce, la fabrication et l'usage des étoffes de soie, d'or et d'argent et autres tissus précieux en Occident, principalement en France, pendant le Moyen Âge*, 2 vols. (Paris: Crapelet, 1852–1854); A. Lecoy de La Marche, *La société au XIIIe siècle* (Paris: V. Palmé, 1880); Louis Petit de Julleville, *Histoire du théâtre en France. La comédie et les mœurs en France au Moyen Âge* (Paris: L. Cerf, 1886); A. Lecoy de la Marche, *L'esprit de nos aïeux: Anecdotes et bons mots tirés des manuscrits du XIIIe siècle* (Paris: C. Marpon et E. Flammarion, 1888).

37. Dr. Cabanès's *Mœurs intimes du passé*, 12 vols. (Paris: Albin Michel, 1908–1936), marks the end of the *histoire des mœurs* in large part replaced by a contemporary reflection on the *mœurs* in the Rétif de la Bretonne tradition; e.g., Octave Uzanne's *Parisiennes de ce temps en leurs divers milieux, états et conditions: Études pour servir à l'histoire des femmes, de la société, de la galanterie française, des mœurs contemporaines et de l'égoïsme masculin* (Paris: Mercure de France, 1910). It is tempting to associate the waning of the *histoire des mœurs* with the trials occurring in the 1850s in France; see Elisabeth Ladenson, *Art for Dirt's Sake: Books on Trial from* Madame Bovary *to* Lolita (Ithaca, N.Y.: Cornell University Press, 2007).

38. "En dressant l'inventaire des vices et des vertus, en rassemblant les principaux faits des passions, en peignant les caractères, en choisissant les événements principaux de la Société, en composant des types par la réunion des traits de plusieurs caractères homogènes, peut-être pouvais-je arriver à écrire l'histoire oubliée par tant d'historiens, celle des mœurs"; Honoré de Balzac, "Avant-propos," *La comédie humaine*, vol. 1 (Paris: Gallimard, Bibliothèque de la Pléiade, 1976 [1842]), p. 11.

39. Jules Michelet, "Préface de 1869," *Œuvres complètes*, vol. 1 (Paris: Flammarion, 1893–1898), pp. i–xlv, at p. iv.

40. "caprices, . . . fantaisie, . . . imagination désordonnée errant à l'aventure à travers la nature et l'histoire, saisissant vivement telle ou telle chose au passage et comme par hasard, sans se faire de règle, ni se proposer de but"; Gabriel Monod, *Les maîtres de l'histoire: Renan, Taine, Michelet*, 2nd ed. (Paris: Calman Lévy, 1894 [1875]), p. 251.

41. "une histoire par tiroirs, le tiroir 'civilisation' ou 'culture' ou 'arts et sciences' prenant la suite, et souvent la fin, des tiroirs 'politique,' 'économique,' 'société'"; Ory, *Histoire culturelle*, p. 32. E.g., Émile de la Bédollière, *Histoire de la mode en France* (Leipzig: A. Dürr, 1858); Charles Louandre, *Les arts somptuaires: Histoire du costume et de l'ameublement et des arts et industries qui s'y rattachent*, 2 vols. (Paris: Hangard-Maugé, 1857–1858); Louis-Nicolas Bescherelle, *Précis de l'histoire générale des arts, des sciences, de l'industrie et des mœurs des nations* (Paris: Breteau et Pichery, 1840).

42. "L'histoire politique et religieuse est rejetée dans une introduction insuffisante. . . . Nul siècle n'y est représenté dans un tableau complet. On a bien l'histoire de chaque chose en particulier; mais une histoire abstraite et dépouillée des rapports qui,

unissant logiquement toutes les diverses manifestations humaines à une époque, concourent à préciser la physionomie de cette époque même"; Larousse, *Grand dictionnaire universel*, "Moyen, -enne," vol. 11 (1874), pp. 656–59, at p. 658.

43. "un ensemble aussi complet que possible," "un morcellement qui empêche le lecteur de se former une idée d'ensemble, et . . . introduit une grande confusion dans l'impression générale" (Larousse, "Moyen, -enne," p. 658).

44. "Ouvrage complétant toutes les Histoires de France"; Émile de la Bédollière, *Histoire des mœurs et de la vie privée des Français. Usages, coutumes, institutions, physionomie de chaque époque, depuis l'origine de la monarchie jusqu'à nos jours*, 3 vols. (Paris: V. Lecou, 1847–1849).

45. "la généalogie des rois, les révolutions des peuples, les changements des dynasties," "Comment vivaient nos ancêtres?" (Bédollière, 1:vii).

46. "Où est la source de tant d'usages vivaces, que nous avons acceptés?" (Bédollière, 1:vii–viii).

47. See Bruce G. Trigger, *A History of Archaeological Thought*, 2nd ed. (Cambridge: Cambridge University Press, 2006 [1989]), pp. 110–11, on the links of the early *histoire des mœurs*, evolutionary anthropology, and antiquarianism in the eighteenth century.

48. "Science qui s'occupe de tout ce qui est relatif aux anciens, spécialement de leurs arts et leurs monuments"; Larousse, *Grand dictionnaire universel*, "Archéologie," vol. 1 (1866), pp. 571–72, at p. 571.

49. "tandis que les historiens nous racontent les faits généraux de la politique, de la guerre, des révolutions, s'attachent aux événements remarquables qui dessinent en grand les figures des peuples, en un mot au côté dramatique et extérieur de la vie des sociétés, l'*archéologie* nous fait pénétrer dans l'intérieur, dans le fond de cette vie; elle nous apprend les mœurs, les usages, les coutumes, les croyances des anciens, nous les rend visibles en transportant notre imagination au milieu de l'antiquité" (Larousse, "Archéologie," p. 571).

50. Michel de Certeau, *The Practice of Everyday Life*, trans. Steven Rendall (Berkeley: University of California Press, 1984) (*L'invention du quotidien* [Paris: Gallimard, 1990 (1980)]).

51. Dominique Poulot, "Alexandre Lenoir et les musées des monuments français," in *Les lieux de mémoire*, ed. Nora, vol. 1, pp. 1515–43, at p. 1534: "Put another way, Lenoir inaugurated the still timid appearance of historical value within an art category—the monumental—which had until then been exclusively reserved for homage and memory" (Autrement dit, Lenoir a signé l'apparition, encore timide, de la valeur historique au sein d'une catégorie artistique—le monumental—jusqu'alors exclusivement dévolue à l'hommage et au souvenir).

52. André Chastel, "The Notion of Patrimony," in *Rethinking France: Les Lieux de Mémoire, Volume 3: Legacies*, ed. Pierre Nora, trans. Nancy Turpin (Chicago: University of Chicago Press, 2009), pp. 1–46, at p. 6, translation modified ("La notion de patrimoine," in *Les lieux de mémoire*, ed. Nora, vol. 1, pp. 1433–69, at p. 1436).

53. "le XIXe siècle français pourra être appelé le *siècle de l'histoire*"; Larousse, *Grand dictionnaire universel*, "Histoire," vol. 9 (1873), pp. 300–14, at p. 301.

54. "on se mit à peindre le passé, tandis qu'auparavant on n'avait encore fait que le juger au point de vue des opinions du moment"; Larousse, *Grand dictionnaire universel*, "France," vol. 8 (1872), pp. 719–51, at p. 747.

55. Trigger, *A History of Archaeological Thought*; Göran Blix, *From Paris to Pompeii: French Romanticism and the Cultural Politics of Archaeology* (Philadelphia: University of Pennsylvania Press, 2009).

56. "Aussi le Moyen Âge est-il, dans cette enquête, un champ d'études privilégié. Là s'est constituée, par rapports successifs, la matrice de la civilisation française, là se sont forgés les systèmes et les instruments du gouvernement, là aussi, en ce temps long, se sont formés et ont émergé la classe moyenne, la bourgeoisie, le peuple, ces forces qui aujourd'hui sont la substance de la nation"; Laurent Theis, "Guizot et les institutions de mémoire," in *Les lieux de mémoire*, ed. Nora, vol. 1, pp. 1575–97, at p. 1583.

57. Nabila Oulebsir, *Les usages du patrimoine: Monuments, musées et politique coloniale en Algérie (1830–1930)* (Paris: Éditions de la Maison des sciences de l'homme, 2004). E.g., for his views on total colonization of Algeria, see Alphonse de Lamartine's *Souvenirs, impressions, pensées et paysages, pendant un voyage en Orient (1832–1833), ou Notes d'un voyageur*, 2 vols. (Paris: C. Gosselin, 1835), *Vues, discours et articles sur la question d'Orient* (Paris: C. Gosselin, 1840), and Dossier "Lamartine, orateur lyrique. Discours parlementaires (1834–1851)," http://www.assemblee-nationale.fr/histoire/lamartine/discours.asp#titre1, accessed February 29, 2012; for writings on the politics of colonization and colonization of Morocco, Algeria, Tahiti, and the Marquises Islands, see François Guizot's *Mémoires pour servir à l'histoire de mon temps* (Paris: Michel-Lévy frères, 1858–1867).

58. Françoise Bercé, "Arcisse de Caumont et les sociétés savantes," in *Les lieux de mémoire*, ed. Nora, vol. 1, pp. 1545–73: Société des antiquaires de Normandie (1824), Antiquaires de la Morinie (1831), Antiquaires de l'Ouest (1834), Antiquaires de Picardie (1836), Société archéologique du midi de la France (1836), and Institut des provinces (1839). This trend will continue in the 1840s.

59. "plusieurs comités historiques du ministère de l'instruction publique" (Larousse, "Lacroix [Paul]," p. 40).

60. "des documents originaux relatifs à notre histoire nationale," "populariser l'étude et le goût de notre histoire nationale dans une voie de saine critique, et surtout par la recherche et l'emploi des documents originaux" (Guizot, qtd. in Theis, pp. 1585, 1585–86).

61. "rechercher et publier tous les documents pouvant contribuer à la connaissance de l'histoire nationale," "la recherche et . . . la publication de documents inédits de la littérature, de la philosophie, des sciences et des arts considérés dans leurs rapports avec l'histoire générale de la France" (Guizot, qtd. in Theis, p. 1594).

62. "The École des chartes thus received a new motivation, and its mission in research, study, and conservation of original monuments of national history, as defined in a notice addressed by the Society of the École des chartes to the minister of Public Instruction in 1839, was precisely established by it" (L'École des chartes reçut ainsi une impulsion nouvelle, et sa mission de recherche, d'étude et de conservation des monuments originaux de l'histoire

nationale, telle que la définit une notice adressée par la Société de l'École des chartes au ministre de l'Instruction publique en 1839, s'en trouva précisément fixée) (Theis, p. 1592).

63. "On a beaucoup parlé . . . de grandeur nationale. Et bien, rien ne contribue davantage à cette grandeur que les monuments qu'on élève, et parmi ces monuments les entreprises littéraires tiennent le premier rang" (Guizot, qtd. in Theis, p. 1590).

64. "Un fait de ce genre, qui n'est ni contestable, ni contesté, c'est que, de toutes les littératures du moyen âge, la française (dans laquelle je comprends celle des Anglo-Normands) est de beaucoup la plus riche en épopées chevaleresques. Il est également certain, également reconu que c'est du français que la plupart de ces épopées ont été traduites ou imitées dans les autres langues de l'Europe"; Claude Fauriel, "Origine de l'épopée chevaleresque du Moyen Âge," *La Revue des deux mondes* 7 (1832): 513–75, 672–710; 8 (1832): 138–94, 268–308, 425–70, at 7:517. See also Ridoux, p. 341, on articles by Edgar Quinet, from 1831 to 1840.

65. Jean-Jacques Ampère, "De la littérature française dans ses rapports avec les littératures étrangères au Moyen Âge," *La Revue des deux mondes* 1 (1833): 22–34, at p. 34.

66. "j'ai voulu montrer déjà vivantes les tendances dont les combinaisons et les luttes formeront, en très grande partie, la vie complexe des siècles modernes. En arrivant à ces siècles plus connus, ou du moins plus étudiés . . . peut-être comprendra-t-on mieux le développement de l'esprit moderne, après en avoir surpris l'embryon dans les flancs vigoureux du moyen-âge"; Jean-Jacques Ampère, "Vue générale de la littérature française au Moyen-Âge," *La Revue des deux mondes* 19 (1839): 179–93, at p. 193. After the 1830s, publications on the literature of the Middle Ages will spike again after 1870, in the aftermath of the Franco-Prussian War; see Elizabeth Emery, "The 'Truth' about the Middle Ages: *La Revue des Deux Mondes* and Late Nineteenth-Century French Medievalism," in *Medievalism and the Quest for the "Real" Middle Ages*, ed. Clare A. Simmons (Portland, Ore.: Frank Cass, 2001), pp. 99–114.

67. "la généalogie de notre idiome national"; Larousse, *Grand dictionnaire universel*, "Français, -aise," vol. 8 (1872), pp. 708–17, at p. 709.

68. "De ce mélange en fermentation va se dégager une forme nouvelle, imparfaite et grossière, mais du moins nationale" (Larousse, "Français, -aise," p. 710).

69. "de la netteté, de la clarté et de la concision" (Larousse, "Français, -aise," p. 712).

70. "ces œuvres . . . avaient . . . jadis joué un grand rôle et exercé une influence étendue. Elles ne demeurèrent pas enserrées en d'étroites limites: l'Europe entière fut leur vaste théâtre. . . . Et c'est là . . . que commence l'usage européen de la langue française"; Émile Littré, *Études sur les barbares et le Moyen Âge* (Paris: Didier, 1869), p. 385.

71. "instrument glorieux de la civilisation du monde," "notre langue n'a pas exercé une moindre influence en Europe que nos mœurs, nos idées et nos armes" (Ampère, "De la littérature française," p. 23).

72. "la France est à elle seule toute une civilisation. À elle surtout, la mission de précéder les autres et de les guider vers de nouvelles destinées" (Larousse, "Moyen, -enne," p. 657).

73. "la force civilisatrice" (Bédollière, 3:312).

74. "trois fois la civilisation française s'est placée à la tête de l'Europe: au moyen âge, par notre litterature, par nos croisades et notre chevalerie; au dix-septième siècle, par le génie de nos écrivains et le règne de Louis XIV; au dix-huitième, par l'ascendant de notre philosophie et les triomphes de notre glorieuse révolution" (Ampère, "De la littérature française," p. 33).

75. "Ce n'est pas l'art qu'il faut chercher dans ces grandes compositions, mais de précieux détails sur l'histoire, sur la géographie des mœurs"; A. Le Roux de Lincy (1841), qtd. in Hans Ulrich Gumbrecht, "'Un souffle d'Allemagne ayant passé': Friedrich Diez, Gaston Paris, and the Genesis of National Philologies," *Romance Philology* 40:1 (1986): 1–37, at p. 21.

76. "Soutenons le pas, messieurs, et pour la quatrième fois reprenons notre poste en tête du mouvement européen. L'Europe nous regarde et nous attend" (Ampère, "De la littérature française," p. 34).

77. "Les grands succès appartiennent moins aux savants qu'aux vulgarisateurs"; Paul Lacroix, *Sciences et lettres au Moyen Âge et à l'époque de la Renaissance* (Paris: Firmin-Didot, 1877), p. iv.

78. "Il a suivi une voie moins critique que littéraire. Il voulait populariser les romans du Moyen Âge"; Henri Wallon, "Notice sur la vie et les œuvres de M. Paulin Paris," *Académie des Inscriptions et Belles-Lettres*, 4e série, t. 10 (1882): 330–88, at p. 338. See also Larousse, *Grand dictionnaire universel*, "Paris (Alexis-Paulin)," vol. 12 (1874), p. 289.

79. "en nouveau langage"; Gaston Paris, "Paulin Paris et la littérature du Moyen Âge," in *La poésie du Moyen Âge*, 2 vols. (Paris: Librairie Hachette, 1885), vol. 1, pp. 211–54, at p. 219.

80. "l'imagination, qui est la qualité maîtresse du romancier ou du poète, a du péril pour l'érudit" (Wallon, pp. 378–79).

81. "un petit nombre d'adeptes qui bornent à des servitudes d'orthographe la *fidèle* reproduction des anciens textes; nous avons appliqué l'orthographe moderne à ces contes du quinzième et du seizième siècle"; Paul Lacroix, *Les vieux conteurs français* (Paris: Société du Panthéon littéraire, 1841), p. vi. His theorization of "archaism" (*archaïsme*) that influenced Balzac dates to 1830 and the publication of *Les deux fous*: "The novelist of the past must, according to Lacroix, adopt a resolutely modern language while allowing his characters to converse in the language of the era that he describes; but it is up to the writer to finesse this archaism a bit 'for the intelligence of the dialogue'" (Le romancier du passé doit, selon Lacroix, adopter un langage résolument moderne tout en laissant ses personnages converser dans le langage de l'époque qu'il décrit; mais libre à l'écrivain d'arranger quelque peu cet archaïsme "pour l'intelligence du dialogue") (Balzac, "Portrait de P.-L. Jacob," pp. 1463–64).

82. "il contemplait [ces antiques volumes] avec un ravissement étonné" (G. Paris, "Paulin Paris," p. 223).

83. "tout imprimer" (Wallon, p. 362). "Reconstituting the original text is assuredly the ideal of truth: but by what sign could the original lesson be recognized?" (Reconstituer le texte original, c'est assurément l'idéal du vrai: mais à quel signe reconnaître la leçon originale?) (Wallon, p. 338). This criticism was in general applied to the entire previous

generation of Romantic medievalists. In 1880, the philological journal *Romania* announced the publication of the first volume of Frédéric Godefroy's *Dictionnaire de l'ancienne langue française*, the most valuable and complete Old French dictionary available to medievalists today, accompanied with this criticism: "We recognize in this work the fruit of a long and immense effort, but we sense also that its author is neither a linguist nor a philologist" (On reconnaît dans cet ouvrage le fruit d'un immense et long travail, mais on sent aussi que son auteur n'est ni linguiste ni philologue) (qtd. in Ridoux, p. 433). A year later, in 1882, this same critique of Godefroy's auto-didacticism is repeated in another medieval journal, the *Revue critique d'histoire et de littérature*, "the chief complaints being that the author had above all accomplished the work of a word collector, and that his work remained that of a self-taught man, rather than an authentic scholar" (les principaux griefs étant que l'auteur avait fait avant tout œuvre de collectionneur de mots, et que son ouvrage demeurait celui d'un autodidacte et non d'un authentique savant) (qtd. in Ridoux, p. 434).

84. "le manque d'une méthode arrêtée et d'un lien philosophique," "par cette lacune dans sa préparation première, il fut empêché, et il le reconnaissait lui-même, d'être aussi bon philologue que littérateur" (G. Paris, "Paulin Paris," p. 251).

85. David F. Hult, "Gaston Paris and the Invention of Courtly Love," in *Medievalism and the Modernist Temper*, ed. R. Howard Bloch and Stephen G. Nichols (Baltimore: Johns Hopkins University Press, 1996), pp. 193–224, at p. 205.

86. For an account, see Paul Lacroix, "Bibliophiles et bibliomanes," *Le Bibliophile, Gazette illustrée des amateurs et bibliophiles des deux mondes* (1883–1885): 49–52, 87–94.

87. "il ne distinguait pas trop entre ses propres livres et ceux dont il avait la garde: procédé qu'il devait sévèrement interdire au public, qu'il aurait dû se refuser aussi peut-être à lui-même, comme conservateur" (Wallon, p. 380).

88. P. Alessandra Maccioni Ruju and Marco Mostert, *The Life and Times of Guglielmo Libri (1802–1869)* (Hilversum: Verloren Publishers, 1995), p. 175.

89. Ridoux, pp. 378–89; cf. Guillaume Libri, *Réponse de M. Libri au rapport de M. Boucly* (Paris, 1848).

90. Ruju and Mostert, p. 274. Cf. Libri, *Réponse*, p. 32, who recalls the libel against François Raynouard: "I was told in the library (at Carcassonne) that this manuscript (of *Flamenca*) was lost, and that Mr. Raynouard, who had borrowed it, had kept it for himself" (je m'entendis dire dans la bibliothèque [à Carcassonne] que ce manuscrit [de *Flamenca*] était perdu, et que M. Raynouard, qui l'avait emprunté, se l'était approprié).

91. Ruju and Mostert, pp. 173, 243, 250, 252–53, 255, 281; Ridoux, p. 383; Libri, pp. 71–74.

92. Ruju and Mostert, p. 244. Cf. Libri, *Réponse*, pp. 41–43, 76.

93. Meyer began examining the Ashburnham library in 1865 and continued to work with Léopold Delisle on establishing the provenance of its manuscripts. Deslile, another *chartiste* who became the director of the Bibliothèque nationale in 1874, succeeded in returning 166 manuscripts from England to France in 1888; Ridoux, pp. 382, 384; see also Mélanie Libri, *Pétition adressée au Sénat sur l'affaire de M. Libri, avec une note à l'appui signée par* M. M. Guizot, le marquis d'Audiffret, Prosper Mérimée, Édouard Laboulaye, Victor

Leclerc, Paulin Paris, Jules Pelletier, Alfred de Wailly, Romain Merlin, Henry Celliez (Paris: C. Lahure, 1861).

94. Guillaume Libri, *Lettre à M. de Falloux, ministre de l'Instruction publique et des cultes, contenant le récit d'une odieuse persécution* (Paris: Paulin, 1849); Lacroix, "Bibliophiles et bibliomanes." See the story of Vrain Lucas in the 1860s and 1870s, a manuscript forger and thief, in Marie-Laure Prevost, "Vrain Lucas, le Balzac du faux," *Revue de la Bibliothèque nationale de France* 13 (2003): 59–69. Medieval forgeries were very much in fashion, as in the case of editions of the supposed fifteenth-century poet Clotilde de Surville; see Laura Kendrick, "The Science of Imposture and the Professionalization of Medieval Occitan Literary Studies," in *Medievalism and the Modernist Temper*, ed. Bloch and Nichols, pp. 95–126.

95. "l'étude des langues, et celle des littératures anciennes," "Il est juste de dire que l'*archéologie*, la *linguistique* et la *philologie* se prêtent un mutuel appui, et que les études archéologiques, linguistiques et philologiques ne peuvent se séparer. Sans la connaissance approfondie des langues anciennes, l'archéologie littéraire (paléographie, épigraphie) est impossible; d'un autre côté, le véritable archéologue ne peut se passer du secours des auteurs anciens pour expliquer les monuments, et à leur tour, les monuments éclaircissent un grand nombre de difficultés, insolubles sans eux, dans les textes des écrivains anciens" (Larousse, "Archéologie," p. 571).

96. "La *philologie*, cette archéologie du langage"; Dr. Cabanès, *L'histoire éclairée par la clinique* (Paris: Albin Michel, 1921), p. 85.

97. "ne doit pas moins que l'art à l'*archéologie*: sans connaissances archéologiques, pas de couleur locale dans les ouvrages dont les sujets sont tirés d'événements anciens" (Larousse, "Archéologie," p. 571).

98. G. Paris, "Paulin Paris," pp. 217, 223. On the position of Gaston Paris as the founder of French medieval studies see two opposing views, R. Howard Bloch and Stephen G. Nichols, eds., *Medievalism and the Modernist Temper* and Ursula Bähler, *Gaston Paris et la philologie romane* (Genève: Droz, 2004); also Ursula Bähler, "Gaston Paris et la philologie romane: Quelques réflexions synthétiques," in *Le Moyen Âge de Gaston Paris. La poésie à l'épreuve de la philologie. Actes du colloque, 27 et 28 mars 2003, Collège de France*, ed. Michel Zink (Paris: Odile Jacob, 2004), pp. 13–40.

99. "nous comprenons aujourd'hui un peu différemment l'étude du moyen âge. Nous nous attachons moins à l'apprécier et à le faire apprécier qu'à le connaître et à le comprendre. Ce que nous cherchons avant tout, c'est de l'histoire. Certes le profit que trouve l'histoire à l'étude des œuvres littéraires du passé préoccupait aussi mon père: il a publié plusieurs textes d'un intérêt surtout historique, et il a insisté à maintes reprises sur les précieux renseignements que les œuvres purement poétiques apportent à l'histoire des mœurs, des institutions, de la civilisation générale" (G. Paris, "Paulin Paris," pp. 219–20).

100. "nous regardons les œuvres poétiques elles-mêmes comme étant avant tout des documents historiques, comme faisant partie de l'histoire prise dans son sens le plus large, comme étant les faits mêmes de l'histoire de la langue, des sentiments et de la pensée" (G. Paris, "Paulin Paris," p. 220).

101. "Les œuvres littéraires, et surtout les œuvres poétiques, sont des produits directs

des âmes, et nous les révèlent, à l'aide du signe transparent des mots, bien plus clairement que ne font les monuments de l'art, plus clairement même et plus complètement que les actes réellement accomplis, dont nous connaissons d'ailleurs un si petit nombre"; G. Paris, "Préface," in *La poésie du Moyen Âge*, vol. 1, pp. v–xiv, at p. xi.

102. Paul Zumthor, *Speaking of the Middle Ages*, trans. Sarah White (Lincoln: University of Nebraska Press, 1986), p. 50 (*Parler du Moyen Âge* [Paris: Éditions de Minuit, 1980], p. 58).

103. "L'histoire de la littérature d'un peuple . . . est l'histoire de sa vie morale, et particulièrement de sa conscience nationale"; G. Paris, "*La Chanson de Roland* et la nationalité française. Leçon d'ouverture au Collège de France, le 8 décembre 1870," in *La poésie du Moyen Âge*, vol. 1, pp. 87–118, at p. 94.

104. "ce sont les époques originales où les nations se sont développées spontanément, sans le trouble qu'apporte trop souvent l'intervention arbitraire des volontés personnelles ou la pression de causes externes"; G. Paris, "La poésie du Moyen Âge," in *La poésie du Moyen Âge*, vol. 1, pp. 1–39, at p. 4. The inadequacy of language as historical document will be criticized first by Paul Zumthor: "Literarity, a modern concept, cannot, without begging the question, found an archaeology"; rather, one needs "to harbor mistrust toward the very materials of these works, toward the language manifested in these texts, refusing to accept it as existing in nature" (Zumthor, pp. 53–54 [pp. 61–62]).

105. As Gumbrecht showed, despite all the political and social changes, unlike Germany, before 1850, "French national identity was never the object of a quest, undertaken between national decadence and national rebirth," which explains "why the discovery of medieval history and literature was never an exclusively academic undetaking" (Gumbrecht, p. 20).

106. Ridoux, pp. 392–408. Cf. the rejection of the philological method in Ferdinand Brunetière, "L'érudition contemporaine et la littérature française au Moyen Âge," *La Revue des deux mondes* 33 (1879): 620–49, at p. 649: "let us not flatter ourselves to acquire these qualities which characterize the German spirit" (ne nous flattons pas d'acquérir ces qualités qui caractérisent l'esprit allemand).

107. Bernard Cerquiglini, *In Praise of the Variant: A Critical History of Philology*, trans. Betsy Wing (Baltimore: Johns Hopkins University Press, 1999), p. 48 (*Éloge de la variante. Histoire critique de la philologie* [Paris: Seuil, 1989], p. 75), translation occasionally modified. This archetype of the "document" parallels the search for the protoype of the "monument," launched by Viollet-le-Duc in 1866: "To restore an edifice is not to maintain it, repair it or remake it, but to reestablish it in a complete state that may never have existed at a given moment" (Restaurer un édifice, ce n'est pas l'entretenir, le réparer ou le refaire, c'est le rétablir dans un état complet qui peut n'avoir jamais existé à un moment donné); Eugène Viollet-le-Duc, "Restauration," *Dictionnaire raisonné de l'architecture française du XIe au XVIe siècle*, vol. 8 (Paris: B. Bance, 1866), pp. 14–34, at p. 14. See also Bruno Foucart, "Viollet-le-Duc et la restauration," in *Les lieux de mémoire*, ed. Nora (Paris: Gallimard/Quarto, 1997 [1986]), vol. 1, pp. 1615–43.

108. "C'est le *stemma codicum*, l'arbre généalogique des manuscrits"; Joseph Bédier,

"La tradition manuscrite du *Lai de l'ombre*. Réflexions sur l'art d'éditer les anciens textes," *Romania* 54 (1928): 160–96, 321–56, at p. 165. See also David F. Hult, "Reading It Right: The Ideology of Text Editing," *Romanic Review* 79:1 (1988): 74–88.

109. This attention to error served as the lance of attack on professional medievalists for Ferdinand Brunetière, who defended the tradition of *belles lettres* against this "evil . . . coming from Germany" (*mal . . . venu d'Allemagne*) (Brunetière, "L'érudition contemporaine," p. 621). He accused medievalists of "pedantry" (*pédantisme*), of "pettiness of scholarship" (*petitesses de l'érudition*), and "the pretension of displacing the center of the history of French literature" (la prétention de déplacer le centre de l'histoire de la littérature française), "since linguistics and philology usurped, in the education of the scholar, this foremost rank that long ago belonged only to the humanities" (depuis que la linguistique et la philologie ont usurpé, dans l'éducation de l'érudit, ce premier rang qui jadis n'appartenait qu'aux seules humanités) (Brunetière, "L'érudition contemporaine," pp. 649, 621, 621–22). They place "at the inception of a language a supposed perfection" (*aux débuts d'une langue une prétendue perfection*), they are "affectionate lawyers of literary barbarity" (*chaleureux avocats de la barbarie littéraire*) (Brunetière, "L'érudition contemporaine," pp. 622, 625). For the same accusation of pedantry, see also Larousse, *Grand dictionnaire universel*, "Philologie," vol. 12 (1874), pp. 823–24, at p. 823. For *La Revue des deux mondes* and Brunetière, see Ridoux, pp. 341–47; Emery.

110. "La littérature du moyen âge . . . est le premier chapitre de nos mémoires de famille" (G. Paris, "Préface," p. xiv).

111. "La littérature est l'expression de la vie nationale" (G. Paris, "*La Chanson de Roland*," p. 99).

112. "la vie de nos pères d'il y a sept ou huit siècles, qui habitaient notre patrie, qui nous ont transmis leur sang, qui parlaient notre langue, et chez lesquels nous trouvons si souvent et notre esprit et notre cœur. Les Français du XIe siècle étaient déjà de vrais Français: ils aimaient la France autant que nous, et s'aimaient entre eux plus peut-être que nous ne faisons; leur génie ressemblait au nôtre, et nous les comprenons pour peu que nous nous en donnions la peine" (G. Paris, "Préface," pp. xiii–xiv).

113. The review of G. Paris's book in the *Grand dictionnaire universel* highlights this national-genealogical aspect: "This volume . . . teaches us to find budding in the literature of the Middle Ages the graces and flaws of our race" (Ce recueil . . . nous apprend à trouver en germe dans la littérature du moyen âge les qualités et les défauts de notre race); Larousse, *Grand dictionnaire universel, Supplément II*, "Moyen Âge (La poésie au)," p. 1617. Note that the title was changed from *La poésie du Moyen Âge*.

114. "Si la philologie érotique pouvait entrer dans une histoire générale de la Prostitution, nous pourrions lui consacrer plusieurs chapitres très-neufs et très-intéressants; car il n'existe pas encore un ouvrage spécial, dans lequel on ait étudié à fond les origines de la langue ou plutôt du jargon des mauvais lieux" (Lacroix, *Histoire*, 5:87–88).

115. "la langue érotique française, qui se dessine déjà très-carrément dès le treizième siècle, procède d'habitude par le pléonasme et la redondance, traduit les mots à son usage dans les langues étrangères, ou se les approprie tels qu'ils sont avec leur consonnance indigène,

recherche les images du style figuré, triomphe dans les équivoques, et obvie sans cesse à la monotonie du discours, au moyen des plus singulières combinaisons philologiques" (Lacroix, *Histoire*, 5:92–93).

116. "évangile de la galanterie française," "*maître-ès-arts* aux écoles de l'argot et de la Prostitution" (Lacroix, *Histoire*, 5:45, 66).

117. "les trouvères, sortis du peuple pour la plupart, conservaient dans leurs œuvres la tache originelle et appliquaient, à des compositions pleines de verve, de gaieté et de malice, la langue crue et grossière qu'ils avaient apprise dans la maison de leurs parents; ils appelaient chaque chose par son nom et ils employaient de préférence l'expression la plus populaire, qui était toujours la plus pittoresque" (Lacroix, *Histoire*, 4:287–88).

118. "c'étaient eux, qui courant le pays, corrompaient à la fois le langage et les mœurs" (Lacroix, *Histoire*, 4:289).

119. "Les trouvères du treizième siècle . . . avaient été les poëtes de la Prostitution" (Lacroix, *Histoire*, 5:41–42).

120. "oublièrent tout à fait les *chansons de geste* et la poésie honnête" (Lacroix, *Histoire*, 4:292).

121. "La langue dut s'assouplir, se déployer, s'enrichir, pour se prêter à ces joyeuses et mordantes exécutions; elle puisa surtout dans le langage familier et même trivial des petites gens de la ville et de la campagne; elle devint plus vivante, plus alerte, plus fine, plus tranchée et plus gaillarde" (Lacroix, *Sciences et lettres*, pp. 389–90).

122. "Le fabliau est ce qu'il y a de plus parfait et de plus scandaleux, hélas! dans le bagage des trouvères et des jongleurs de la langue d'oil. Ce sont parfois des chefs-d'œuvre . . . et ce ne sont pas les moins immoraux" (Lacroix, *Sciences et lettres*, p. 466).

123. "On prouverait aisément, dans une étude philologique sur le jargon de la Prostitution, que ce jargon est contemporain de la langue vulgaire, et qu'il s'est formé d'un mélange confus de tous les idiomes et de tous les dialectes, comme s'il eût la prétention de représenter une langue universelle" (Lacroix, *Histoire*, 5:89).

124. "une foule de mots, qui n'ont pas pris la peine de quitter leur caractère national, et qui se sont faits français, en restant grecs, latins, italiens, allemands ou espagnols," "un langage de convention, qu'on parlait et qu'on entendait également dans les différentes provinces de la France, à une époque où deux villes voisines étaient souvent étrangères l'une à l'autre à cause de leurs patois" (Lacroix, *Histoire*, 5:89–90).

125. Lionel Gossman, *Medievalism and the Ideologies of the Enlightenment: The World and Work of LaCurne de Sainte-Palaye* (Baltimore: Johns Hopkins University Press, 1968), pp. 153–71.

126. "il n'y a aucune partie nécessaire à la perfection d'un Ouvrage de cette nature qui n'ait été rendue dans les Fabliaux de nos anciens Poëtes d'une maniere [*sic*] à servir de modèle"; qtd. in Barbazan, *Fabliaux et contes des poëtes françois des XII, XIII, XIV & XVe siècles, tirés des meilleurs auteurs*, 3 vols. (1756), p. xxxiii. See Kris Peeters, "La découverte littéraire du Fabliau au XVIIIe siècle: le Comte de Caylus dans l'histoire d'un genre médiéval," *Revue d'histoire littéraire de la France* 4:106 (2006): 827–42; Nicholas Cronk and Kris Peeters, eds., *Le Comte de Caylus: Les arts et les lettres. Actes du colloque international*

Université d'Anvers (UFSIA) et Voltaire Foundation, Oxford, 26–27 mai 2000 (Amsterdam: Rodopi, 2004).

127. Harry F. Williams, "French Fabliau Scholarship," *South Atlantic Review* 46:1 (1981): 76–82, at p. 77. Legrand d'Aussy used them (see his vol. 1, p. 64), as did Méon in his 1823 edition of Barbazan (but not in the 1808 edition); Dominique Martin Méon, ed., *Nouveau recueil de fabliaux et contes inédits, des poètes français des XIIe, XIIIe, XIVe et XVe siècles*, 2 vols. (Paris: Chasseriau, 1823), vol. 1, p. v.

128. "les mœurs et les usages de nos pères"; Jean-Baptiste de La Curne de Sainte-Palaye, *Mémoires sur l'ancienne chevalerie*, ed. Charles Nodier, 2 vols. (Paris: Girard, 1826 [1759–1781]), vol. 1, p. i. On La Curne de Sainte-Palaye, see Chapter 6.

129. "C'est surtout par ce tableau si intéressant des mœurs et du costume de leur temps . . . que pourront plaire les fabliaux. . . . Faits par leur nature . . . pour peindre les actions ordinaires de la vie privée, ils montrent la Nation en deshabillé, s'il est permis de parler ainsi. Opinions, préjugés, superstitions, coutumes, ton de conversation, manière de faire l'amour, tout se trouve là; & beaucoup de choses ne se trouvent que là. J'ose même croire que quand on les aura lus, on connaîtra mieux les Français du treizième siècle, que si on lisait toutes nos histoires modernes"; Pierre Jean-Baptiste Legrand d'Aussy, *Fabliaux ou contes, fables et romans du XIIe et du XIIIe siècle*, ed. A. A. Renouard, 5 vols. (Paris: J. Renouard, 1829 [1781]), vol. 1, p. 49.

130. "siècles d'ignorance," "anciens rimeurs"; Geoffrey Wilson, *A Medievalist in the Eighteenth Century: Le Grand d'Aussy and the Fabliaux ou Contes* (The Hague: Nijhoff, 1975), pp. 7, 70.

131. "le commencement de notre poésie"; M. A***, ed., *Recueil de fabliaux, précédé d'une introduction* (Paris: Bureau de la bibliothèque choisie, 1829), p. vii.

132. "trivialité," "des évènements [*sic*] domestiques" (Legrand d'Aussy, 1:61).

133. "accuse[] moins la licence de l'époque que sa naïveté. Le Français d'alors *bravait l'honnêteté dans les mots*; il ignorait les périphrases ou circonlocutions imposées par le rigorisme moderne, et appelait toutes choses par leur nom, sans penser à mal" (Bédollière, 3:337–38).

134. "un chat chez les fabliers est appelé un chat, et rien n'y est nommé que par son nom" (Legrand d'Aussy, 1:50), "on ne se scandalisoit que du mauvais usage que l'on en faisoit [des mots]" (Barbazan, p. xxxix).

135. "on ne peut pas blâmer un Auteur de se servir des termes que cet usage ou que la politesse n'ont point bannis du langage" (Barbazan, p. xliv), "plutôt ceux de leur siècle que les leurs" (Legrand d'Aussy, 1:45).

136. "Quelques étranges que soient les mœurs des fabliaux, il est de mon devoir de les représenter telles qu'elles sont, puisqu'elles peignent leur siècle" (Legrand d'Aussy, 1:52).

137. "négligées & même méprisées, parce que l'on s'est persuadé qu'elles étoient grossieres [*sic*], sans inventions, sans imagination & sans conduite," "On y trouve une heureuse simplicité, des narrés intéressans [*sic*], des images vives, des pensées fines, des réflexions justes, des expressions énergiques, une agréable variété de la conduite & de l'ordonnance" (Barbazan, pp. xxvii, xxxiii).

138. "ni si barbares, ni si obscurs" (Barbazan, p. xlvii), "une manière de narrer simple, claire et naïve," "nulle affectation" (Legrand d'Aussy, 1:47).

139. In reference to *Les amours du bon vieux tems. On n'aime plus comme on aimoit jadis* (1756). The full bibliographic reference is to "'Romance d'Aucassin et de Nicolette,' en prose avec des chansons; et 'La Châtelaine de Saint-Gilles,' en vers, adaptés par Jean-Baptiste de La Curne de Saint-Palaye, d'après Barbier" [The 'Romance of Aucassin and Nicolette' in prose with songs; and 'The Lady of the Manor of Saint-Gilles,' in verse, adapted by Jean-Baptiste of La Curne de Saint-Palaye, according to Barbier]. This will be the standard expression in the nineteenth century to refer to the unadulterated naiveté of the Middle Ages. See the following editions: *Recueil de fabliaux* (Paris: La renaissance du livre, 1910), p. 112: "Gallic humor" (la gauloiserie), but also "true models of narrative art" (de vrais modèles de l'art narratif); Maurice Allem, *Les conteurs français des origines au XXe siècle* (Paris: Librairie des Annales, n.d. [1912?]); M. L. Tarsot, *Fabliaux et contes du Moyen Âge* (Paris: Renouard, 1913), p. iii: "Our good ancestors were people without malice; a bold word, a story a bit daring had nothing to scare them. They laughed about them unpretentiously and without a second thought. Civilization has made us more scrupulous" (Nos bons aïeux étaient gens sans malice; un mot gras, un récit un peu leste n'étaient pas pour les effrayer. Ils en riaient ingénument et sans arrière-pensée. La civilisation nous a rendus plus scrupuleux).

140. "Il est des contes licencieux que je supprimerai en entier" (Legrand d'Aussy, 1:52).

141. "Ce sont des textes non-seulement utiles, mais même nécessaires pour l'histoire de la langue et pour l'histoire littéraire. . . . Nous n'avons pas à les juger"; Anatole de Montaiglon, Gaston Raynaud, eds., *Recueil général et complet des fabliaux des XIIIe et XIVe siècles*, 6 vols. (Paris: Librairie des bibliophiles, 1872–1890), vol. 1, p. x. See also Jubinal, *Nouveau recueil de contes, dits, fabliaux*, vol. 1, p. iii: "to reveal to us the greatest number of local traditions, to give us the key to the greatest number of rare expressions and special usages" (à nous révéler le plus de traditions locales, à nous donner la clef du plus grand nombre de locutions obscures et d'usages singuliers).

142. "graveleux répertoire. Il abonde en saillies piquantes, en inventions drolatiques, en traits d'une gaîté communicative; mais il est souvent d'une dégoûtante obscénité" (Bédollière, 3:316).

143. "barbare, rude comme les mœurs et grossière comme les appétits," "cette impuissance de l'esprit et . . . cette stérilité de la pensée" (Brunetière, "L'érudition contemporaine," pp. 626, 630).

144. "la plupart de nos fabliaux sont littéralement illisibles . . . quel intérêt il pouvait y avoir, non pas même pour l'histoire littéraire, mais pour l'histoire de la langue, à tirer de l'obscurité qui nous les cachait jusqu'ici, ces hideux modèles de la brutalité dans les mœurs, de la grossièreté dans la plaisanterie, de la naïveté dans l'impudeur" (Brunetière, "L'érudition contemporaine," p. 634). See also Brunetière's later article, "Les fabliaux du Moyen Âge et l'origine des contes," *La Revue des deux mondes* 119 (1893): 189–213.

145. "dévergondage complet" (Bédollière, 3:317).

146. "Il est du moins un fait qu'il seroit difficile de nier, c'est que la Chevalerie est née sur le sol de la France. . . . Nous ne connoissons rien dans les souvenirs de la France de plus

essentiellement françois"; Charles Nodier, "Introduction," in La Curne de Sainte-Palaye, *Mémoires sur l'ancienne chevalerie*, vol. 1, pp. i–xxv, at p. xxi.

147. "un rempart contre la violence," "de tout temps la nation avoit été brave" (Legrand d'Aussy, 1:56).

148. "Pour peu qu'on étudie la littérature du moyen âge, on est stupéfait du mouvement intellectuel qui suivit les Croisades. . . . Nos aïeux, dont nous avons trop longtemps denigré la mémoire, nous apparaîtraient entourés d'une splendide auréole; nous verrions le travail fécond, la lumière, la force civilisatrice, là où nous n'avons aperçu que la stérilité, les ténèbres et la barbarie" (Bédollière, 3:312–13).

149. "vinrent les croisades, où les Français parurent aux premiers rangs, car ce n'est pas sans motif que le nom de Franc fut donné à tous ces guerriers aventureux qui allaient combattre pour la cause de la civilisation, en croyant ne suivre que l'étendard de la foi" (Ampère, "De la littérature," p. 26).

150. "la langue nationale de la France s'était répandue dans l'Europe entière" (Lacroix, *Sciences et lettres*, p. 470).

151. "Au moyen-âge, nous avons beaucoup donné et très peu reçu. . . . Nous avons reçu beaucoup de contes de l'Orient, nous comme tous les autres peuples de l'Europe, peut-être plus qu'aucun autre, et en outre c'est très souvent pour [*sic*] nous que la transmission s'est opérée. . . . Nous avons été constamment le véhicule par lequel les contes orientaux, transformés par nous en fabliaux, ont été disséminés dans le reste de l'Europe" (Ampère, "Vue générale," p. 185). In this 1839 article Ampère comes down on the side of the Oriental origin of the fabliaux imported by the crusading French. In his 1833 article he spoke both of oriental legends, "these fables . . . of which France was among the first to gather the heritage" (ces fables . . . dont la France, une des premières, a recueilli l'héritage) and of the fabliaux, "a genre of literature whose origin belongs to us more completely" (un genre de littérature dont l'origine nous appartient plus complètement) (Ampère, "De la littérature," pp. 27, 29). See also Legrand d'Aussy, 1:31; Montaiglon, pp. xvi–xvii: "Many tales, assuredly, all of them, perhaps, come from the Orient, and they are more or less [still] found there; but assuredly also the authors of our Fabliaux did not take them directly from the Orient. . . . What must be the origin of the Fabliaux are the volumes of short stories written in Latin" (Assurément beaucoup de contes, tous les contes peut-être, viennent de l'Orient, et on les y retrouve plus ou moins; mais assurément aussi les auteurs de nos Fabliaux ne les ont pas pris directement à l'Orient. . . . Ce qui doit être l'origine des Fabliaux, ce sont des recueils de petits contes écrits en latin).

152. "Dans cette histoire de l'esprit français, il y a deux choses à noter, l'antériorité et le renom" (Littré, p. 390).

153. "Ainsi se forma peu à peu cette littérature des *fableaux* [*sic*], qui . . . a fini par être le plus vraiment populaire de nos anciens genres poétiques, bien qu'elle ait sa cause et ses racines, à l'extrémité de l'Orient"; Gaston Paris, *Les contes orientaux dans la littérature française du Moyen Âge* (Paris: Librairie A. Franck, 1875), p. 21. G. Paris inherits Legrand d'Aussy's eighteenth-century ideas on the origins of the fabliaux, though he is also

influenced by Theodor Benfey's *Panschatantra, fünf Bücher indischer Fabeln, Märchen und Erzählungen* (1859).

154. "d'approprier les contes orientaux aux mœurs européennes" (G. Paris, *Contes orientaux*, p. 21).

155. "quand on aborde cette littérature avec les idées courantes sur la galanterie délicate et passionnée et le culte de la femme qu'on attribue aux temps chevalresques" (G. Paris, *Contes orientaux*, pp. 21–22).

156. "Quant aux contes innombrables, presque toujours plaisants, trop souvent grossiers, qui ont pour sujet les ruses et les perfidies des femmes, ils ne sont pas nés spontanément de la société du moyen âge: ils proviennent de l'Inde et ils ont leur raison d'être dans le milieu qui les a produits" (G. Paris, *Contes orientaux*, p. 23).

157. "Nos épopées chevaleresques, provençales et françaises, ont été le type des épopées chevalresques de l'Angleterre et de l'Allemagne, qui n'en sont en général que des traductions, tout au plus des reproductions un peu modifiées; et il en a été ainsi non-seulement pour notre héros national, Charlemagne, mais même pour des héros qui ne nous appartiennent pas par droit de naissance, comme Arthur ou Tristan" (Ampère, "Vue générale," pp. 186–87). On the Provençal origins of chivalric epic in Fauriel and Ampère versus the argument for Celtic origins in Guizot, P. Paris, and E. Renan, and also on the hierarchical relationship of the Carolingian *chanson de geste* and the Arthurian chivalric romances, central to the work of G. Paris's father, Paulin Paris, see Michael Glencross, *Reconstructing Camelot: French Romantic Medievalism and the Arthurian Tradition* (Cambridge: D. S. Brewer, 1995).

158. "le plus ancien de nos romans français" (Lacroix, *Sciences et lettres*, p. 412). See G. Paris, "*La Chanson de Roland*," pp. 117–18; Luke Sunderland, "Entre résistance et révolte: les enjeux du territoire," *Revue des langues romanes* 112 (2008): 465–90.

159. "Car dès lors . . . la littérature de la France était la reine et l'initiatrice des littératures voisines; aucune œuvre de notre époque classique n'a été traduite en plus de langues, n'a exercé autour de nous une influence plus étendue et plus durable" (G. Paris, "*La Chanson de Roland*," pp. 117–18).

160. "quand nous revenons écouter ce langage naïf . . . il nous semble entendre le bégaiement de l'enfance. . . . Cet enfant qui est déjà le grand peuple français parle aussi la grande langue française" (G. Paris, "*La Chanson de Roland*," p. 117).

161. Gustave Lanson, *Histoire de la littérature française*, 5th ed. (Paris: Hachette, 1898 [1895]), pp. 82–88; Achille Luchaire, *Les premiers Capétiens (987–1137)*, in Ernest Lavisse, ed., *Histoire de France depuis les origines jusqu'à la Révolution*, vol. 2, part 2 (Paris: Hachette, 1911), p. 378.

162. Joseph Bédier, *Les fabliaux: Études de littérature populaire et d'histoire littéraire du Moyen Âge* (Paris: E. Bouillon, 1893). It is fascinating that Bédier, who comes from the French colony La Réunion, refuted the Oriental origin of the fabliaux and established them as universal, popular narratives, well grounded in the European continent and metropole. Bédier's view of the origin of the *chanson de geste* as collective oral narratives along pilgrimage routes is very similar to the degree in which it displaces the national question toward universal, albeit European origins. On Bédier, see Michelle R. Warren, *Creole Medievalism:*

Colonial France and Joseph Bédier's Middle Ages (Minneapolis: University of Minnesota Press, 2011).

163. "l'une [forme] qui est centrée sur les nationalités, et qui va se trouver essentiellement en continuité, d'une part, avec les phénomènes de la langue, et par conséquent avec la philologie," "Philologie, économie politique, biologie. Parler, travailler, vivre"; Michel Foucault, *"Il faut défendre la société," Cours au Collège de France (1975–1976)* (Paris: Gallimard; Seuil, 1997), p. 170.

164. For a first such groundbreaking study, see James A. Schultz's *Courtly Love, the Love of Courtliness, and the History of Sexuality* (Chicago: University of Chicago Press, 2006).

CHAPTER 6

1. "religion de l'amour, arrivée à son exaltation suprême"; Henri Martin, *Histoire de France depuis 1789 jusqu'à nos jours*, 17 vols. (Paris: Furne, 1855–1860 [1838–1853]), vol. 3, p. 383.

2. "une théorie très-répandue et très-funeste à la fidélité conjugale"; Émile de la Bédollière, *Histoire des mœurs et de la vie privée des Français. Usages, coutumes, institutions, physionomie de chaque époque, depuis l'origine de la monarchie jusqu'à nos jours*, 3 vols. (Paris: Victor Lecou, 1847–1849), vol. 3, pp. 324, 335. See also "the law of love" (*la loi de l'amour*) (Martin, 3:383).

3. "De là les adultères et les vengeances sanglantes" (Martin, 3:386).

4. Gaston Paris, "Études sur les romans de la table ronde: *Lancelot du Lac*," *Romania* 12 (1883): 459–534; see also Gaston Paris, "Études sur les romans de la table ronde," *Romania* 10 (1881): 465–96, esp. pp. 468–69. Depending on the interpretation, the year could have been 1881. In 1881, G. Paris speaks of the concept for the first time, though he does not name it; see Henry Ansgar Kelly, "The Varieties of Love in Medieval Literature," *Romance Philology* 40:3 (1987): 301–27; David F. Hult, "Gaston Paris and the Invention of Courtly Love," in *Medievalism and the Modernist Temper*, ed. R. Howard Bloch and Stephen G. Nichols (Baltimore: Johns Hopkins University Press, 1996), pp. 193–224; Ji-hyun Philippa Kim, "For a Modern Medieval Literature: Gaston Paris, *Courtly Love,* and the Demands of Modernity" (Ph.D. diss., Harvard University, 2005).

5. "de jolis pages, d'écuyers magnifiques, de fiers chevaliers et de troubadours langoureux, jongleurs, chanteurs et ménestrels," "animés des sentiments les plus chevaleresques," "des *cours d'amour* où l'on professait des principes à faire dresser les cheveux sur la tête d'une honnête femme moderne, sous prétexte d''amour épuré'"; Dr. Th.-F. Debray, *Histoire de la prostitution et de la débauche chez tous les peuples du globe depuis l'antiquité jusqu'à nos jours* (Paris: S. Lambert, 1879), p. 230.

6. Jean-Baptiste de La Curne de Sainte-Palaye, *Mémoires sur l'ancienne chevalerie*, ed. Charles Nodier, 2 vols. (Paris: Girard, 1826 [1759–1781]). La Curne de Sainte-Palaye saw chivalry "as a creation of the early monarchy, rather than an instrument of government by the nobility"; see Lionel Gossman, *Medievalism and the Ideologies of the Enlightenment: The*

World and Work of LaCurne de Sainte-Palaye (Baltimore: Johns Hopkins University Press, 1968), p. 282. Cf. Voltaire, "De la chevalerie," in *Essai sur les mœurs et l'esprit des nations, et sur les principaux faits de l'histoire, depuis Charlemagne jusqu'à Louis XIII*, in *Œuvres*, ed. M. Beuchot, vols. 15–18 (Paris: Werdet et Lequien, 1829 [1756–1763]), vol. 17, pp. 1–6, at p. 1: "This chivalry was a military establishment which had created itself among the lords" (Cette chevalerie était un établissement guerrier qui s'était fait de lui-même parmi les seigneurs).

7. "*courtoisie*, c'est-à-dire, civilité ou honnêteté," "une douceur, une modestie, une politesse que le nom de courtoisie exprimoit parfaitement, et dont on ne trouve dans aucunes autres lois des préceptes aussi formels que dans celles de la Chevalerie" (Sainte-Palaye, 1:69, 67).

8. "le gouvernement françois sortit du chaos où l'avoient plongé les troubles qui suivirent l'extinction de la seconde race de nos rois [les Carolingiens]. Déjà l'autorité royale commençoit à se faire respecter; tout reprenoit une nouvelle face; les loix se formèrent; les communes et les bourgeoisies furent institutées; les fiefs acquirent une forme et une discipline plus régulières" (Sainte-Palaye, 1:58).

9. "Un juge de la dispute qui répondoit à ce qu'on appeloit *prince d'amour*, ou *prince de Puy dans les cours d'amour*, juridictions établies dans quelques contrées, pour connoître de ces importantes matières, un juge, dis-je, prononçoit des sentences presque toujours équivoques, obscures et souvent énigmatiques, auxquelles les parties se soumettoient avec une respectueuse docilité" (Sainte-Palaye, 1:348–49).

10. "un idiôme [*sic*] intermédiaire qui, par la décomposition de la langue des Romains, et l'établissement d'un nouveau système grammatical, a fourni le type commun d'après lequel se sont successivement modifiés les divers idiômes de l'Europe latine"; Just Marie François Raynouard, *Choix des poésies originales des troubadours*, 6 vols. (Paris: F. Didot, 1816–1821), vol. 2 (1817), p. cxxxvii.

11. "une nouvelle littérature," "la sévère franchise de leurs opinions morales et politiques, leur enthousiasme pour les exploits honorables et pour les illustres personnages qui les exécutaient, leur juste et courageuse indignation contre les erreurs et les fautes de leurs contemporains" (Raynouard, 2:i).

12. "institution remarquable," "l'un des points les plus intéressants de l'histoire des mœurs, des usages et de l'esprit du moyen âge" (Raynouard, 2:lxxviii).

13. Raynouard was subsequently criticized and marginalized by Gaston Paris, Paul Meyer, and the Romance philologists, especially for his rendering of Romance linguistics and the historical interpretation of the courts of love; see Charles Ridoux, *Évolution des études médiévales en France de 1860 à 1914* (Paris: Honoré Champion, 2001), pp. 21–24, 704, 936.

14. "n'a pas été l'ouvrage du législateur, mais l'effet de la civilisation, des mœurs, des usages, et des préjugés de la chevalerie" (Raynouard, 2:lxxx).

15. "épurait et ennoblissait, au profit de la civilisation, des mœurs, de l'enthousiasme chevaleresque, ce sentiment impétueux et tendre" (Raynouard, 2:lxxix).

16. "Un des traits qui ont caractérisé davantage le François dans les temps de la Chevalerie, et dont il conserve encore des traces qui le distinguent de toute autre nation" (Sainte-Palaye, 1:ii).

17. See Noémi Hepp, "Gallantry," in *Rethinking France: Les Lieux de Mémoire, Volume 3: Legacies*, ed. Pierre Nora, trans. Deke Dunisberre (Chicago: University of Chicago Press, 2009), pp. 233–73 ("Galanterie," *Les lieux de mémoire*, ed. Pierre Nora, 3 vols. [Paris: Gallimard/Quarto, 1997 (1986)], vol. 3, pp. 3677–710). See also Jean Frappier, "'Amour courtois,'" in *Amour Courtois et Table Ronde* (Genève: Droz, 1973), pp. 33–41.

18. "l'âge d'or de la galanterie" (Sainte-Palaye, 1:349; and elsewhere).

19. "les usages galants de la chevalerie," "mœurs chevaleresques," "esprit chevaleresque," "passion chevaleresque," "humeur sévère, chevaleresque et galante des troubadours" (Raynouard, 2:lxxxv, ii, iii, v, 215).

20. "le code d'amour," "code amoureux" (Raynouard, 2:ciii, ci; and elsewhere).

21. Quinet contrasted it to "an ecstasy known by the name of mystic love" (une extase connue sous le nom d'amour mystique), "the thought of God [which] alone remained for oppressed peoples" (la pensée de Dieu [qui] restait seule aux peuples opprimés); Edgar Quinet, "Études sur le Moyen Âge," in *La pensée d'Edgar Quinet: Étude sur la formation de ses idées avec essais de jeunesse et documents inédits*, ed. Willy Aeschimann (Paris: Éditions Anthropos; Genève: Georg, 1986), pp. 469–509, at p. 478.

22. Claude Fauriel, *Histoire de la poésie provençale*, 3 vols. (Genève: Slatkine, 1969 [1846]), vol. 1, pp. 496, 497, and elsewhere. Fauriel believed that chivalric love and Arthurian and Carolingian romances all originated in southern France in Provençal poetry, "the intimate union of chivalry and poetry, the poetic and chivalric spirit" (l'union intime de la chevalerie et de la poésie, de l'esprit poétique et de l'esprit chevaleresque) (1:529; also 1:478). On Fauriel, see Michael Glencross, *Reconstructing Camelot: French Romantic Medievalism and the Arthurian Tradition* (Cambridge: D. S. Brewer, 1995); Simone Bernard-Griffiths, Pierre Glaudes, and Bertrand Vibert, eds., *La fabrique du Moyen Âge au XIXe siècle: Représentations du Moyen Âge dans la culture et la littérature françaises du XIXe siècle* (Paris: Champion, 2006), pp. 205–11, 227–29.

23. "Les préceptes que donne [maître André] sont de ceux qui portent encore de nos jours la qualification de chevaleresques, et pour qu'ils ne fussent pas enfreints impunément, des tribunaux spéciaux, les *cours d'amour*, en surveillaient l'application" (Bédollière, 3:328).

24. "l'idéal du pur amour," "C'est une révolution générale et simultanée [des deux côtés de la Loire] que celle qui introduit, dans les usages comme dans les idées de la société du moyen âge, un élément nouveau, et presque aussitôt dominateur, à côté de l'élément purement guerrier et de l'élément religieux" (Martin, 3:386, 376).

25. "La chevalerie ne se contente pas d'une morale enseignée par la poésie, et propagée par l'opinion: elle crée une institution qui concentre la force de l'opinion et qui donne une sanction à cette morale, aussi différente de l'enseignement ecclésiastique que des maximes féodales" (Martin, 3:380).

26. "Il est du moins un fait qu'il seroit difficile de nier, c'est que la Chevalerie est née sur le sol de la France. . . . Nous ne connoissons rien dans les souvenirs de la France de plus essentiellement françois"; Charles Nodier, "Introduction," in La Curne de Sainte-Palaye, *Mémoires sur l'ancienne chevalerie*, vol. 1, pp. i–xxv, at p. xxi. In contrast, see Mme de Staël's views on chivalry as a European phenomenon: "chivalry, which was the warrior religion of

Europe" (la chevalerie, qui fut la religion guerrière de l'Europe); Mme de Staël, chapter 4, "De l'influence de l'esprit de chevalerie sur l'amour et l'honneur" in *De l'Allemagne* (Paris: Garnier Flammarion, 1968 [1813]), pp. 69–73, at p. 71. In the same year, a member of Mme de Staël's *groupe de Coppet*, Simonde de Sismondi, argued that chivalry was a fictional ideal and a moral agent; see Jean Charles Léonard Simonde de Sismondi, *De la littérature du Midi de l'Europe* (Paris: Treuttel et Würtz, 1813).

27. This was an eighteenth-century dispute represented in the views of Henri de Boulainvilliers's monarchist Frankish theory, abbé du Bos's pro-third estate Romanist version of French history, and Montesquieu's *Esprit des lois*: after the Germanic invasions and the fall of the Roman Empire, did the Franks melt into the population under the influence of the Gallo-Romans, or did the Franks colonize and subjugate the Gallo-Romans? The theory of the three races was reignited with Augustin Thierry's *Histoire de la conquête d'Angleterre par les Normands* (1825) and then again in the *Récits des temps mérovingiens* (1840 [1833–1837]), where Thierry defended the vanquished Gauls. On Augustin Thierry, see *La fabrique du Moyen Âge*, ed. Bernard-Griffiths, Glaudes, and Vibert, pp. 347–61. On the theory of the three races, see Michel Foucault, *"Il faut défendre la société," Cours au Collège de France, 1975* (Paris: Gallimard/Seuil, 1997); Krzysztof Pomian, "Franks and Gauls," in *Realms of Memory: Rethinking the French Past, Volume 1: Conflicts and Divisions*, ed. Pierre Nora, trans. Arthur Goldhammer (New York: Columbia University Press, 1992), pp. 25–76 ("Francs et Gaulois," in *Les lieux de mémoire*, ed. Nora, vol. 2, pp. 2245–300).

28. "une sorte de patriotisme européen," "elle s'est embellie par le charme de la poésie et de l'amour" (Mme de Staël, pp. 70, 69).

29. "fureur des généalogies" (Ch. Nodier, qtd. in Glencross, p. 19).

30. "la morale . . . basée sur le caractère national des François" (Nodier, p. xvi), "une des grandes phases du développement de l'âme humaine" (Martin 3:389), "un champ ouvert à tous ces sentiments, à tous ces principes de civilisation" (Fauriel 1:484).

31. "l'une des institutions les plus originales, . . . les plus franchement civilisatrices de la vie au Moyen-Âge"; Antony Méray, *La vie au temps des cours d'amour* (Paris: Claudin, 1876), p. 107.

32. "C'est d'elle [la poésie chevaleresque] que provient tout ce qui subsiste de délicatesse morale dans le monde moderne" (Martin, 3:391).

33. "il fallait, pour exciter à de bonnes actions sociales, un mobile plus immédiat, plus sensible, et pour tout dire, moins relevé que celui de la religion. Ce mobile, la chevalerie le trouve dans l'amour" (Fauriel, 1:496).

34. "Une autre marque que ces mœurs étaient fort avancées sur la route de la véritable civilisation, c'est qu'à peine sorties des horreurs du Moyen Âge, et de la féodalité où la force était tout, nous voyons le sexe le plus faible moins tyrannisé qu'il ne l'est *légalement* aujourd'hui"; Stendhal, *De l'amour* (Paris: Gallimard/Folio, 1980 [1822]), p. 186. It is important to note that Stendhal, who did not write about chivalry but only about love, considered chivalric love a purely Provençal phenomenon that was eradicted during the Albigensian crusade by the French north: "A wild rage animated them against all that carried any trace of civilization. . . . All was finished for them [the Provençals]: no more

love, no more gaity, no more poetry, less than twenty years after the Conquest (1235), they were almost as barbarian and uncouth as the French, as our fathers" (une rage sauvage les animait contre tout ce qui portait quelque trace de civilisation. . . . Tout fut fini pour ceux-ci [les Provençaux]: plus d'amour, plus de gaieté, plus de poésie; moins de vingt ans après la conquête [1235], ils étaient presque aussi barbares et aussi grossiers que les Français, que nos pères) (Stendhal, p. 190). In this, Stendhal differed from most authors who thought that chivalry was born in the North or that it was imported in the North from the South.

35. "des seigneurs galants et magnifiques, dont l'urbanité supérieure fut unanimement reconnue par les nations européennes" (Bédollière, 3:164).

36. "nous verrions le travail fécond, la lumière, la force civilisatrice, là où nous n'avons aperçu que la stérilité, les ténèbres et la barbarie" (Bédollière, 3:312–13).

37. "en lisant tout ce que nous avons dit à l'honneur de la Chevalerie, on se sera rappelé que les siècles dans lesquels elle étoit la plus florissante, furent des siècles de débauche, de brigandage, de barbarie et d'horreur; et que souvent tous les vices et tous les crimes se trouvoient réunis dans les mêmes chevaliers qu'alors on érigeoit en héros. . . . Néanmoins ces deux vérités, si contraires en apparence, sont également constatées" (Sainte-Palaye, 1:337–38). On the difficulty of reconciling the relations between chivalry and feudalism for the liberal view of the Middle Ages, see Glencross, pp. 1–26.

38. "Le régime féodal, cette institution politique triste et sévère, mais qui consolidait, à quelques égards, l'esprit de la chevalerie en le transformant en lois" (Mme de Staël, p. 70).

39. "quel contraste entre cette couronne de vertus que la chevalerie veut poser sur son front, et la brutalité, l'avidité, la tyrannie, la versatilité déloyale que le monde féodal a héritées des barbares" (Martin, 3:383).

40. "un rempart contre la violence," "les malheurs de l'état et l'anarchie du gouvernement"; Pierre Jean-Baptiste Legrand d'Aussy, *Fabliaux ou contes, fables et romans du XIIe et du XIIIe siècle*, ed. A. A. Renouard, 5 vols. (Paris: J. Renouard, 1829 [1781]), vol. 1, p. 56.

41. "Elle ne pouvoit vivre que dans l'état social où elle étoit née. La confusion des pouvoirs, l'absence de la justice..., l'inflexibilité des coutumes féodales" (Nodier, p. xxii).

42. "Notre chevalerie" (Sainte-Palaye, 1:337).

43. "la discipline primitive de l'ancienne Chevalerie étoit tombée dans le relâchement dès ce temps-là," "Les plus sages réglements [*sic*] ne furent pas capables d'arrêter les progrès de la corruption" (Sainte-Palaye, 1:71).

44. "Les divisions, survenues entre les princes du sang royal pendant les accès de la maladie du roi Charles VI, [qui] causèrent dans toutes les parties du gouvernement une infinité de désordres; et ceux qui s'introduisirent dans la Chevalerie ne furent pas les moins pernicieux" (Sainte-Palaye, 1:359).

45. "l'ignorance et la barbarie . . . principalement depuis qu'ils eurent abandonné les glorieuses fonctions de la justice," "l'oisiveté," "la lecture des ouvrages de leurs trouvères et jongleurs, gens grossiers et libertins" (Sainte-Palaye, 3:368, 369).

46. "Aussi devons-nous ce qu'il y a de noble dans nos mœurs à ces croisades et aux Maures d'Espagne" (Stendhal, p. 198).

47. "la plus execrable dépravation," "la plus honteuse de toutes les prostitutions," "Un horrible libertinage s'était glissé dans toutes les classes de la société depuis les croisades, et le vice contre nature, que le séjour des Français en Palestine avait acclimaté en France, menaçait encore, en dépit de la chevalerie, d'infecter les mœurs et de corrompre la population toute entière"; Pierre Dufour (Paul Lacroix), *Histoire de la prostitution, chez tous les peuples du monde depuis l'antiquité la plus reculée jusqu'à nos jours* [History of prostitution among all of the peoples of the world from the earliest antiquity to the current day], 6 vols. (Paris: Seré; Martinon, 1851–1853), vol. 4, pp. 59, 64, 58. Sainte-Palaye was one of Lacroix's main sources (Lacroix, 4:272).

48. "au même degré de décadence que la chevalerie, et la société tout entière semblait aller à sa dissolution," "l'on ne saurait dire si c'était l'Église qui démoralisait la chevalerie, ou la chevalerie qui démoralisait l'Église" (Lacroix, 4:314, 315).

49. "la prostitution du langage littéraire" (Lacroix, 5:53), "degré de perversité auquel étaient parvenus quelques nobles, quelques grands seigneurs, qui s'abandonnaient à toutes les aberrations de la débauche" (Lacroix, 4:324).

50. "si quelque chose peut aider à concevoir les prétentions exagérées, les raffinements, les subtilités de cet amour, ce sont les motifs intéressés et précaires du mariage féodal" (Fauriel, 1:499).

51. "sa nécessité la plus directe et la plus grossière" (Fauriel, 1:504).

52. "les idées chevaleresques attribuaient aux femmes une véritable suprématie sur les hommes," "Son service était un culte" (Fauriel, 1:513). Stendhal believed that "love-passion . . . was founded on justice and the equality of rights between the two sexes" (amour-passion . . . était fondé sur la justice et sur l'égalité de droits entre les deux sexes) (Stendhal, p. 189).

53. "un piédestal de tendre admiration et un trône d'honneur," "C'est là évidemment le principe de la chevalerie: elle affranchit un sexe que la Prostitution avait soumis à la plus dégradante servitude. Ici, la femme était esclave et humiliée de son rôle indigne; là, elle est reine, et sa souveraineté repose encore sur l'amour; mais ce n'est plus l'amour charnel, dont les coupables jouissances étouffent l'instinct du bien et prédisposent le cœur à tous les vices; c'est l'amour parfait, c'est l'amour héroïque, qui prend sa source dans les plus beaux sentimens [*sic*] et qui s'exalte par l'imagination en se dégageant des entraves de la nature physique" (Lacroix, 4:271).

54. "C'était pour se faire distinguer par elle et pour se faire aimer aussi, qu'il se montrait preux et vaillant, honnête et courtois, loyal et vertueux. Le nom et les couleurs de cette dame lui tenaient lieu de talisman dans les circonstances les plus difficiles de sa vie. . . . Rien ne ressemblait moins à l'amour matériel, que cette profonde et délicate dévotion amoureuse à l'égard d'une seule dame, qui souvent ne recompensait pas même d'un chaste baiser un sentiment si exalté; mais ce sentiment, pur et ardent à la fois, trouvait en soi une force invincible qui s'augmentait sans cesse par l'idée fixe et par l'extase" (Lacroix, 4:276–77). Already in the 1830s Lacroix was concerned with the status of women: "divorce, this crucial corrective to marriage, this first declaration of sexual equality" (divorce, cet indispensable correctif du mariage, cette première déclaration de l'égalité sexuelle), "I have

no fear to avow my conviction: marriage without divorce is absurd and immoral" (je ne crains pas de vous avouer ma conviction: le mariage sans le divorce est absurde et immoral); P.-L. Jacob (Paul Lacroix), *Un divorce. Histoire du temps de l'Empire, 1812–1814* (Paris: E. Renduel, 1832), pp. 14–15. Marriage without divorce is immoral for Lacroix because it foments adultery and prostitution.

55. "l'incompatibilité de l'amour et du mariage" (Fauriel, 1:506).

56. "On eût dit que les amants prenaient plaisir à surexciter leurs desirs, afin de prouver jusqu'à quel point ils pouvaient les combattre ensuite et les vaincre; en cherchant le péril et en s'y exposant avec une sorte d'orgueil, on peut croire qu'ils y succombaient quelquefois" (Lacroix, 4:278).

57. "Cet amour presque mystique," "l'amour métaphysique" (Lacroix, 4:278, 271).

58. "métaphysique d'amour, ce vaste champ où s'exerçoient les plus beaux esprits qui brilloient parmi nos respectueux serviteurs des dames" (Sainte-Palaye, 1:350).

59. "l'essence et le caractère du parfait et véritable amour," "fausses subtilités," "Un vain cérémonial de révérences, de génuflexions, de prosternations jusqu'à terre" (Sainte-Palaye, 1:348, 352).

60. "l'innocence prétendue de leurs amusements avec les dames et les demoiselles, de leurs conversations, des récits," "Jamais on ne vit les mœurs plus corrompues que du temps de nos chevaliers, et jamais le règne de la débauche ne fut plus universel" (Sainte-Palaye, 1:347, 351).

61. "La noblesse française eut seulement l'avantage d'un extérieur plus brillant que les autres nations" (Voltaire, 16:437).

62. "grossières et avilissantes tyrannies de l'amour matériel," "Ce fut sous le règne de Charles VI que commença cette décadence des mœurs chevaleresques" (Lacroix, 4:271, 313).

63. "amour exalté et presque mystique" (G. Paris, "*Lancelot du Lac*," p. 519).

64. I do not take up the issue of whether "courtly love" is or is not a medieval term, whether we should use it or not, and whether it applies to only the Arthurian romances or the whole romance corpus. The exhaustive bibliography of this quarrel is too long to list; however, we should note: F. X. Newman, ed., *The Meaning of Courtly Love* (Albany: State University of New York Press, 1968); Jean Frappier, "Sur un procès fait à l'amour courtois," *Romania* 93 (1972): 145–93, reprinted in Frappier, *Amour Courtois et Table Ronde*, pp. 61–96; Roger Boase, *The Origin and Meaning of Courtly Love: A Critical Study of European Scholarship* (Manchester: Manchester University Press, 1977); Joan M. Ferrante, "*Cortes'Amor* in Medieval Texts," *Speculum* 55:4 (1980): 686–95; Henry Ansgar Kelly, "Gaston Paris's Courteous and Horsely Love," in *The Spirit of the Court: Selected Proceedings of the Fourth Congress of the International Courtly Literature Society (Toronto 1983)*, ed. Glyn S. Burgess and Robert A. Taylor (Woodbridge, Suffolk: D. S. Brewer, 1985), pp. 217–23; Kelly, "The Varieties of Love"; Hult, "Gaston Paris and the Invention of Courtly Love"; E. Jane Burns, "Courtly Love: Who Needs It? Recent Feminist Work in the Medieval French Tradition," *Signs* 27:1 (2001): 23–57.

65. "illégitime, furtif," "la tendresse qu'il souhaite ou qu'il a déjà obtenue," "l'amour est un art, une science, une vertu, qui a ses règles tout comme la chevalerie ou la courtoisie,

règles qu'on possède et qu'on applique mieux à mesure qu'on a fait plus de progrès, et auxquelles on ne doit pas manquer sous peine d'être jugé indigne" (G. Paris, "*Lancelot du Lac*," pp. 518–19).

66. "les règles d'un art raffiné," "une science, . . . un code" (G. Paris, "*Lancelot du Lac*," pp. 518, 519).

67. "tendresse," "nullement platonique toutefois, et fondé sur la pleine possession, mais ne laissant aux sens qu'une part secondaire" (G. Paris, "*Lancelot du Lac*," pp. 529–30).

68. "la création de la société courtoise par l'établissement dans l'aristocratie . . . des règles d'une étiquette subtile et souvent bizarre dont l'observation rigoureuse était une science" (G. Paris, "*Lancelot du Lac*," p. 520).

69. "un effort pour créer et faire accepter aux hommes un amour idéal et raffiné . . . étroitement lié à la pratique et l'accroissement des vertus sociales" (G. Paris, "*Lancelot du Lac*," pp. 529–30).

70. "La forme des jugements et des arrêts était la même que dans les tribunaux de justice royale et seigneuriale" (Lacroix, 4:280).

71. "des jeux d'esprit, . . . des amusements de société," "est inconnu au moyen âge"; Gaston Paris, "Les cours d'amour du Moyen Âge," in *Mélanges de littérature française du Moyen Âge*, ed. Mario Roques (Paris: Honoré Champion, 1966 [1888]), pp. 473–97, at p. 494; also G. Paris, "*Lancelot du Lac*," p. 529.

72. The search for sociohistoric reasons of these courtly rituals will be revived by Georges Duby, who will attribute it to the management of erotic desires of young knights (*les puinés, juvenes*), pushed out of inheritance by the rule of primogeniture: "What, therefore, the love songs of the second half of the twelfth century suggest is a new kind of erotic relationship, better suited to the position of the *juvenes*'—that husbands should no longer pay court to ladies, and that they should no longer prevent their wives from receiving 'youths' and accepting their services of love," thus forming the erotic triangle "husband-lady-young courtly servant"; Georges Duby, "Youth in Aristocratic Society: Northwestern France in the Twelfth Century," in *The Chivalrous Society*, trans. Cynthia Postan (Berkeley: University of California Press, 1977), pp. 112–22, at p. 122.

73. Francis Ronsin, *Les divorciaires: Affrontements politiques et conceptions du mariage dans la France du XIXe siècle* (Paris: Aubier, 1992); Alfred Naquet, *Le divorce*, 2nd ed. (Paris: E. Dentu, 1881 [1877]). See also Alfred Naquet, *La loi du divorce* (1903) and *Vers l'union libre* (1908).

74. See Theresa McBride, "Divorce and the Republican Family," in *Gender and the Politics of Social Reform in France, 1870–1914*, ed. Elinor A. Accampo, Rachel G. Fuchs, and Mary Lynn Stewart (Baltimore: Johns Hopkins University Press, 1995), pp. 59–81; Jean Elisabeth Pedersen, *Legislating the French Family: Feminism, Theater, and Republican Politics, 1870–1920* (New Brunswick, N.J.: Rutgers University Press, 2003), pp. 13–41.

75. "le seul grand argument des législateurs de 1816, c'était l'argument religieux"; Alfred Naquet, *Le divorce, Chambre des Députés (séance du 27 mai 1879), Extrait du Journal officiel* (Paris: Bureaux de l'École laïque, 1879), p. 28. See also Francis Ronsin, *Le contrat sentimental: Débats sur le mariage, l'amour, le divorce, de l'Ancien Régime à la Restauration* (Paris: Aubier, 1990).

76. See Johann Jakob Bachofen, *Das Mutterrecht* (1861); Henry Sumner Maine, *Ancient Law* (1861); Numa Denis Fustel de Coulanges, *La cité antique* (1864); John Ferguson McLennan, *Primitive Marriage* (1865); Henry Lewis Morgan, *System of Consanguinity and Affinity of the Human Family* (1871) and *Ancient Society* (1881). See also Alfred Naquet, *Religion, propriété, famille* (1869); Alexis Giraud-Teulon, *La mère chez certains peuples de l'antiquité* (1867), *Les Origines de la famille. Questions sur les antécédents des sociétés patriarcales* (1874), and *Les origines du mariage et de la famille* (1884); Charles Letourneau, *L'évolution du mariage et de la famille* (1888) and *L'évolution de la propriété* (1889).

77. "théorie de l'adultère" (Ronsin, *Les divorciaires*, p. 72).

78. Laure Adler, *Secrets d'alcôve: Histoire du couple de 1830 à 1930* (Paris: Hachette, 1983), p. 166.

79. "Qu'il soit commis par l'homme ou par la femme, l'adultère est toujours un crime énorme dans ses conséquences, qu'une bonne législation devrait sinon effacer, du moins rendre de plus en plus rare en moralisant le mariage par le divorce"; Auguste Debay, *Philosophie du mariage. Histoire de l'homme et de la femme mariés, dans leurs rapports physiques et moraux* (Paris: Moquet, 1849), p. 142.

80. "Il est évident que dans la société actuelle, le divorce, dans une multitude de cas, est devenu pour ainsi dire indispensable, et qu'il vaut mieux rétablir le divorce dans la loi que d'être forcé d'admettre le meurtre dans les mœurs" (Dumas *fils*, qtd. in Ronsin, *Les divorciaires*, p. 170).

81. "Mieux vaut le divorce que la fornication, que la prostitution, que l'adultère, qui sont devenus l'élément le plus dissolvant du mariage et de la famille; mieux vaut le divorce que la séparation de corps qui est un mensonge et qui est devenue la plus radicale expression de la liberté du célibat dans le mariage"; Armand Hayem, *Le mariage* (Paris: Didier, 1872), p. 276. The 1876 reprinting of this work calls attention to the fact that it received honorable mention from the Académie des sciences morales et politiques.

82. "des ennemis de la famille"; Victor Poupin, "Préface," in Armand Hayem, *Le mariage*, 2 vols. (Paris: Librairie de la bibliothèque démocratique, 1876), pp. 6–7, at p. 6.

83. "Le rétablissement du divorce, loin de nuire au mariage, le moraliserait, au contraire, et diminuerait le nombre si considérable des célibataires. Aujourd'hui, on le sait, la plupart des mariages sont un commerce de supercherie; on prend un masque avant de se marier, on déguise ses défauts, on fait taire, pour quelques jours, ses vices, et on leur lâche la bride après que le serment indissoluble a été prononcé. Le divorce mettrait fin à ce honteux artifice" (Debay, p. 147).

84. "En demandant le divorce, réduit à des cas justement limités, nous demandons un encouragement au mariage, une contre-assurance de bonheur conjugal: nous plaidons surtout la cause du plus faible, de la femme dont le divorce deviendrait la plus ferme garantie, la plus sure protection, le plus honorable abri" (Hayem, p. 281).

85. "Depuis l'institution du divorce et son annexion à notre code, le mariage a été mis à la portée de tout le monde, il n'y a pas de mérite à s'y soustraire"; Dr. Henry Deville, *L'amour dans le mariage, Le médecin populaire*, vol. 2 (Paris: L. Boulanger, n.d. [1893]), p. 26.

86. "la durée et [le] bonheur des sociétés humaines"; Dr. Henry Deville, *L'amour libre, Le médecin populaire*, vol. 3 (Paris: L. Boulanger, n.d. [1894]), p. 31.

87. "le salon du 11, rue de Varenne . . . se tenait entre 1880 et 1885, et . . . a été fréquenté par Renan, Taine, Paul Bourget, Sully Prudhomme, Hérédia, Alexandre Dumas fils, Sorel, Berthelot, Pasteur, et le vicomte de Vogüé. D'ailleurs, le début des années 1880 est la période où Gaston Paris travaillait sur *l'amour courtois* [et] où le mouvement de la Décadence s'épanouissait" (Kim, pp. 56–57, see also pp. 43–57). On Gaston Paris's salon, see also Ursula Bähler, *Gaston Paris et la philologie romane* (Genève: Droz, 2004), pp. 180–84.

88. "I will affirm that prostitution is a necessary evil to save both the honor and the well being of many people" (Je dirai que la prostitution est un mal nécessaire pour sauvegarder et l'honneur et le bien-être de beaucoup de gens); Dr. Bouglé, *Les vices du peuple, suivis de l'histoire des maladies vénériennes* (Paris: Georges Carré, 1888), p. 107.

89. "[notre poésie la plus ancienne] nous apprend à nous refaire enfants pour goûter les joies naïves de l'enfance; elle nous rouvre les trésors de l'imagination de nos pères, et fait jaillir de nouveau, dans nos intelligences desséchées, les sources vives de la joyeuse et jeune poésie"; Gaston Paris, "La poésie du Moyen Âge," in *La poésie du Moyen Âge*, 2 vols. (Paris: Librairie Hachette, 1885), vol. 1, pp. 1–39, at p. 9; see Ridoux, p. 64: "urgency of national renewal" (*l'urgence d'une restauration nationale*).

90. "On a vu de nos jours des littératures créer des nations, c'est-à-dire que la conscience nationale, presque complètement éteinte . . . a retrouvé . . . la plénitude de sa force et de sa vie"; G. Paris, "*La Chanson de Roland* et la nationalité française. Leçon d'ouverture au Collège de France, le 8 décembre 1870," in *La poésie du Moyen Âge*, vol. 1, pp. 87–118, at p. 100.

91. "tout fait présager encore [à notre littérature] de longs siècles de prospérité, car elle semble ne traverser une phase nouvelle que pour s'enrichir d'un nouveau perfectionnement, de ressources nouvelles"; Pierre Larousse, *Grand dictionnaire universel du XIXe siècle* (Paris: Administration du Grand dictionnaire universel, 1866–1877), "Français, -aise," vol. 8 (1872), pp. 708–17, at p. 711.

92. "une passion simple, ardente, naturelle" (G. Paris, "*Lancelot du Lac*," p. 519).

93. "les subtilités et les raffinements" (G. Paris, "*Lancelot du Lac*," p. 519). See also Kelly, "Varieties of Love."

94. Kim, p. 121. Bähler argues that for Paris in his study of *Tristan et Iseut* (1894) their love was "already 'courtly love,' conventional and ritualized love that will find its ultimate expression in the relationship of Lancelot and Guinevere" (déjà "l'amour 'courtois,'" l'amour conventionnel et réglementé qui trouvera son expression complète dans la liaison de Lancelot et de Guenièvre) (Bähler, p. 614). But Paris would have changed his opinion in 1901 and returned to the original idea: "Just as with the *chansons de geste*, barbarism and savagery are . . . entirely positive terms, connected again to sincerity and truthfulness" (Tout comme c'est le cas pour les chansons de geste, la barbarie et la sauvagerie sont . . . des termes tout à fait positifs, et se voient à nouveau rapprochés de la sincérité et de la véracité) (Bähler, p. 634).

95. "à partir du onzième siècle, une amélioration heureuse se fit sentir dans les mœurs

publiques et privées"; Dr. Jaf (Dr. Fauconney), *Physiologie du vice, son histoire à travers les âges* (Paris: C. Offenstadt, 1903), p. 62.

96. "avec le XIIe siècle les mœurs changent soudainement de caractère. Les sentiments les plus épurés d'amour divin et terrestre se réveillent dans tous les cœurs et prennent un caractère d'exaltation incroyable; on court en Orient pour défendre la cause de Dieu, on traverse les villes et les royaumes pour se montrer en aide aux veuves, aux orphelins, aux dames belles, innocentes et persécutées"; Dr. Michel Villemont, "Cours d'amour," in *Dictionnaire historique et scientifique de l'Amour et du mariage* (Paris: A. Lévy et Cie, 1886), pp. 112–16.

97. "Il y eut là, dans tous les cas, une tentative louable pour réformer et discipliner ces deux grands mobiles de l'action humaine: la passion érotique et l'amour des combats" (Dr. Debray, p. 229).

98. "Le moyen âge avait fait de l'amour le principe de la chevalerie. Il en fit même une sorte d'institution contre le mariage. Les cours d'amour n'admettaient pas la possibilité de l'amour dans le mariage"; Dr. Jaf (Dr. Fauconney), *L'amour secret* (Paris: Charles Offenstadt, 1903), p. 45. Also qtd. in Dr. Raoul Désormaux, *L'amour conjugal, Bibliothèque sexuelle du Dr Désormeaux*, vol. 11 (Paris: Librairie P. Fort, n.d. [1905]), p. 5. See also Drs. Caufeynon and Jaf (Dr. Fauconney), *Les messes noires. Le culte de Satan-Dieu* (Paris: Pancier, 1905), pp. 202–3; Dr. Caufeynon (Dr. Fauconney), *La prostitution à travers les âges, depuis les temps les plus reculés* (Paris: Librairie P. Fort, n.d. [1903]), pp. 167–68. Dr. Jean Fauconney was a great fin-de-siècle popularizer.

99. "À quelque point de vue que l'on se place, il est impossible de concevoir un état social dont la base fondamentale ne serait pas le mariage. . . . Oui, sans mariage, aucune société n'est possible; la famille est un mot vide de sens et la propriété une chose aléatoire qui ne se fixe nulle part"; Félix Roubaud, *Le livre des époux. Guide pour la guérison de l'impuissance, de la stérilité, et de toutes les maladies des organes génitaux* (Paris: L'auteur, 1852), p. 1. See also G. Tourdes, "Mariage (Médecine légale)," in *Dictionnaire encyclopédique des sciences médicales*, ed. Amédée Dechambre (Paris: G. Masson, 1864–1888), 2e série, vol. 5 (1874), pp. 83–104.

100. "Le Mariage et ses lois constituent la base de tout l'édifice social"; Dr. Georges Morache, *Le mariage: Étude de socio-biologie et de médecine légale* (Paris: Felix Alcan, 1902), p. ii.

101. "Le mariage, c'est la famille, et la famille constitue les assises, la raison d'être de notre société" (Dr. Deville, *L'amour dans le mariage*, p. 27).

102. "Le mariage, avons-nous dit, est l'union légale de l'homme et de la femme, instituée pour la satisfaction des besoins physiques, pour perpétuer l'espèce, pour l'éducation des enfants et pour s'aider, par des secours mutuels, pendant le cours de la vie" (Dr. Deville, *L'amour dans le mariage*, p. 22), "l'union de l'homme et de la femme, commandée par la nature pour perpétuer leur espèce" (Debay, p. 45).

103. "qui convient le mieux à l'état de l'homme"; Dr. Michel Villemont, *L'amour conjugal* (Paris: A. Lévy, 1885), p. 135.

104. "diminue les chances de mortalité, d'aliénation et surtout de criminalité; il est

favorable à la fécondité et plus encore à la vitalité des nouveau-nés"; Émile Littré and Charles Robin, *Dictionnaire de médecine, de chirurgie, de pharmacie, de l'art vétérinaire et des sciences qui s'y rapportent*, 14th rev. ed. (Paris: J.-B. Baillière, 1878), "Mariage," pp. 940–41, at p. 940.

105. "Plus vous aurez d'hommes mariés, moins il y aura de crimes. . . . Le père de famille ne veut pas rougir devant ses enfants"; Dr. Michel Villemont, *Hygiène de l'amour et du mariage* (Paris: Librairie des publications nouvelles, n.d. [1883]), p. 401.

106. "la solution la plus naturelle et la plus morale, en même temps qu'elle est la plus favorable à la société et à l'individu," Dr. Raoul Désormaux, *Le mariage, Bibliothèque seuxelle du Dr Désormeaux*, vol. 5 (Paris: Librairie P. Fort, n.d [1905]), p. 7; also qtd. in Dr. Jaf (Dr. Fauconney), *Le mariage et son hygiène* (Paris: Librairie des connaissances médicales, 1905), p. 7.

107. "le mariage devient aussi pour le corps le plus précieux gage de santé, assurant la vitalité et la longévité des époux"; Jacques Bertillon, "Mariage (Démographie; Hygiène matrimoniale)," in *Dictionnaire encyclopédique des sciences médicales*, ed. Dechambre, 2e série, t. 5 (1874), pp. 7–83, at p. 53.

108. "Cette union légale a une influence marquée sur la santé physique et morale des populations" (Littré and Robin, p. 940).

109. "Quel est le but du mariage? de perpétuer la société, non seulement par le nombre des individus, mais plus encore par la beauté, par la vigueur et l'organisation parfaite, autant que possible, des races"; Dr. Serrurier, *Du mariage considéré dans ses rapports physiques et moraux. Inconvénients des unions entre des individus entachés d'un principe constitutif héréditaire* (Paris: J.-B. Baillière, 1845), p. 35.

110. "les mystères de l'amour conjugal et de l'hygiène du mariage, comme moyen de perfectionner l'espèce humaine"; Parent-Aubert, *Almanach des mystères de l'amour conjugal et de l'hygiène du mariage* (Paris, 1852), p. 7.

111. "Il faut cependant protéger toujours, et à certaines heures plus que jamais, la famille et la race"; Dr. Henry Cazalis, *La science et le mariage* (Paris: Octave Doin, 1900), p. 6.

112. "La mortalité extrême des enfants, en France, et notre faible *natalité* met en péril l'existence même de notre nation"; Dr. A. Coriveaud, *Le lendemain du mariage, étude d'hygiène* (Paris: J.-B. Baillière, 1884), p. 267.

113. "Si les armées allemandes n'avaient remporté sur nous que des victoires militaires, la fortune de la France serait vite réparée. Mais c'est la victoire d'une société sur une autre, d'une civilisation sur une autre, d'une grande famille humaine sur une autre grande famille humaine plus divisée, moins nombreuse, moins laborieuse. Le mal n'est pas seulement dans les rangs des soldats, il est dans les rangs de la société même, il est dans la famille. . . . En France, la population est à peu près stationnaire; en Allemagne elle double en moins de soixante ans" (Hayem, p. b).

114. "Mais il n'est pas contestable que la crainte de se trouver retenus dans des liens indissolubles doit arrêter beaucoup de jeunes gens malgré leur goût pour le mariage; et qu'ainsi le régime de la séparation de corps est peut-être un des plus graves obstacles au plus grand nombre des mariages" (Hayem, pp. 280–81).

115. Jacques Bertillon, *Étude démographique du divorce et de la séparation de corps dans les différents pays de l'Europe* (Paris: G. Masson, 1883) and *Le divorce et la séparation de corps dans les différents pays de l'Europe, étude démographique* (Nancy: Berger-Levrault, 1884).

116. Paul Leroy-Beaulieu, *La question de la population* (Paris: F. Alcan, 1913), p. i.

117. "Cette excessive prévoyance du malthusianisme médical, contre le danger imaginaire d'une natalité exagérée, est la destruction même du mariage et sa négation"; Dr. Pierre Garnier, *Le mariage dans ses devoirs, ses rapports et ses effets conjugaux au point de vue légal, hygiénique, physiologique et moral* [A free and expanded translation of F. Monlau, *Higiene del matrimonio*] (Paris: Garnier Frères, 1879), p. 361.

118. "La théorie de Malthus est une absurdité économique, et sa mise en pratique, un crime de lèse-nation" (Dr. Coriveaud, p. 267). See Francis Ronsin, *La grève des ventres: Propagande néo-malthusienne et baisse de la natalité française, XIXe–XXe siècles* (Paris: Aubier Montaigne, 1980).

119. Dr. Brennus, *Amour et securité*, 108th ed. (Paris: André Hal, 1908 [1893]). See Dr. Brennus, *Histoire du célèbre ouvrage "Amour et sécurité" par Doctor Brennus, poursuivi en cour d'assises à Paris le 29 août 1895* (Paris: F. Montel, 1896).

120. "Plus d'hésitation ni de déception en amour, grâce aux secrets dévoilés par les Docteurs Jaf et Raphaël Saldo. C'est le Guide des jeunes gens et des jeunes filles, c'est le Bréviaire des Amoureux par excellence"; Dr. Riolan, *Chasteté, virginité et célibat, Nouvelle collection exclusive d'hygiène et de médecine*, vol. 3 (Paris: F. Pierre, 1909), n.p.

121. "un ouvrage de médecine," "un guide, un manuel des époux, tant au point de vue moral que d'hygiène. . . . Contrairement à la doctrine courante, le devoir capital d'un mari intelligent et sage consiste à rechercher son bonheur dans celui de sa femme"; Dr. Riolan, *Pédérastie et homosexualité, Nouvelle collection exclusive d'hygiène et de médecine*, vol. 10 (Paris: F. Pierre, 1909), n.p.

122. "Préservatifs pour hommes," "*le Chaste, le Mignon, le Favori,*" "Préservatifs pour Dames," "*le Simplex, le Discret*"; Dr. Riolan, *Accouchement* and *Les maladies honteuses, Nouvelle collection exclusive d'hygiène et de médecine*, vols. 7 and 12 (Paris: F. Pierre, 1909), n.p.

123. "On ne connaît plus l'exaltation d'amour dans nos siècles; ils ne sont plus ces temps de la chevalerie et des *cours d'amour*, où les femmes dispensaient la gloire, devenaient les arbitres de la courtoisie et de la prouesse des paladins; elles régnaient par les seuls regards, et leur doux empire se perpétuait par la vertu la plus pure et l'attachement le plus fidèle. Tels étaient aussi ces *galois* et ces *galoises* [*sic*], sorte de confrérie dans le moyen âge, qui faisaient vœu de souffrir et l'ardeur des étés et le froid de l'hiver, et tous les tourments, s'il le fallait, pour une personne adorée"; J. J. Virey, *De la femme sous ses rapports physiologique, moral et littéraire* (Paris: Crochard, 1823), pp. 382–83.

124. "Ce que nous appelons *amour* n'est que la manifestation extérieure de ce mouvement vital qui tend à se répandre dans d'autres êtres, pour leur communiquer la vie" (Virey, p. 358).

125. "Aimer, c'est vivre pour son espèce" (Virey, p. 208).

126. "La cause primordiale de l'amour est sans contredit dans l'instinct de reproduction"; J.-B.-F. Descuret, *La médecine des passions, ou Les passions considérées dans leurs*

rapports avec les maladies, les lois et la religion, 2 vols., 3rd rev. ed. (Paris: Labé, 1860 [1841]), vol. 2, p. 156.

127. "L'amour est le principe de la vie, celle-ci se caractèrise par l'amour, c'est-à-dire la vigueur réproductive"; Dr. Caufeynon (Dr. Fauconney), *L'amour et l'accouplement, Bibliothèque populaire des connaissances médicales*, vol. 6 (Paris: Nouvelle librairie médicale, 1902), p. 6.

128. "communion intime de deux êtres semblant faits l'un pour l'autre et qui réalisera pour chacun d'eux les conditions du bonheur le plus complet" (Dr. Coriveaud, p. 2).

129. "L'amour dans le mariage n'est pas seulement une condition de bonheur domestique que chacun devrait rechercher, de préférence à tous les autres éléments qui entrent d'ordinaire dans les combinaisons matrimoniales"; Dr. Jaf (Dr. Fauconney), *Le mariage et son hygiène* (Paris: Librairie des connaissances médicales, 1905), pp. 19–20.

130. Robert A. Nye, "Honor, Impotence, and Male Sexuality in Nineteenth-Century French Medicine," *French Historical Studies* 16:1 (1989): 48–71, at p. 52.

131. "Beaucoup de moralistes semblent avoir pris à tâche de confondre la galanterie avec l'amour. Aussi doit-on à cette confusion le désaccord qui règne dans un certain nombre d'écrits sur cette dernière passion. Et cependant quelle différence! Moins vive, moins sérieuse, mais plus clairvoyante et plus sensuelle que l'amour, la galanterie recherche plutôt la beauté physique que la beauté morale.... L'amour nous attache uniquement, généreusement et sans réserve à l'objet de notre affection"; Dr. Alibert, *L'amour et l'accouplement, Bibliothèque médicale populaire et d'hygiène*, vol. 3 (Paris: Librairie de la nouvelle France, n.d. [19??]), pp. 80–81.

132. "Dans l'intérêt de l'hygiène corporelle et intellectuelle de l'humanité, il faut pousser au mariage et donner aux jeunes hommes les préceptes anatomiques de l'amour"; Dr. Jules Guyot, *Bréviaire de l'amour expérimental, méditations sur le mariage selon la physiologie du genre humain* (Paris: Marpon et Flammarion, 1882), p. 31.

133. Dr. Paul Belouino, *La femme: Physiologie, histoire, morale* (Paris: Waille, 1845), p. 384.

134. "Lorsque l'amour, quelle que soit sa violence, n'a pour base que les attraits physiques de la jeunesse et de la beauté, il est rare que la passion et surtout que l'abus des plaisirs ne finissent par amener peu à peu l'indifférence et même le dégoût. Aussi est-ce en parlant des unions de cette nature qu'on a dit avec raison que l'hymen est le tombeau de l'amour" (Dr. Alibert, p. 13).

135. "Ce ne peut donc pas être l'amour qui soit le principe d'une union durable entre l'homme et la femme. L'amour détermine l'union, mais il est impuissant à la maintenir, et, comme un dieu enfant qui ne sait pas user de son pouvoir, il fait et défait son ouvrage" (Hayem, p. 10).

136. "L'amour est le mobile de toutes choses: nous parlons ici de cet amour tendre, honnête que le cœur inspire et que la raison dirige, de cet amour tranquille qu'on prendrait pour de *l'amitié*, et non de cet amour fougueux, enfanté par l'ardeur des sens, qui disparaît aussitôt que les désirs sont satisfaits" (Debay, pp. 57–58, emphasis added).

137. "De cette première nuit dependent souvent le bonheur et l'avenir d'un ménage" (Dr. Désormaux, *L'amour conjugal*, p. 23). See Adler, p. 31–65.

138. "la cause la plus commune d'un grand nombre de ruptures conjugales" (Dr. Désormaux, *L'amour conjugal*, p. 27).

139. "Quelle émotion pénible doit éprouver la malheureuse, lorsque sans préliminaires, sans préparation, sous les secousses de celui dont elle espérait les plus délicieuses caresses, elle éprouve de douloureuses meurtrissures de la chair et des sens. Quelques paroles d'amour et d'affection doivent, au contraire, précéder et préparer, d'abord l'entrée au lit, ensuite le premier contact charnel. L'homme s'avancera donc avec précaution, tentera des essais préparatoires d'entière pénétration, lesquels exciteront les organes extérieurs et feront préalablement entrer la femme dans les spasmes de la volupté" (Dr. Brennus, *Amour et sécurité*, pp. 106–8).

140. "L'homme ne ménage pas seulement les organes, il ménage aussi la susceptiblité de la femme" (Dr. Fauconney, *Le mariage et son hygiène*, p. 29).

141. Dr. Caufeynon (Dr. Fauconney), *La procréation à volonté des filles et des garçons, suivie de La fécondation artificielle, et de L'ami des jeunes femmes* (Paris: P. Fort, n.d. [1903]), pp. 109–204.

142. "C'est un temps d'ivresse sans limites, donné par la nature aux époux pour se comprendre et s'accorder sur la satisfaction normale et complète des besoins et du sens génésiques" (Dr. Guyot, p. 104).

143. "il faut user, et non abuser. . . . La modération dans les plaisirs est une sage économie qui donnera un charme toujours nouveau" (Dr. Deville, *L'amour dans le mariage*, pp. 5–6).

144. "Le mariage, dit-on, est le tombeau de l'amour . . . entièrement physique, qui, suivant le proverbe, s'éteint dans le mariage. . . . Le véritable amour commence à cette mort, que c'est un effet nécessaire du mariage que *la galanterie se change en culte*, que tout mari, quelque mine qu'il fasse, est, au fond de l'âme, *idolâtre*; que s'il y a conspiration ostensible entre les hommes pour secouer le joug du sexe, il y a convention tacite pour l'*adorer*; que la faiblesse seule de la femme oblige de temps à autre l'homme à ressaisir l'empire; que sauf ces rares exceptions, la femme est *souveraine*, et que là est le principe de la tendresse et de l'harmonie conjugale" (Dr. Désormaux, *Le Mariage*, pp. 11–13, emphasis added). Copied in Dr. Fauconney, *Mariage et son hygiène*, p. 19.

145. "un mariage d'inclination"; Dr. Gébé, *L'amour conjugal: Conseils* (Paris: F. Pigeon, 1891), pp. 35–36.

146. "Libre choix dans les deux sexes; pas de contrat imposé par les parents et subi par les enfants"; Paolo Mantegazza, *L'amour dans l'humanité: Essai d'une ethnologie de l'amour*, trans. Emilien Chesneau (Paris: F. Fetscherin et Chuit, 1886 [1885]), p. 402.

147. "À plus forte raison pour ceux qui se sont choisis, aimés librement et qu'un mutuel amour réunit. Libres et confondus, ils n'ont eu qu'à se dire et se prouver réciproquement tout le bonheur qu'ils en éprouvent"; Dr. Pierre Garnier, *La génération universelle*, 4th rev. ed. (Paris: Garnier Frères, 1887), p. 279.

148. "Il importe de ne pas gêner la liberté naturelle du choix dans le ménage, et les parents sont coupables qui exercent un empire tyrannique sur la volonté de leurs enfants,"

"le mariage . . . est l'union d'abord libre de deux cœurs qui se fait sanctionner par les lois pour constituer la famille" (Dr. Désormaux, *L'amour conjugal*, pp. 19, 21).

149. "Il est même permis d'affirmer que l'acte est toujours imparfait sans cette union des cœurs avec la cooperation des sens" (Dr. Garnier, *La génération universelle*, p. 271).

150. "Dans l'amour vrai, le plaisir des sens y est pour beaucoup"; also "This book is not properly speaking a book of medicine; it is an apologia for true love and for its laws, for *secret love*, or, better yet, for a rational nomenclature of the secrets of love, meaning what must be done to inspire true love, to bolster it and perpetuate it between two lovers" (Ce livre n'est pas un livre de médecine proprement dite, c'est une apologie de l'amour vrai et de ses lois, de l'*amour secret*, ou mieux une nomenclature raisonnée des secrets d'amour, c'est-à-dire de ce qui doit être fait pour faire naître l'amour vrai, pour le consolider et le perpétuer entre deux amants) (Dr. Fauconney, *L'amour secret*, p. 10).

151. "Dans l'accouplement amoureux, il ne devrait y avoir en réalité que jouissance commune et réciproque" (Dr. Désormaux, *L'amour conjugal*, p. 25).

152. "le *mariage parfait*. Illuminée et réchauffée par l'amour,—né d'elle ou avant elle,—une telle union donne aux deux *amis* qu'elle lie pour toujours une quiétude morale, un équilibre spirituel, une énergie dans la volonté qui décuplent leur valeur et leurs forces. Pour nous, en un mot, dans le *mariage parfait*, au-dessus de la forme sociale et purement conventionnelle, plane et rayonne un sentiment que, faute d'autre mot, nous appelons l'*amour*" (Dr. Coriveaud, p. 2).

153. Jacques Lacan, *The Seminar of Jacques Lacan, Book VII: The Ethics of Psychoanalysis (1959–1960)*, ed. Jacques-Alain Miller, trans. Dennis Porter (New York: W. W. Norton & Co., 1992), pp. 153, 148 (*Le Séminaire 7 [1959–1960]: L'éthique de la psychanalyse*, ed. Jacques-Alain Miller [Paris: Éditions du Seuil, 1986], pp. 178, 183); Denis de Rougemont, *L'amour et l'Occident* (Paris: Plon, 1939).

EPILOGUE

1. James A. Schultz's *Courtly Love, the Love of Courtliness, and the History of Sexuality* (Chicago: University of Chicago Press, 2006) is a groundbreaking study that analyzes courtly love as a discourse of sexuality.

2. "cette réunion unique et sans rivale de petits monuments du moyen âge destinés à faire l'ornement d'un des musées de la capitale," "jeter un nouveau jour sur cette époque ténébreuse et reculé [*sic*]"; Letter from Arthur Forgeais to Edmond du Sommerard, dated March 5/6, 1861, Archives du Musée national du Moyen Âge (dossiers d'acquisition-1861). The date of March 5 appears to have been corrected to March 6. I am grateful to Jean-Christophe Ton-That for providing me with a copy of this letter.

3. Arthur Forgeais, *Collection de plombs historiés trouvés dans la Seine et recueillis par Arthur Forgeais, I. Méreaux des corporations de métiers; II. Enseignes de pélerinages; III. Variétés numismatiques; IV. Imagerie religieuse; V. Numismatique populaire* (Paris: Aubry, 1862–1866).

4. "À cause de leur contenu et malgré leur présence dès les origines au sein de la collection du musée, elles [les enseignes érotiques] n'avaient guère été commentées ni publiées dans le passé"; Elisabeth Taburet-Delahaye, ed., with Jean-Christophe Ton-That and Marie-Alice Virlouvet, *Thermes et hôtel de Cluny, Musée national du Moyen Âge: Œuvres nouvelles, 1995–2005* (Paris: Réunion des musées nationaux, 2006), p. 58.

5. "Il est des turpitudes que l'on ne doit pas réunir sans motif sérieux, et j'ai cette excuse que mon ouvrage ne sera probablement pas feuilleté hors du centre des gens graves. La forme, d'ailleurs, y est toujours assez grossière pour ne pas enflammer les imaginations, qui savent se garder quand le cœur ne se livre pas de plein gré à l'ivresse des sens"; Arthur Forgeais, *Priapées* (n.p.: n.d.), p. 3.

6. "Quant à l'usage que l'on pouvait faire . . . c'est ce qu'il est difficile de déterminer" (Forgeais, p. 4).

7. "petits monuments priapiques" (Forgeais, p. 4).

8. "tirons du moins de ces bouffoneries érotiques ce qu'elles peuvent offrir en moralité profonde, bien que l'objet du dessinateur pût être tout autre; car il ne semble pas que l'on ait prétendu moraliser à l'aide de formes aussi grossières, tant les sens sont prêts à prendre le dessus si on les dépouille absolument de leurs voiles" (Forgeais, p. 12).

9. "monté sur deux jambes, comme si tout l'animal était complet ainsi, sauf les ailes (qui peuvent toutefois désigner l'empressement et l'ardeur inquiète). Ce n'est pas un portrait fort absurde de certaines natures où la lubricité s'est enflammée par l'habitude, jusqu'à devenir monomanie constante" (Forgeais, p. 10).

10. "Ainsi la femme chevauchant un *phallus* peut susciter cette idée: *l'humanité qui s'abandonne à la pente bestiale finit par n'être plus maîtresse d'elle-même*" (Forgeais, p. 4).

11. "On retrouverait encore une leçon analogue dans la Figure de l'homme porteur d'un organe plus gros que le reste de son corps. Suivant l'expression de Balzar . . . cet homme sera devenu un tempérament" (Forgeais, pp. 4–5).

12. "Mais pourquoi ne serait-ce pas tantôt une sorte d'*enseigne* adoptée par les femmes de mauvaise vie pour se faire entendre sans pourparlers, tantôt une note d'infamie qu'on leur imposait pour les classer nettement, ou, peut-être, quelque signe d'abonnement à un mauvais lieu?" (Forgeais, p. 4). The sheer number of surviving artifacts and the cheapness of their production suggest that they were very important in medieval popular culture, but there is still no agreement among scholars as to their meaning and use. Denis Bruna asserts that religious badges, whose purpose was clearly to mark pilgrimage, should not be opposed to the secular badges, if for no other reason than the fact that they were found together in the riverbed of the Seine. He eludes any further interpretation, however: "If the pilgrim badges lend themselves easily to relatively coherent classification as to their use (identity, religious devotion, and protection), the secular badges are, as for them, too diverse for one to be able to undertake such an exercise" (Si les enseignes de pèlerinage se prêtent aisément à une classification relativement cohérente de leur usage [identité, dévotion et protection], les enseignes profanes sont, quant à elles, trop diverses pour que l'on puisse se livrer à un tel exercice); Denis Bruna, *Enseignes de pèlerinages et enseignes profanes: Musée national du Moyen âge-Thermes de Cluny* (Paris: Réunion des musées nationaux, 1996), p. 20. Authors

of the entry for Figure 11 offer this explanation in the 2006 catalogue of new acquisitions at the Cluny Museum: "Two clearly opposed interpretations are possible for these objects. They can be taken, despite appearances, for religious badges, in the same manner as comparable objects in antiquity; these badges would then be tied to sanctuaries visited to obtain fertility. But erotic badges can also be considered as secular ones, even of the most profane kind, in contrast to the courtly badges or surpassing them, revealing themselves to be the vulgar and brutal underbelly of the enchanted world of medieval love" (Deux interprétations sont possibles pour ces objets, nettement opposées. On peut les tenir malgré les apparences, pour des enseignes religieuses, au même titre que bien des objets comparables dans l'antiquité; ces enseignes seraient alors liées à des sanctuaires visités pour obtenir fertilité et fécondité. Mais on peut aussi considérer les enseignes érotiques comme des enseignes profanes, voire pour les plus profanes des enseignes, s'opposant aux enseignes courtoises ou les dépassant, se révélant comme un envers grossier et brutal du monde enchanté de l'amour médiéval) (Taburet-Delahaye et al., p. 58). Malcolm Jones and Ruth Melinkoff have argued in favor of their apotropaic significance, that is, to avert bad and bring good luck; Malcolm Jones, *The Secret Middle Ages* (Stroud, Gloucestershire: Sutton, 2002), esp. pp. 248–73; Ruth Mellinkoff, *Averting Demons: The Protective Power of Medieval Visual Motifs and Themes*, 2 vols. (Los Angeles: Ruth Mellinkoff Publications, 2004). For an insightful reading of the cultural code contained in the sexual badges, both as entertainment and moral instruction, see Stefanie Stockhorst, "Passionate Pilgrims: Secular Lead Badges as Precursors of *Emblemata Amatoria*," in *Profane Images in Marginal Arts of the Middle Ages*, ed. Elaine C. Block, with Frédéric Billiet, Sylvie Bethmont-Gallerand, and Paul Hardwick (Turnhout, Belgium: Brepols, 2009), pp. 157–81.

13. "Conformément à l'étymologie grecque du mot, c'est surtout aux livres relatifs à la prostitution que convient le titre d'ouvrage pornographique. Tel est le livre de Restif de la Bretonne, intitulé: *Le Pornographe ou Idées d'un honnête homme sur un projet de règlement pour les prostituées, propre à prévenir les malheurs qu'occasionne le publicisme des femmes*"; Pierre Larousse, *Grand dictionnaire universel du XIXe siècle* (Paris: Administration du Grand dictionnaire universel, 1866–1877), "Pornographie," vol. 12 (1874), pp. 1438–39, at p. 1438; Nicolas Edme Rétif de la Bretonne, *Le pornographe ou Idées d'un honnête homme sur un projet de règlement pour les prostituées, propre à prévenir les malheurs qu'occasionne le publicisme des femmes* (London; Paris, 1769).

14. "modifications survenues dans les sens de certains mots . . . on n'oserait plus donner au livre de Parent-Duchâtelet le nom de pornographique" (Larousse, "Pornographie," p. 1439); Alexandre-Jean-Baptiste Parent-Duchâtelet, *De la prostitution dans la ville de Paris, considérée sous le rapport de l'hygiène publique, de la morale et de l'administration: Ouvrage appuyé de documens statistiques puisés dans les archives de la Préfecture de police*, 2 vols. (Paris: J.-B. Baillière, 1836).

15. "dont l'immoralité fait le fond et le but," "un dessein plus suivi d'immoralité, sont plus complétement [*sic*] faits dans le but de servir les appétits lubriques, de réveiller les sens blasés" (Larousse, "Pornographie," p. 1439).

16. Lynn Hunt, "Introduction: Obscenity and the Origins of Modernity," in *The*

Invention of Pornography: Obscenity and the Origins of Modernity, 1500–1800, ed. Lynn Hunt (New York: Zone Books, 1993), pp. 9–45, at p. 13. For a thorough exploration of the emergence of modern pornography, see Joan DeJean, *The Reinvention of Obscenity: Sex, Lies, and Tabloids in Early Modern France* (Chicago: University of Chicago Press, 2002). For a take on pornography in the Middle Ages, see Bill Burgwinkle and Cary Howie, *Sanctity and Pornography in Medieval Culture: On the Verge* (Manchester: Manchester University Press, 2010).

17. "De fait, qu'est-ce que la pudeur, sinon le soin habituel de ne réveiller, ni en nous, ni dans les autres, des instincts périlleux toujours prêts à faire explosion comme la bête farouche qui ne rêve que de proie?" (Forgeais, p. 4).

18. "à notre plume une réserve d'autant plus grande et plus facile que la vue seule de tels tableaux donne une suffisante leçon de moralité. D'un autre côté, c'est avec sévérité, c'est avec la plus extrême rigueur que nous choisirons nos citations, que nous rapellerons les faits historiques"; Larousse, *Grand dictionnaire universel*, "Prostitution," vol. 13 (1875), pp. 289–98, at p. 289.

19. "À cause d'un parti-pris de pudeur mal placée, ces questions [de pédérastie] n'ont été abordées au point de vue scientifique qu'avec une certaine timidité. Les auteurs, même ceux qui se croient le plus émancipés de toute idée religieuse, craignent d'être soupçonnés de pornographie ou d'inconvenance scientifique, se répandent avec la fougue d'un prédicateur en épithètes variées sur un *vice* qu'ils qualifient d'abomination, de monstruosité, d'infamie, etc., etc., comme si le caractère véritablement extraordinaire de ces faits, dans toutes les sociétés, à toutes les époques de l'histoire, ne devait pas attirer l'attention du psychologue et du médecin légiste, afin d'élucider ces complexes et savoir si . . . il faut traiter"; Alexandre Lacassagne, "Pédérastie," in *Dictionnaire encyclopédique des sciences médicales*, ed. Amédée Dechambre (Paris: G. Masson, 1864–1888), 3e série, t. 22 (1886), pp. 239–59, at p. 240. Repeated in Lacassagne's "Préface" to Marc-André Raffalovich, *Uranisme et unisexualité: Étude sur différentes manifestations de l'instinct sexuel* (Lyon: A. Storck, 1896), n.p.

20. "Il ne convient guère, croyons-nous, d'exciter le public, par l'attrait de gravures d'un goût douteux, à acheter un livre qui, par le soin apporté à sa composition et par la nature même du sujet qu'il traite, doit rester aux mains seules des moralistes et des érudits" (Larousse, "Prostitution," p. 298).

21. "L'éditeur n'a rien négligé pour donner au livre de M. Dufour tous les dehors séduisants propres à attirer l'attention des lycéens et des vieillards désœuvrés et libertins. En agissant de cette façon, peut-être, le succès de scandale a été obtenu, mais cet ouvrage n'est pas digne de prendre rang dans les bibliothèques à côté des auteurs sérieux que nous venons de nommer" (Larousse, "Prostitution," p. 298).

22. Walter Kendrick, *The Secret Museum: Pornography in Modern Culture* (Berkeley: University of California Press, 1996 [1987]).

23. "Les archéologues seuls . . . ont osé l'aborder," "délicat et suspect à la fois"; Pierre Dufour (Paul Lacroix), *Histoire de la prostitution, chez tous les peuples du monde depuis l'antiquité la plus reculée jusqu'à nos jours*, 6 vols. (Paris: Seré; Martinon, 1851–1853), vol. 1, p. 9.

24. "Nous ne perdrons jamais de vue que ce livre, préparé lentement au profit de la science, doit servir à la morale et qu'il a pour principal objet de faire détester le vice en dévoilant ses turpitudes. . . . C'est là, surtout, que nous nous distinguerons des archéologues et des savants proprement dits, qui ne se préoccupent pas de la moralité des faits et qui ne se soucient pas d'en tirer des conséquences philosophiques. Ils dissertent longuement, . . . ils en dévoilent les monstruosités, ils en retracent les infâmies, mais ils oublient ensuite de nous purifier la pensée et de nous tranquiliser l'esprit, en opposant à ces images impures et dégradantes les chastes leçons de la philosophie et l'action bienfaisante du christianisme" (Lacroix, 1:9).

25. "Quant à nous, tout archéologue que nous sommes aussi, nous n'oublierons pas que . . . nous nous adressons à un public français qui veut être instruit, mais qui en même temps veut être respecté" (Lacroix, 1:9).

26. "souvent d'une dégoutante obscénité. Les mots les plus sales de la langue française y semblent prodigues à plaisir; les fonctions les plus vulgaires de la machine humaine y sont le sujet de grossières plaisanteries, les parties les plus secrètes du corps y sont nommées en termes dont rougiraient les prostituées d'aujourd'hui"; Émile de la Bédollière, *Histoire des mœurs et de la vie privée des Français. Usages, coutumes, institutions, physionomie de chaque époque, depuis l'origine de la monarchie jusqu'à nos jours*, 3 vols. (Paris: V. Lecou, 1847–1849), vol. 3, p. 316.

27. "un langage scientifique, clair, net, compris et admis par tous" (Lacassagne, p. 239).

28. As Kathleen Biddick did by looking at what still remains on the exterior of the medieval discipline, because new medievalism failed "to address others excluded by such methods," though it expelled "the now suspect 'fathers' to the exterior of the discipline"; Kathleen Biddick, *The Shock of Medievalism* (Durham, N.C.: Duke University Press, 1998), p. 2.

Bibliography

ARCHIVAL SOURCES

Forgeais, Arthur. Letter to Edmond du Sommerard, March 5/6, 1861. Archives du Musée national du Moyen Âge-Thermes de Cluny (dossiers d'acquisition-1861).

Gide, André. Letter to Georges Hérelle, July 14, 1934. Transcribed by Georges Hérelle. Bibliothèque municipale de Troyes, MS 3188, fol. 359.

Lacroix, Paul. "Discours pour sa reception [*sic*] supposée à l'Académie Francaise [*sic*]." Bibliothèque de l'Arsenal, MS 13427, 1942, no. 1.253, fol. 16.

PRINT MATERIALS

A***, M., ed. *Recueil de fabliaux, précédé d'une introduction*. Paris: Bureau de la bibliothèque choisie, 1829.

Adler, Laure. *Secrets d'alcôve: Histoire du couple de 1830 à 1930*. Paris: Hachette, 1983.

Agamben, Giorgio. *Signatura rerum. Sur la méthode*. Trans. Joël Gayraud. Paris: Vrin, 2009.

Aldrich, Robert. *Colonialism and Homosexuality*. New York: Routledge, 2003.

Dr. Alibert. *L'amour et l'accouplement. Bibliothèque médicale populaire et d'hygiène*, vol. 3. Paris: Librairie de la nouvelle France, n.d. (19??).

Dr. Alibert. *Pédérastie et aberrations sexuelles. Bibliothèque médicale populaire et d'hygiène*, vol. 11. Paris: Librairie de la nouvelle France, n.d. (1908).

Allem, Maurice. *Les conteurs français des origines au XXe siècle*. Paris: Librairie des Annales, n.d. (1912?).

Altschul, Nadia R. *Geographies of Philological Knowledge: Postcoloniality and the Transatlantic National Epic*. Chicago: University of Chicago Press, 2012.

Ampère, Jean-Jacques. "De la littérature française dans ses rapports avec les littératures étrangères au Moyen Âge." *La Revue des deux mondes* 1 (1833): 22–34.

Ampère, Jean-Jacques. "Vue générale de la littérature française au Moyen-Âge." *La Revue des deux mondes* 19 (1839): 179–93.

Arnold, David, ed. *Imperial Medicine and Indigenous Societies*. Manchester: Manchester University Press, 1988.

Audiffret, H. "Retz, Gilles de Laval, seigneur de." In *Biographie universelle, ancienne et moderne*, vol. 37. Paris: L. G. Michaud, 1824, pp. 398–401.

Bähler, Ursula. *Gaston Paris et la philologie romane*. Genève: Droz, 2004.

Bähler, Ursula. "Gaston Paris et la philologie romane. Quelques réflexions synthétiques." In *Le Moyen Âge de Gaston Paris. La poésie à l'épreuve de la philologie*, ed. Michel Zink. *Actes du colloque, 27 et 28 mars 2003, Collège de France*. Paris: O. Jacob, 2004, pp. 13–40.

Baillarger, J. "Des hallucinations, des causes qui les produisent, et des maladies qui les caractérisent." *Mémoires de l'Académie royale de médecine*, vol. 12. Paris: J.-B. Baillière, 1846, pp. 273–475.

Ballet, Gilbert. "Préface." In Dr. H.-M. Fay, *Histoire de la lèpre en France. Lépreux et cagots du sud-ouest*. Paris: Honoré Champion, 1910, pp. v–xvi.

Ballet, Gilbert, ed. *Traité de pathologie mentale*, by D. Anglade, F.-L. Arnaud, H. Colin, E. Dupré, A. Dutil, J. Roubinovitch, J. Seglas, and Ch. Vallon. Paris: Octave Doin, 1903.

Balzac, Honoré de. "Avant-propos." *La comédie humaine*, vol. 1. Paris: Gallimard, Bibliothèque de la Pléiade, 1976 (1842).

Balzac, Honoré de. "Portrait de P.-L. Jacob." In *Œuvres diverses 2*, ed. Pierre-Georges Castex, Roland Chollet, and René Guise. Paris: Gallimard, Bibliothèque de la Pléiade, 1996, pp. 654–57, 1462–64.

Barbazan. *Fabliaux et contes des poëtes françois des XII, XIII, XIV & XVe siècles, tirés des meilleurs auteurs*. 3 vols. 1756.

Barber, Malcolm. *The Trial of the Templars*. Cambridge: Cambridge University Press, 1978.

Barthes, Roland. *Michelet*. Trans. Richard Howard. New York: Hill and Wang, 1987. (*Michelet*. Paris: Seuil, 1995 [1954].)

Barthes, Roland. "Michelet, l'Histoire et la Mort." *Esprit* 178 (1951): 497–511.

Bartlett, Robert. *The Making of Europe: Conquest, Colonization, and Cultural Change, 950–1350*. Princeton, N.J.: Princeton University Press, 1993.

Bartlett, Robert. "Medieval and Modern Concepts of Race and Ethnicity." *Journal of Medieval and Early Modern Studies* 31:1 (2001): 39–56.

Bataille, Georges. *The Trial of Gilles de Rais*. Trans. Richard Robinson. Los Angeles: Amok Books, 2004. (*Le procès de Gilles de Rais*. Paris: Éditions Pauvert, 1959.)

Beaujeu, Maurice. *Psychologie des premiers Césars. Une étude de médecine légale dans l'histoire. Thèse pour obtenir le grade de docteur en médecine*. Faculté de médecine et de pharmacie de Lyon, no. 796. Lyon: A. Storck, 1893.

Beaumont, G. *Discours de distribution des prix au Collège de Saint-Dié, le 12 juilliet 1919*. Saint-Dié: Imprimerie G. Freisz, 1919.

Beaune, Colette. *Jeanne d'Arc*. Paris: Perrin, 2004.

Beaune, Colette. *Jeanne d'Arc, vérités et légendes*. Paris: Perrin, 2008.

Bédier, Joseph. "La tradition manuscrite du *Lai de l'ombre*. Réflexions sur l'art d'éditer les anciens textes." *Romania* 54 (1928): 160–96, 321–56.

Bédier, Joseph. *Les fabliaux: Études de littérature populaire et d'histoire littéraire du Moyen Âge*. Paris: E. Bouillon, 1893.

Bédollière, Émile de la. *Histoire de la mode en France*. Leipzig: A. Dürr, 1858.

Bédollière, Émile de la. *Histoire des mœurs et de la vie privée des Français. Usages, coutumes, institutions, physionomie de chaque époque, depuis l'origine de la monarchie jusqu'à nos jours*. 3 vols. Paris: Victor Lecou, 1847–1849.

Belouino, Dr. Paul. *La femme: Physiologie, histoire, morale*. Paris: Waille, 1845.

Benoiston de Châteauneuf, Louis-François. *Mémoire sur la durée des familles nobles de France* (Extrait du compte-rendu des séances et travaux de l'Académie des sciences morales et politiques, 1845). *Annales d'hygiène publique et de médecine légale* 35 (janvier 1846): 27–58.

Bercé, Françoise. "Arcisse de Caumont et les sociétés savantes." In *Les lieux de mémoire*, ed. Nora. vol. 1, pp. 1545–73.

Bernard-Griffiths, Simone, Pierre Glaudes, and Bertrand Vibert, eds. *La fabrique du Moyen Âge au XIXe siècle: Représentations du Moyen Âge dans la culture et la littérature françaises du XIXe siècle*. Paris: Champion, 2006.

Bernelle, Frédéric-Henri. *La psychose de Gilles de Rais, Sire de Laval, Maréchal de France, 1404–1440. Thèse pour le doctorat en médecine*, no. 218. Paris, 1910.

Bertillon, Jacques. *Étude démographique du divorce et de la séparation de corps dans les différents pays de l'Europe*. Paris: G. Masson, 1883.

Bertillon, Jacques. *Le divorce et la séparation de corps dans les différents pays de l'Europe, étude démographique*. Nancy: Berger-Levrault, 1884.

Bertillon, Jacques. "Mariage (Démographie; Hygiène matrimoniale)." In *Dictionnaire encyclopédique des sciences médicales*, ed. Dechambre, 2e série, t. 5 (1874), pp. 7–83.

Bertrand, Alexandre. "De l'état d'extase. Jeanne d'Arc." *Le Globe, Journal littéraire* (April 23, 1825): 489–91.

Bescherelle, Louis-Nicolas. *Précis de l'histoire générale des arts, des sciences, de l'industrie et des mœurs des nations*. Paris: Breteau et Pichery, 1840.

Bhabha, Homi K. *The Location of Culture*. New York: Routledge, 1994.

Biddick, Kathleen. "The Cut of Genealogy: Pedagogy in the Blood." *Journal of Medieval and Early Modern Studies* 30:3 (2000): 449–62.

Biddick, Kathleen. *The Shock of Medievalism*. Durham, N.C.: Duke University Press, 1998.

Biddick, Kathleen. *The Typological Imaginary: Circumcision, Technology, History*. Philadelphia: University of Pennsylvania Press, 2003.

Bildhauer, Bettina. *Medieval Blood*. Cardiff: University of Wales Press, 2006.

Bleys, Rudi. *The Geography of Perversion: Male-to-Male Sexual Behaviour Outside the West and the Ethnographic Imagination, 1750–1918*. London: Cassell, 1996.

Blix, Göran. *From Paris to Pompeii: French Romanticism and the Cultural Politics of Archaeology*. Philadelphia: University of Pennsylvania Press, 2009.

Bloch, Marc. *Feudal Society*. Trans. L. A. Manyon. London: Routledge & Kegan Paul, 1961. (*La société féodale*. Paris: Albin Michel, 1982 [1939].)

Bloch, Marc. *Les rois thaumaturges*. London: H. Milford, Oxford University Press, 1924.

Bloch, R. Howard. "842: The Birth of Medieval Studies." In *A New History of French Literature*, ed. Denis Hollier. Cambridge, Mass.: Harvard University Press, 1994 (1989), pp. 6–13.

Bloch, R. Howard. *Etymologies and Genealogies: A Literary Anthropology of the French Middle Ages.* Chicago: University of Chicago Press, 1983.

Bloch, R. Howard. "Genealogy as a Medieval Mental Structure and Textual Form." In *La littérature historiographique des origines à 1500, Grundriss der romanischen Literaturen des Mittelalters*, ed. Hans Ulrich Gumbrecht, Ursula Link-Heer, and Peter-Michael Spangenberg, vol. 11, part 1. Heidelberg: Carl Winter, 1986, pp. 135–56.

Bloch, R. Howard. "The Medieval Text—'Guigemar'—as a Provocation to the Discipline of Medieval Studies." In *The New Medievalism*, ed. Marina Brownlee, Kevin Brownlee, and Stephen G. Nichols. Baltimore: Johns Hopkins University Press, 1991, pp. 99–112.

Bloch, R. Howard, and Stephen G. Nichols. "Introduction." In *Medievalism and the Modernist Temper*, ed. Bloch and Nichols, pp. 1–22.

Bloch, R. Howard, and Stephen G. Nichols, eds. *Medievalism and the Modernist Temper.* Baltimore: Johns Hopkins University Press, 1996.

Boase, Roger. *The Origin and Meaning of Courtly Love: A Critical Study of European Scholarship.* Manchester: Manchester University Press, 1977.

Bossard, Eugène. *Gilles de Rais, Maréchal de France, dit Barbe-Bleue, 1404–1440, d'après des documents inédits.* Grenoble: Jérôme Millon, 1992 (1885).

Boswell, John. *Christianity, Social Tolerance, and Homosexuality: Gay People in Western Europe from the Beginning of the Christian Era to the Fourteenth Century.* Chicago: University of Chicago Press, 1980.

Bouchard, Constance Brittain. *Those of My Blood: Constructing Noble Families in Medieval Francia.* Philadelphia: University of Pennsylvania Press, 2001.

Boucher, Dr. Louis. *Jeanne d'Arc au point de vue médical.* N.p.: Union typographique, n.d.

Dr. Bouglé. *Les vices du peuple, suivis de l'histoire des maladies vénériennes.* Paris: Georges Carré, 1888.

Bourdeault, A. "Chantoncé: Gilles de Rais et les ducs de Bretagne." *Mémoires de la Société d'histoire et d'archéologie de Bretagne* 5 (1924): 41–150.

Boureau, Alain. *The Lord's First Night: The Myth of the* Droit de Cuissage. Trans. Lydia G. Cochrane. Chicago: University of Chicago Press, 1998. (*Le droit de cuissage: La fabrication d'un mythe, XIIIe–XXe siècle.* Paris: Albin Michel, 1995.)

Bourgeois, Alfred. *Quelle est l'influence des mariages consanguins sur les générations? Thèse pour le doctorat en médecine.* Paris: Rignoux, 1859.

Bowler, Peter J. *The Mendelian Revolution: The Emergence of Hereditarian Concepts in Modern Science and Society.* Baltimore: Johns Hopkins University Press, 1989.

Boyarin, Daniel, and Elizabeth A. Castelli. "Introduction: Foucault's *The History of Sexuality*: The Fourth Volume, or, a Field Left Fallow for Others to Till." *Journal of the History of Sexuality* 10:3–4 (2001): 357–74.

Brachet, Auguste. *Pathologie mentale des rois de France. Louis XI et ses ascendants. Une vie humaine étudiée à travers six siècles d'hérédité, 852–1483.* Paris: Librairie Hachette, 1903 (1896).

Dr. Brennus. *Amour et securité.* 108th ed. Paris: André Hal, 1908 (1893).

Dr. Brennus. *Histoire du célèbre ouvrage "Amour et sécurité" par Doctor Brennus, poursuivi en cour d'assises à Paris le 29 août 1895*. Paris: F. Montel, 1896.

Brierre de Boismont, Alexandre. *De l'hallucination historique ou Étude médico-psychologique sur les voix et les révélations de Jeanne d'Arc*. Paris: Germer Baillière, 1861.

Brierre de Boismont, Alexandre. *Des hallucinations ou Histoire raisonnée des apparitions, des visions, des songes, de l'extase, du magnétisme et du somnambulisme*. Paris: Germer Baillière, 1845.

Brierre de Boismont, Alexandre. "Réponse à l'article d'Alfred Maury." *Annales médico-psychologiques. Journal de l'anatomie, de la physiologie et de la pathologie du système nerveux* 5 (mai 1845): 339–41.

Broca, Paul. *Recherches sur l'hybridité animale en général et sur l'hybridité humaine en particulier*. Paris: J. Claye, 1860.

Brownlee, Marina, Kevin Brownlee, and Stephen G. Nichols, eds. *The New Medievalism*. Baltimore: Johns Hopkins University Press, 1991.

Bruna, Denis. *Enseignes de pèlerinages et enseignes profanes: Musée national du Moyen Âge-Thermes de Cluny*. Paris: Réunion des musées nationaux, 1996.

Brunet, Gustave. "Tribunaux secrets." In *Le Moyen Âge et la Renaissance*, ed. Lacroix and Seré, vol. 3, n.p.

Brunetière, Ferdinand. "L'érudition contemporaine de la littérature française au Moyen Âge." *La Revue des deux mondes* 33 (1879): 620–49.

Brunetière, Ferdinand. "Les fabliaux du Moyen Âge et l'origine des contes." *La Revue des deux mondes* 119 (1893): 189–213.

Bryder, Linda. "Sex, Race, and Colonialism: An Historiographical Review." *International History Review* 20:4 (1998): 806–22.

Bullough, Vern L., and James A. Brundage, eds. *Handbook of Medieval Sexuality*. New York: Garland, 1996.

Burgwinkle, Bill, and Cary Howie. *Sanctity and Pornography in Medieval Culture: On the Verge*. Manchester: Manchester University Press, 2010.

Burns, E. Jane. "Courtly Love: Who Needs It? Recent Feminist Work in the Medieval French Tradition." *Signs* 27:1 (2001): 23–57.

Burns, E. Jane, et al. "Feminism and the Discipline of Old French Studies: *Une belle disjointure*." In *Medievalism and the Modernist Temper*, ed. Bloch and Nichols, pp. 224–66.

Buzareingues, Charles Girou de. *Philosophie physiologique, politique et morale*. Paris: Firmin Didot, 1828.

Bynum, Caroline Walker. *Wonderful Blood: Theology and Practice in Late Medieval Northern Germany and Beyond*. Philadelphia: University of Pennsylvania Press, 2007.

Dr. Cabanès. *Dans les coulisses de l'Histoire*, vol. 2. Paris: Albin Michel, 1937.

Dr. Cabanès. *La flagellation dans l'histoire et la littérature*. Clermont: Imprimerie Daix Frères, 1899.

Dr. Cabanès. *Le cabinet secret de l'Histoire*, vol. 1. Paris: Albin Michel, 1929 (1895–1898).

Dr. Cabanès. "Le légendaire Barbe-Bleue." In *Légendes et curiosités de l'Histoire*. Paris, Albin Michel, 1922 (1912), vol. 1, pp. 45–147.

Dr. Cabanès. *Le mal héréditaire I. Les descendants de Charles-Quint. Ouvrage orné de 63 illustrations, Le mal héréditaire II. Les Bourbons d'Espagne. Ouvrage orné de 93 illustrations.* Paris: Albin Michel, 1926–1927.

Dr. Cabanès. *Les indiscrétions de l'Histoire.* 6 vols. Paris: Albin Michel, 1903.

Dr. Cabanès. *Les morts mystérieuses de l'Histoire. Souverains et princes français de Charlemagne à Louis XVII,* vol. 1. Paris: A. Maloine, 1901.

Dr. Cabanès. *L'histoire éclairée par la clinique.* Paris, Albin Michel, 1921.

Dr. Cabanès. *Mœurs intimes du passé.* 12 vols. Paris: Albin Michel, 1908–1936.

Caillot, Antoine. *Mémoires pour servir à l'histoire des mœurs et usages des Français, depuis les plus hautes conditions jusqu'aux classes inférieures de la société, pendant le règne de Louis XVI, sous le Directoire exécutif, sous Napoléon Bonaparte, et jusqu'à nos jours.* 2 vols. Paris: Dauvin, 1827.

Calmeil, Louis-Florentin. *De la folie considérée sous le point de vue pathologique, philosophique, historique et judiciaire: Depuis la renaissance des sciences en Europe jusqu'au dix-neuvième siècle.* 2 vols. Paris: J.-B. Baillière, 1845.

Carlson, Eric T. "Medicine and Degeneration: Theory and Praxis." In *Degeneration: The Dark Side of Progress,* ed. Chamberlin and Gilman, pp. 120–44.

Caro, Elme. *M. Littré et le positivisme.* Paris: Libraire Hachette, 1883.

Cartron, Laure. "Degeneration and 'Alienism' in Early Nineteenth-Century France." In *Heredity Produced,* ed. Müller-Wille and Rheinberger, pp. 155–74.

Cauzons, Thomas de. *La magie et la sorcellerie en France.* 3 vols. Paris: Dorbon ainé, 1910–1912.

Cazacu, Matei. *Gilles de Rais.* Paris: Tallandier, 2005.

Cazalis, Dr. Henry. *La science et le mariage. Paris:* Octave Doin, 1900.

Cerquiglini, Bernard. *In Praise of the Variant: A Critical History of Philology.* Trans. Betsy Wing. Baltimore: Johns Hopkins University Press, 1999. (*Éloge de la variante. Histoire critique de la philologie.* Paris: Seuil, 1989.)

Certeau, Michel de. *The Practice of Everyday Life.* Trans. Steven Rendall. Berkeley: University of California Press, 1984. (*L'invention du quotidien.* Paris: Gallimard, 1990 [1980].)

Certeau, Michel de. "Walking in the City." In *The Cultural Studies Reader,* ed. Simon During. New York: Routledge, 1993, pp. 126–33. ("*Marcher dans la ville.*" In *Arts de faire. L'invention du quotidien,* vol. 1. Paris: Gallimard, 1990 [1980], pp. 139–64.)

Chamberlin, J. Edward, and Sander L. Gilman, eds. *Degeneration: The Dark Side of Progress.* New York: Columbia University Press, 1985.

Chambure, Charles Hippolyte Maillard de. *Règle et statuts secrets des Templiers, précédés de l'Histoire de l'établissement, de la destruction et de la continuation moderne de L'ordre du Temple.* Paris: Brockaus et Avenarius; Dijon: Lamarche, 1840.

Champagny, Franz de. *Les Césars.* Paris: Renouard, 1841.

Charcot, Jean-Martin, and Paul Richer. *Les démoniaques dans l'art.* Paris: A. Delahaye et E. Lecrosnier, 1887.

Charcot, Jean-Martin, and Paul Richer. *Les difformes et les malades dans l'art.* Paris: Lecrosnier et Babé, 1889.

Charlton, Donald Geoffrey. *Positivist Thought in France during the Second Empire, 1852–1870*. Oxford: Clarendon Press, 1959.

Chastel, André. "The Notion of Patrimony." In *Rethinking France: Les Lieux de Mémoire, Volume 3: Legacies*, ed. Pierre Nora. Trans. Nancy Turpin. Chicago: University of Chicago Press, 2009, pp. 1–46. ("La notion de patrimoine." In *Les lieux de mémoire*, ed. Nora, vol. 1, pp. 1433–69.)

Cherveix, Jean de. *Deux cents viols par un Maréchal de France*. Paris: Offenstadt, 1903.

Chevalier, Julien. *Une maladie de la personnalité: L'inversion sexuelle*. Paris: G. Masson, 1893.

Christian, J. "Hallucination." In *Dictionnaire encyclopédique des sciences médicales*, ed. Dechambre, 4e série, t. 12 (1886), pp. 77–121.

Cohen, Jeffrey Jerome, ed. *The Postcolonial Middle Ages*. New York: St. Martin's Press, 2000.

Coriveaud, Dr. A. *Le lendemain du mariage, étude d'hygiène*. Paris: J.-B. Baillière, 1884.

Corlieu, Dr. Auguste. *Étude médicale sur la dynastie des Valois*. Paris: Adrien Delahaye, 1872.

Corlieu, Dr. Auguste. *La fistule de Louis XIV*. Paris: Librairie Germer-Baillière, 1874.

Corlieu, Dr. Auguste. *La médecine de l'histoire. Étude médicale sur la dynastie des Valois*. Paris: Adrien Delahaye, 1872.

Coubard, Dr. Ch. "Gilles de Rais devant la médecine et devant la mystique." *Aesculape, Revue mensuelle illustrée des lettres et des arts dans leurs rapports avec les sciences et la médecine* 1 (janvier 1931): 9–16, 40–47, 148–52.

Cronk, Nicholas, and Kris Peeters, eds. *Le Comte de Caylus: Les arts et les lettres. Actes du colloque international Université d'Anvers, UFSIA et Voltaire Foundation, Oxford, 26–27 mai 2000*. Amsterdam: Rodopi, 2004.

Cullerre, Alexandre. *Les frontières de la folie*. Paris: J.-B. Baillière, 1888.

Cuvier, Georges. *Histoire des sciences naturelles*. Paris: Fortin, Masson et Cie, 1841.

Dakyns, Janine R. *The Middle Ages in French Literature, 1851–1900*. London: Oxford University Press, 1973.

Dally, Eugène. *Des dangers attribués aux mariages entre consanguins*. Extrait de la *Gazette hebdomadaire de médecine et de chirurgie* 9 (1862): 1–45.

d'Arc, P. L. "Gilles de Rais, maréchal de France, dit *Barbe-Bleue,* (1404–1440), par l'abbé Eugène Bossard." *Revue catholique des institutions et du droit* 30 (1888): 463–67.

Daremberg, Charles. *Faculté de médecine de Paris. Cours sur l'histoire de la médecine et de la chirurgie, par le professeur Ch. Daremberg. Leçon d'ouverture, le 11 novembre 1871*. Paris: F. Malteste, 1871.

Daremberg, Charles. "Préface" and "Introduction." In *Histoire des sciences médicales*, vol. 1. Paris: J.-B. Baillière, 1870, pp. vii–xxviii, 1–66.

Davis, Kathleen, and Nadia Altschul, eds. *Medievalisms in the Postcolonial World: The Idea of "the Middle Ages" Outside Europe: Rethinking Theory*. Baltimore: Johns Hopkins University Press, 2010.

Debay, Auguste. *Philosophie du mariage. Histoire de l'homme et de la femme mariés, dans leurs rapports physiques et moraux*. Paris: Moquet, 1849.

Debray, Dr. Th.-F. *Histoire de la prostitution et de la débauche chez tous les peuples du globe depuis l'antiquité jusqu'à nos jours*. Paris: S. Lambert, 1879.

Debret, Dr. Fernand. *La sélection naturelle dans l'espèce humaine. Contribution à l'étude de l'hérédité convergente*. Paris: G. Steinheil, 1901.

Dechambre, Amédée, Jacques Raige-Delorme, Léon Lereboullet, and François Louis Hahn. *Dictionnaire encyclopédique des sciences médicales*. 100 vols. Paris: Masson; Asselin et Houzeau, 1864–1889.

Dehérain, Henri. *Silvestre de Sacy, ses contemporains et ses disciples. Bibliothèque archéologique et historique*, vol. 27. Paris: Librairie orientaliste Paul Geuthner, 1938.

DeJean, Joan E. *The Reinvention of Obscenity: Sex, Lies, and Tabloids in Early Modern France*. Chicago: University of Chicago Press, 2002.

Dejerine, Jules. *L'hérédité dans les maladies du système nerveux*. Paris: Anselin et Huzeau, 1886.

Delage, Yves. *L'hérédité et les grands problèmes de la biologie générale*. 2nd ed. Paris: Librairie C. Reinwald, 1903 (1895).

Derennes, Charles. *Amours et crimes*. Paris: Éditions de France, 1927.

Descuret, J.-B.-F. *La médecine des passions, ou Les passions considérées dans leurs rapports avec les maladies, les lois et la religion*. 2 vols. 3rd rev. ed. Paris: Labé, 1860 (1841).

Désormaux, Dr. Raoul. *L'amour conjugal. Bibliothèque sexuelle du Dr Désormeaux*, vol. 11. Paris: Librairie P. Fort, n.d. (1905).

Désormaux, Dr. Raoul. *Le mariage. Bibliothèque seuxelle du Dr Désormeaux*, vol. 5. Paris: Librairie P. Fort, n.d. (1905).

Devay, Dr. Francis. *Du danger des mariages consanguins au point de vue sanitaire*. Paris: Labé; Lyon: M. Savy, 1857.

Devay, Francis. *Du danger des mariages consanguins sous le rapport sanitaire*. 2nd rev. ed. Paris: Victor Masson, 1862.

Devay, Dr. Francis. *Hygiène des familles ou perfectionnement physique et moral de l'homme considéré particulièrement dans ses rapports avec l'éducation et les besoins de la civilisation moderne*. 2 vols. Lyon: Dorier, 1846.

Devay, Dr. Francis. *Nouvelles observations sur les dangers des mariages entre consanguins au point de vue sanitaire*. Paris: Victor Masson, 1860.

Devay, Dr. Francis. *Traité spécial d'hygiène des familles, particulièrement dans ses rapports avec le mariage, au physique et au moral et les maladies héréditaires*. 2nd rev. ed. Paris: Labé, 1858.

Devay, Dr. Francis. *Un mot sur le danger des mariages consanguins, réponse à une attaque. État de la question*. Paris: Victor Masson, 1863.

Deville, Dr. Henry. *L'amour dans le mariage. Le médecin populaire*, vol. 2. Paris: L. Boulanger, n.d. (1893).

Deville, Dr. Henry. *L'amour libre. Le médecin populaire*, vol. 3. Paris: L. Boulanger, n.d. (1894).

Dorlin, Elsa. *La matrice de la race: Généalogie sexuelle et coloniale de la nation française*. Paris: Éditions la Découverte, 2006.

Dowbiggin, Ian R. *Inheriting Madness: Professionalization and Psychiatric Knowledge in Nineteenth-Century France*. Berkeley: University of California Press, 1991.

Dromard, Dr. Gabriel. *Les fausses 'messes noires'. Causeries psychologiques et médico-légales sur quelques scandales modernes*. Paris: Éditions de la Revue libre, n.d. (1904).

Dubuisson, Paul, and A. Vigouroux. *Responsabilité pénale et folie, étude médico-légale*. Paris: Librairie Felix Alcan, 1911.

Duby, Georges. *The Chivalrous Society*. Trans. Cynthia Postan. Berkeley: University of California Press, 1980.

Duby, Georges. "French Genealogical Literature." In *The Chivalrous Society*, pp. 149–57.

Duby, Georges. *La société aux XIe et XIIe siècles dans la région mâconnaise*. Paris: A. Colin, 1953.

Duby, Georges. "Lineage, Nobility and Knighthood. The Mâconnais in the Twelfth Century: A Revision." In *The Chivalrous Society*, pp. 59–80.

Duby, Georges. "The Nobility in Medieval France." In *The Chivalrous Society*, pp. 94–111.

Duby, Georges. "The Structure of Kinship and Nobility." In *The Chivalrous Society*, pp. 134–48.

Duby, Georges. "Youth in Aristocratic Society: Northwestern France in the Twelfth Century." In *The Chivalrous Society*, pp. 112–22.

Dulaure, J.-A. *Histoire physique, civile et morale de Paris*. 7th ed. Paris: Dutertre, 1867 (1821–1825).

Dusolier, Émile. *Psychologie des derniers Valois (François II, Charles IX, Henri III). Thèse présentée à la Faculté de médecine et de pharmacie de Lyon, et soutenue publiquement le mercredi 27 novembre 1895*. Lyon: Imprimerie de A.-H. Storck, 1895.

Eigen Figal, Sara. *Heredity, Race, and the Birth of the Modern*. New York: Routledge, 2008.

Emery, Elizabeth. *Romancing the Cathedral: Gothic Architecture in Fin-De-Siècle French Culture*. Albany: State University of New York Press, 2001.

Emery, Elizabeth. "The 'Truth' about the Middle Ages: *La Revue des Deux Mondes* and Late Nineteenth-Century French Medievalism." In *Medievalism and the Quest for the "Real" Middle Ages*, ed. Clare A. Simmons. Portland, Ore.: Frank Cass, 2001, pp. 99–114.

Emery, Elizabeth, and Laura Morowitz. *Consuming the Past: The Medieval Revival in Fin-De-Siècle France*. Aldershot, UK: Ashgate, 2003.

Encyclopaedia Britannica, 11th ed. London: Encyclopaedia Britannica, 1910–1911.

Erber, Nancy. "Queer Follies: Effeminacy and Aestheticism in Fin-De-Siècle France, the Case of Baron d'Adelsward Fersen and Count de Warren." In *Disorder in the Court: Trials and Sexual Conflict at the Turn of the Century*, ed. George Robb and Nancy Erber. New York: New York University Press, 1999, pp. 186–208.

Esquirol, Étienne. "Folie." In *Dictionnaire des sciences médicales*, ed. Charles Louis Fleury Pancoucke, vol. 16. Paris, 1816, pp. 151–240.

Esquirol, Étienne. *Mental Maladies: A Treatise on Insanity*. Trans. E. K. Hunt. New York: Hafner Publishing Co., 1965 (facsimile of 1845 English ed.). (*Des maladies mentales, considérées sous les rapports médical, hygiénique et médico-légal*. Paris: J.-B. Baillière, 1838 [1817].)

Evergates, Theodore. *The Aristocracy in the County of Champagne, 1100–1300*. Philadelphia: University of Pennsylvania Press, 2007.

Everson, Ted. *The Gene: A Historical Perspective*. Westport, Conn.: Greenwood Press, 2007.

Fauconney, Dr. Jean (pseud. Dr. Caufeynon). *L'amour et l'accouplement. Bibliothèque populaire des connaissances médicales*, vol. 6. Paris: Nouvelle librairie médicale, 1902.

Fauconney, Dr. Jean (pseud. Dr. Caufeynon). *L'amour secret*. Paris: Charles Offenstadt, 1903.

Fauconney, Dr. Jean (pseud. Dr. Caufeynon). *La procréation à volonté des filles et des garçons, suivie de La fécondation artificielle, et de L'ami des jeunes femmes*. Paris: P. Fort, n.d. (1903).

Fauconney, Dr. Jean (pseud. Dr. Caufeynon). *La prostitution à travers les âges, depuis les temps les plus reculés*. Paris: Librairie P. Fort, n.d. (1903).

Fauconney, Dr. Jean (pseud. Dr. Caufeynon). *Le mariage et son hygiène*. Paris: Librairie des connaissances médicales, 1905.

Fauconney, Dr. Jean (pseud. Drs. Caufeynon and Jaf). *Les messes noires. Le culte de Satan-Dieu*. Paris: Pancier, 1905.

Fauconney, Dr. Jean (pseud. Dr. Jaf). *Physiologie du vice, son histoire à travers les âges*. Paris: C. Offenstadt, 1903.

Fauriel, Claude. *Histoire de la poésie provençale*. 3 vols. Genève: Slatkine, 1969 (1846).

Fauriel, Claude. "Origine de l'épopée chevaleresque du Moyen Âge." *La Revue des deux mondes* 7 (1832): 513–75, 672–710; 8 (1832): 138–94, 268–308, 425–70.

Féré, Charles. *Dégénérescence et criminalité: Essai physiologique*. Paris: Félix Alcan, 1888.

Féré, Charles. *La famille névropathique: Théorie tératologique de l'hérédité et de la prédisposition morbides et de la dégénérescence*. 2nd ed. Paris: Félix Alcan, 1898 (1894).

Ferlampin-Acher, Christine, and Denis Hüe, eds. *Lignes et lignages dans la littérature arthurienne. Actes du 3e colloque arthurien organisé à l'université de Haute-Bretagne, 13–14 octobre 2005*. Rennes: Presses Universitaires de Rennes, 2007.

Ferrante, Joan M. "*Cortes'Amor* in Medieval Texts." *Speculum* 55:4 (1980): 686–95.

Fleuret, Fernand. *De Gilles de Rais à Guillaume Apollinaire*. Paris: Mercure de France, 1933.

Fleuret, Fernand (pseud. Ludovico Hernandez). *Le procès inquisitorial de Gilles de Rais, maréchal de France avec un essai de réhabilitation*. Paris: Bibliothèque des curieux, 1921.

Forgeais, Arthur. *Collection de plombs historiés trouvés dans la Seine et recueillis par Arthur Forgeais, I. Méreaux des corporations de métiers; II. Enseignes de pélerinages; III. Variétés numismatiques; IV. Imagerie religieuse; V. Numismatique populaire*. Paris: Aubry, 1862–1866.

Forgeais, Arthur. *Priapées*. N.p.: n.d.

Foucart, Bruno. "Viollet-le-Duc et la restauration." In *Les lieux de mémoire*, ed. Nora, vol. 1, pp. 1615–43.

Foucault, Michel. *The Archaeology of Knowledge*. Trans. A. M. Sheridan Smith. New York: Pantheon Books, 1972. (*L'Archéologie du savoir*. Paris: Gallimard, 1969.)

Foucault, Michel. *The Birth of the Clinic: An Archaeology of Medical Perception*. Trans. A. M. Sheridan Smith. New York: Vintage Books, 1994. (*Naissance de la clinique*. Paris: Quadrige/Presses Universitaires de France, 1997 [1963].)

Foucault, Michel. "The Confession of the Flesh." In *Power/Knowledge: Selected Interviews and Other Writings, 1972–1977*, ed. Colin Gordon. Trans. Colin Gordon, Leo Marshall, John Mepham, and Kate Soper. New York: Pantheon Books, 1980, pp. 194–228. ("Le jeu de Michel Foucault." *Dits et écrits II, 1976–1988*, ed. Daniel Defert and François Ewald. Paris: Quarto/Gallimard, 2001 [1977], pp. 298–329.)

Foucault, Michel. *The History of Sexuality, Volume I: An Introduction*. Trans. Robert Hurley. New York: Vintage, 1990. (*Histoire de la sexualité I: La Volonté de savoir*. Paris: Gallimard/Tel 2005 [1976]).

Foucault, Michel. *"Il faut défendre la société." Cours au Collège de France, 1975*. Paris: Gallimard/Seuil, 1997.

Foucault, Michel. *Naissance de la biopolitique, Cours au Collège de France, 1978–1979*. Paris: Gallimard/Seuil, 2004.

Foucault, Michel. "Nietzsche, Genealogy, History." In *Language, Counter-Memory, Practice: Selected Essays and Interviews*, ed. Donald F. Bouchard. Trans. Donald F. Bouchard and Sherry Simon. Ithaca, N.Y.: Cornell University Press, 1977, pp. 139–64. ("Nietzsche, la généalogie, l'histoire." *Dits et écrits I, 1954–1975*, ed. Daniel Defert and François Ewald. Paris: Quarto Gallimard, 2001 [1971], pp. 1004–24.)

Fradenburg, Louise, and Carla Freccero, eds. *Premodern Sexualities*. New York: Routledge, 1996.

Frappier, Jean. "'Amour courtois.'" In *Amour Courtois et Table Ronde*. Genève: Droz, 1973, pp. 33–41.

Frappier, Jean. "Sur un procès fait à l'amour courtois." *Romania* 93 (1972): 145–93. Reprinted in *Amour Courtois et Table Ronde*. Genève: Droz, 1973, pp. 61–96.

Freedman, Paul, and Gabrielle M. Spiegel. "Medievalisms Old and New: The Rediscovery of Alterity in North American Medieval Studies." *American Historical Review* 103:3 (1998): 677–704.

Gabory, Émile. *Alias Bluebeard: The Life and Death of Gilles de Raiz*. Trans. Alvah C. Bessie. New York: Brewer & Warren, 1930. (*La vie et la mort de Gilles de Raiz, dit - à tort - Barbebleue*. Paris: Perrin, 1926.)

Gabory, Émile. *La psychologie de Gilles de Rais, extrait de la Revue du Bas-Poitou*. Fontenay-le-Comte: H. Lussaud, 1925.

Galippe, Dr. Victor. *L'hérédité des stigmates de dégénérescence et les familles souveraines*. Paris: Masson, 1905.

Galton, Sir Francis. *Hereditary Genius: An Inquiry into Its Laws and Consequences*. London: Macmillan, 1869.

Ganim, John M. *Medievalism and Orientalism*. New York: Palgrave Macmillan, 2005.

Garnier, Dr. Pierre. *La génération universelle*. 4th rev. ed. Paris: Garnier Frères, 1887.

Garnier, Dr. Pierre. *Le mariage dans ses devoirs, ses rapports et ses effets conjugaux au point de vue légal, hygiénique, physiologique et moral*. Paris: Garnier Frères, 1879.

Dr. Gébé. *L'amour conjugal: Conseils*. Paris: F. Pigeon, 1891.

Genil-Perrin, Georges. *Histoire des origines et l'évolution de l'idée de dégénérescence en médecine mentale*. Paris: Leclerc, 1913.

Gillispie, Charles Coulston. *The Edge of Objectivity: An Essay in the History of Scientific Ideas*. Princeton, N.J.: Princeton University Press, 1960.

Gillispie, Charles Coulston. *Science and Polity in France: The Revolutionary and Napoleonic Years*. Princeton, N.J.: Princeton University Press, 2004.

Giron, Aimé, and Albert Tozza. *La bête de luxure. Gilles de Rais*. 4th ed. Paris: Ambert, 1907.

Glencross, Michael. *Reconstructing Camelot: French Romantic Medievalism and the Arthurian Tradition*. Cambridge: D. S. Brewer, 1995.

Gobineau, Arthur de. *Essai sur l'inégalité des races humaines*. 2 vols. 2nd ed. Paris: Firmin-Didot, 1884 (1853–1855).

Goldstein, Jan. *Console and Classify: The French Psychiatric Profession in the Nineteenth Century*. Cambridge: Cambridge University Press, 1987.

Gossman, Lionel. *Medievalism and the Ideologies of the Enlightenment: The World and Work of LaCurne de Sainte-Palaye*. Baltimore: Johns Hopkins University Press, 1968.

Gourevitch, Danielle. "Charles Daremberg, His Friend Émile Littré, and Positivist Medical History." In *Locating Medical History: The Stories and Their Meanings*, ed. Frank Huisman and John Harley Warner. Baltimore: Johns Hopkins University Press, 2004, pp. 53–73.

Gourevitch, Danielle, ed. *Médecins érudits de Coray à Sigerist. Actes du colloque de Saint-Julien-en-Beaujolais, juin 1994*. Paris: De Broccard, 1995.

Grmek, Mirko D., ed. *Histoire de la pensée médicale en Occident*, vol. 3. Trans. Louise Lambrichs. Paris: Seuil, 1999.

Grmek, Mirko D. *Les maladies à l'aube de la civilisation occidentale*. Paris: Payot, 1983.

Guardia, José Miguel (Joseph-Michel). *La médecine à travers les siècles*. Paris: J.-B. Baillière, 1865.

Guéraud, Armand. *Notice sur Gilles de Rais*. Rennes: Imprimerie de A. Marteville et Oberthur, 1855, pp. 3–74. (Reprinted, with minor revisions, as A. Guéraud, "Rais, Gilles, baron de." In *Biographie bretonne*, ed. P. Levot, vol. 2. Vannes: Cauderan; Paris: Dumoulin, 1857, pp. 677–91.)

Guerra, Francisco. "The Description of Syphilis in Avicenna." *27th International Congress of the History of Medicine, 1980*. Barcelona: Academia de Ciencies Mediques de Catalunya i Balears, 1981, pp. 731–33.

Guerra, Francisco. "The Dispute over Syphilis: Europe versus America." *Clio Medica* 13:1 (1978): 39–61.

Guerreau-Jalabert, Anita. "La désignation des relations et groupes de parenté en latin médiéval." *Bulletin du Cange* 46–47 (1988): 65–108.

Guerreau-Jalabert, Anita. "La parenté dans l'Europe médiévale et moderne: À propos d'une synthèse récente." *L'homme* 110 (1989): 69–93.

Guerreau-Jalabert, Anita. "Sang." In *Dictionnaire du Moyen Âge*, ed. Claude Gauvard, Alain de Libera, and Michel Zink. Paris: Presses universitaires de France, 2002, pp. 1280–81.

Guerreau-Jalabert, Anita. "*Spiritus* et *caritas*. Le baptême dans la société médiévale." In *La parenté spirituelle*, ed. Françoise Héritier-Augé and Elisabeth Copet-Rougier. Paris: Éditions des archives contemporaines, 1995, pp. 133–203.

Guizot, François. *Mémoires pour servir à l'histoire de mon temps.* Paris: Michel-Lévy frères, 1858–1867.

Gumbrecht, Hans Ulrich. "'Un souffle d'Allemagne ayant passé': Friedrich Diez, Gaston Paris, and the Genesis of National Philologies." *Romance Philology* 40:1 (1986): 1–37.

Guyot, Dr. Jules. *Bréviaire de l'amour expérimental, méditations sur le mariage selon la physiologie du genre humain.* Paris: Marpon et Flammarion, 1882.

Hahn, Thomas, ed. *Race and Ethnicity in the Middle Ages. Journal of Medieval and Early Modern Studies* 31:1 (2001).

Hamblin-Smith, Dr. M. "Introduction." In A. L. Vincent and Clare Binns, *Gilles de Rais: The Original Bluebeard.* London: A.M. Philpot, 1926, pp. 7–35.

Hammer-Purgstall, Joseph von, introduction and trans. *Ancient Alphabets and Hieroglyphic Characters Explained: With an Account of the Egyptian Priests, their Classes, Initiation, and Sacrifices.* London, 1806.

Hammer-Purgstall, Joseph von. *Mémoire des deux coffrets gnostiques du Moyen Âge, du cabinet de M. le Duc de Blacas* (Paris: Librairie orientale de Dondey-Dupré père et fils, 1832).

Hammer-Purgstall, Joseph von. "Mysterium Baphometis revelatum." In *Fundgruben des Orients, bearbeitet durch eine Gesellschaft von Liebhabern*, vol. 6. Wien: Anton Schmid, 1818, pp. 3–120, 445–92.

d'Hancarville, Pierre-François Hugues. *Monumens de la vie privée des douze Césars, d'après une suite de pierres gravées sous leur règne.* Rome, 1780.

Harsin, Jill. *Policing Prostitution in Nineteenth-Century Paris.* Princeton, N.J.: Princeton University Press, 1985.

Hayem, Armand. *Le mariage.* Paris: Didier, 1872.

Dr. Hayès. *La pédérastie, historique, conséquences funestes de ce vice honteux. Bibliothèque d'hygiène des deux sexes*, vol. 7. Paris: Librairie des publications modernes, 1891.

Hebert, J. "Une cause célèbre au XVe siècle: Gilles de Rais." *Bulletin de la Société académique de Brest* 2:18 (1892–1893): 183–303.

Heers, Jacques. *Gilles de Rais.* Paris: Perrin, 1994.

Hepp, Noémi. "Gallantry." In *Rethinking France: Les Lieux de Mémoire, Volume 3: Legacies*, ed. Pierre Nora. Trans. Deke Dunisberre. Chicago: University of Chicago Press, 2009, pp. 233–73. ("Galanterie." *Les lieux de mémoire*, ed. Nora, vol. 3, pp. 3677–710.)

Holsinger, Bruce. *The Premodern Condition: Medievalism and the Making of Theory.* Chicago: University of Chicago Press, 2005.

Hugo, Victor. "Préface." *Les Orientales.* Paris: Librairie de L. Hachette, 1858 (1829).

Hult, David F. "Gaston Paris and the Invention of Courtly Love." In *Medievalism and the Modernist Temper*, ed. Bloch and Nichols, pp. 193–224.

Hult, David F. "Reading It Right: The Ideology of Text Editing." *Romanic Review* 79:1 (1988): 74–88.

Hunt, Lynn. "Introduction: Obscenity and the Origins of Modernity." In *The Invention of Pornography: Obscenity and the Origins of Modernity, 1500–1800*, ed. Lynn Hunt. New York: Zone Books, 1993, pp. 9–45.

Hunt, Lynn. "Pornography and the French Revolution." In *The Invention of Pornography: Obscenity and the Origins of Modernity, 1500–1800*, ed. Lynn Hunt. New York: Zone Books, 1993, pp. 301–39.

Hunt, Lynn, Margaret C. Jacob, and Wijnand Mijnhardt. *The Book That Changed Europe: Picart & Bernard's Religious Ceremonies of the World.* Cambridge, Mass.: Belknap Press of Harvard University Press, 2010.

Huysmans, J.-K. *Là-Bas: A Journey into the Self.* Trans. Brendan King. Cambs, UK: Dedalus, 2001. (*Là-bas.* Paris: Gallimard/Folio, 1985 [1891].)

Huysmans, J.-K. *La magie en Poitou. Gilles de Rais.* Ligugé, 1899.

Jacob, François. *The Logic of Life: A History of Heredity.* Princeton, N.J.: Princeton University Press, 1993 (1970).

Jacoby, Paul. *Études sur la sélection chez l'homme.* 2nd ed. Paris: Felix Alcan, 1904 (1881).

Jones, Malcolm. *The Secret Middle Ages.* Stroud, Gloucestershire, UK: Sutton, 2002.

Jubinal, Achille. *Jongleurs et trouvères, ou Choix de saluts, épîtres, rêveries et autres pièces légères des XIIIe et XIVe siècles.* Paris: J. A. Merklein, 1835.

Jubinal, Achille. *Nouveau recueil de contes, dits, fabliaux et autres pièces inédites des XIIIe, XIVe et XVe siècles, pour faire suite aux collections de Legrand d'Aussy, Barbazan et Méon.* 2 vols. Paris: E. Pannier, 1839–1842.

Jubinal, Achille. *Recherches sur l'usage et l'origine des tapisseries à personnages dites historiées, depuis l'antiquité jusqu'au XVIe siècle inclusivement.* Paris: Challamel, 1840.

Keller, Barbara G. *The Middle Ages Reconsidered: Attitudes in France from the Eighteenth Century through the Romantic Movement.* New York: Peter Lang, 1994.

Kelly, Henry Ansgar. "Gaston Paris's Courteous and Horsely Love." In *The Spirit of the Court: Selected Proceedings of the Fourth Congress of the International Courtly Literature Society, Toronto 1983*, ed. Glyn S. Burgess and Robert A. Taylor. Woodbridge, Suffolk: D. S. Brewer, 1985, pp. 217–23.

Kelly, Henry Ansgar. "The Varieties of Love in Medieval Literature." *Romance Philology* 40:3 (1987): 301–27.

Kendrick, Laura. "The Science of Imposture and the Professionalization of Medieval Occitan Literary Studies." In *Medievalism and the Modernist Temper*, ed. Bloch and Nichols, pp. 95–126.

Kendrick, Walter. *The Secret Museum: Pornography in Modern Culture.* Berkeley: University of California Press, 1996 (1987).

Kim, Ji-hyun Philippa. "For a Modern Medieval Literature: Gaston Paris, *Courtly Love* and the Demands of Modernity." Ph.D. diss., Harvard University, 2005.

Kinoshita, Sharon. "Deprovincializing the Middle Ages." In *The Worlding Project: Doing Cultural Studies in the Era of Globalization*, ed. Rob Wilson and Christopher Leigh Connery. Santa Cruz, Calif.: New Pacific Press, 2007, pp. 75–89.

Kinoshita, Sharon. *Medieval Boundaries: Rethinking Difference in Old French Literature.* Philadelphia: University of Pennsylvania Press, 2006.

Klapisch-Zuber, Christiane. "La genèse de l'arbre généalogique." In *L'arbre. Histoire naturelle et symbolique de l'arbre, du bois et du fruit au Moyen Âge*, ed. Michel Pastoureau. Paris: Le Léopard d'Or, 1993, pp. 41–81.

Klapisch-Zuber, Christiane. "Le corps de la parenté." *Micrologus* 1 (1993): 43–60.

Klapisch-Zuber, Christiane. *L'ombre des ancêtres: Essai sur l'imaginaire médiéval de la parenté*. Paris: Fayard, 2000.

Kottek, Samuel, and Danielle Gourevitch. "Un correspondant allemand, malchanceux de Daremberg, Julius *Rosenbaum*, 1807–1874." In *Médecins érudits de Coray à Sigerist*, ed. Gourevitch, pp. 71–87.

Krafft-Ebing, Richard von. *La responsabilité criminelle et la capacité civile dans les états de trouble intellectuel*. Trans. Dr. Chatelain. Paris: G. Masson, 1875.

Krafft-Ebing, Richard von. *Psychopathia Sexualis: A Clinical-Forensic Study*. Trans. F. J. Rebman from the twelfth and final edition, complete and unabridged. Burbank, Calif.: Bloat, 1999 (1886).

Lacan, Jacques. *The Seminar of Jacques Lacan, Book VII: The Ethics of Psychoanalysis (1959–1960)*. Ed. Jacques-Alain Miller. Trans. Dennis Porter. New York: W. W. Norton & Co., 1992. (*Le Séminaire 7 [1959–1960]: L'éthique de la psychanalyse*, ed. Jacques-Alain Miller. Paris: Éditions du Seuil, 1986.)

Lacassagne, Alexandre. "Consanguinité." In *Dictionnaire encyclopédique des sciences médicales*, ed. Dechambre, 1ère série, t. 19 (1876), pp. 652–716.

Lacassagne, Alexandre. "Pédérastie." In *Dictionnaire encyclopédique des sciences médicales*, ed. Dechambre, 3e série, t. 22 (1886), pp. 239–59.

Lacassagne, Alexandre. "Préface." In Dr. Cabanès, *Les morts mystérieuses de l'Histoire*, pp. v–xi.

Lacassagne, Alexandre. "Préface." In Chevalier, *Une maladie de la personnalité*, pp. i–vi.

Lacassagne, Alexandre. "Préface." In Raffalovich, *Uranisme et unisexualité*, n.p.

Lacroix, Paul. "Bibliophiles et bibliomanes." *Le Bibliophile, Gazette illustrée des amateurs et bibliophiles des deux mondes* (1883–1885): 49–52, 87–94.

Lacroix, Paul (pseud. Pierre Dufour). *Histoire de la prostitution, chez tous les peuples du monde depuis l'antiquité la plus reculée jusqu'à nos jours*. 6 vols. Paris: Seré; Martinon, 1851–1853.

Lacroix, Paul. *La bibliothèque de Jules Janin*. Paris: Librairie des bibliophiles, 1877.

Lacroix, Paul (pseud. P.-L. Jacob, bibliophile). "Le Maréchal de Rais." In *Curiosités de l'histoire de France, procès célèbres*. Paris: Adolphe Delahays, 1858, pp. 1–119.

Lacroix, Paul and Ferdinand Seré, eds. *Le Moyen Âge et la Renaissance: Histoire et description des mœurs et usages, du commerce et l'industrie, des sciences, des arts, des littératures et des beaux-arts en Europe*. 5 vols. Paris, 1848–1851.

Lacroix, Paul. *Les vieux conteurs français*. Paris: Société du Panthéon littéraire, 1841.

Lacroix, Paul (pseud. P.-L. Jacob, bibliophile). "Origines du Mal de Naples." In *Curiosités de l'histoire des croyances populaires au Moyen Âge*. Paris: Adolphe Delahays, 1859, pp. 239–323.

Lacroix, Paul (pseud. Bibliophile Jacob). *Physiologie des rues de Paris. Suivie de notes et renseignements et d'un plan de Paris et de ses faubourgs*, by Ch. Piquet. Paris: Martinon; Bureau du Musée des familles, 1842. (Reprinted in *Curiosités de l'histoire du vieux Paris*. Paris: Adolphe Delahays, 1858, pp. 5–34.)

Lacroix, Paul (pseud. P.-L. Jacob, bibliophile). *Recherches historiques sur les Maladies de Vénus, dans l'antiquité et le Moyen Âge*. Bruxelles: Brancart, 1883.

Lacroix, Paul. *Sciences et lettres au Moyen Âge et à l'époque de la Renaissance*. Paris: Firmin-Didot, 1877.

Lacroix, Paul (pseud. P.-L. Jacob). *Un divorce. Histoire du temps de l'Empire, 1812–1814*. Paris: E. Renduel, 1832.

La Curne de Sainte-Palaye, Jean-Baptiste de. *Mémoires sur l'ancienne chevalerie*. Ed. Charles Nodier. 2 vols. Paris: Girard, 1826 (1759–1781).

Ladenson, Elisabeth. *Art for Dirt's Sake: Books on Trial from* Madame Bovary *to* Lolita. Ithaca, N.Y.: Cornell University Press, 2007.

La Marche, A. Lecoy de. *La société au XIIIe siècle*. Paris: V. Palmé, 1880.

La Marche, A. Lecoy de. *L'esprit de nos aïeux: Anecdotes et bons mots tirés des manuscrits du XIIIe siècle*. Paris: C. Marpon et E. Flammarion, 1888.

Lamartine, Alphonse de. "Joan of Arc." In *Memoirs of Celebrated Characters*, vol. 2. New York: Harper, 1854, pp. 49–118. ("Jeanne d'Arc." In *Vies de quelques hommes illustres. Œuvres complètes de Lamartine publiées et inédites*, vol. 35. Paris: L'auteur, 1863 [1855–1856], pp. 121–95.)

Lamartine, Alphonse de. *Souvenirs, impressions, pensées et paysages, pendant un voyage en Orient (1832–1833), ou Notes d'un voyageur*. 2 vols. Paris: C. Gosselin, 1835.

Lamartine, Alphonse de. *Vues, discours et articles sur la question d'Orient*. Paris: C. Gosselin, 1840.

Langlois, Charles-Victor. *Saint Louis, Philippe le Bel, les derniers Capétiens directs, 1226–1328*. In *Histoire de France depuis les origines jusqu'à la Révolution*, ed. Ernest Lavisse, vol. 3, part 2. Paris: Hachette, 1901.

Lanson, Gustave. *Histoire de la littérature française*. 5th ed. Paris: Hachette, 1898 (1895).

Larger, Dr. René. *Théorie de la contre-évolution ou dégénérescence par l'hérédité pathologique*. Paris: Félix Alcan, 1917.

Larousse, Pierre. *Grand dictionnaire universel du XIXe siècle*. 17 vols. Paris: Administration du Grand dictionnaire universel, 1866–1877.

Larousse, Pierre. *Grand dictionnaire universel du XIXe siècle, Supplément II*. Paris: Administration du Grand dictionnaire universel, 1890.

Dr. Laupts. *Perversion et perversité sexuelles*. Paris: Georges Carré, 1896.

Lea, Henry Charles. "Gilles de Rais." In *History of the Inquisition of the Middle Ages*, vol. 3. New York: Harper, 1888, pp. 468–88.

Le Brun, Annie. *Les châteaux de la subversion*. Paris: J.-J. Pauvert, 1982.

Le Goff, Jacques. "Métiers licites et métiers illicites dans l'Occident médiéval." In *Pour un autre Moyen Âge*. Paris: Gallimard, 1977, pp. 91–107.

Le Goff, Jacques, with Nicolas Truong. *Une histoire du corps au Moyen Âge*. Paris: Éditions Liana Levi, 2003.

Legrand d'Aussy, Pierre Jean-Baptiste. *Fabliaux ou contes, fables et romans du XIIe et du XIIIe siècle.* Ed. A. A. Renouard. 5 vols. Paris: J. Renouard, 1829 (1781).

Lélut, Louis-Francisque. *Du démon de Socrate: Spécimen d'une application de la science psychologique à celle de l'histoire.* Paris: Trinquart, 1836.

Lélut, Louis-Francisque. *L'amulette de Pascal: Pour servir à l'histoire des hallucinations.* Paris: J.-B. Baillière, 1846.

Léonard, Jacques. *La médecine entre les savoirs et les pouvoirs: Histoire intellectuelle et politique de la médecine française au XIXe siècle.* Paris: Aubier Montaigne, 1981.

Leroy-Beaulieu, Paul. *La question de la population.* Paris: F. Alcan, 1913.

Letourneau, Charles. "Hérédité." In *Dictionnaire encyclopédique des sciences médicales,* ed. Dechambre, 4e série, t. 13 (1888), pp. 588–605.

Leuret, François. *Fragmens psychologiques sur la folie.* Paris: Crochard, 1834.

Lévi, Éliphas (Alphonse Constant). *Dogme et rituel de la haute magie.* 2 vols. Paris: G. Baillière, 1856.

Libri, Guillaume. *Lettre à M. de Falloux, ministre de l'Instruction publique et des cultes, contenant le récit d'une odieuse persécution.* Paris: Paulin, 1849.

Libri, Guillaume. *Réponse de M. Libri au rapport de M. Boucly.* Paris: 1848.

Libri, Mélanie. *Pétition adressée au Sénat sur l'affaire de M. Libri, avec une note à l'appui signée par* MM. Guizot, le marquis d'Audiffret, Prosper Mérimée, Édouard Laboulaye, Victor Leclerc, Paulin Paris, Jules Pelletier, Alfred de Wailly, Romain Merlin, Henry Celliez. Paris: C. Lahure, 1861.

Littré, Émile. *Auguste Comte et la philosophie positive.* Paris: Librairie Hachette, 1863.

Littré, Émile. *Conservation, révolution et positivisme.* Paris: Ladrange, 1852.

Littré, Émile. *De la philosophie positive.* Paris: Ladrange, 1845.

Littré, Émile. *Études sur les barbares et le Moyen Âge.* Paris: Didier, 1869.

Littré, Émile. *Médecine et médecins.* 2nd ed. Paris: Didier, 1872.

Littré, Émile, and Charles Robin. *Dictionnaire de médecine, de chirurgie, de pharmacie, de l'art vétérinaire et des sciences qui s'y rapportent.* 14th rev. ed. Paris: J.-B. Baillière, 1878.

Littré, Émile, and Charles Robin. *Dictionnaire de médecine, de chirurgie, de pharmacie, des sciences accessoires et de l'art vétérinaire de P.-H. Nysten.* 10th ed. Paris: J.-B. Baillière, 1855.

Livingstone, Amy. *Out of Love for My Kin: Aristocratic Family Life in the Lands of the Loire, 1000–1200.* Ithaca, N.Y.: Cornell University Press, 2010.

Lochrie, Karma. "Desiring Foucault." *Journal of Medieval and Early Modern Studies* 27:1 (1997): 3–16.

Lochrie, Karma. *Heterosyncrasies: Female Sexuality When Normal Wasn't.* Minneapolis: University of Minnesota Press, 2005.

Lochrie, Karma, Peggy McCracken, and James A. Schultz, eds. *Constructing Medieval Sexuality.* Minneapolis: University of Minnesota Press, 1997.

Loiseleur, Jules. *La doctrine secrète des Templiers.* Cazilhac: Belisane, 1999 (1873).

Lombroso, Cesare. *L'uomo delinquente, studiato in rapporto alla antropologia, alla medicina legale ed alle discipline carcerarie.* Milan: U. Hoepli, 1876.

Lombroso, Cesare. *L'uomo di genio in rapporto alla psichiatria, alla storia ed all'estetica.* Turin: Fratelli Bocca, 1888.

López-Beltrán, Carlos. "In the Cradle of Heredity: French Physicians and 'l'hérédité naturelle' in the Early Nineteenth Century." *Journal of the History of Biology* 37 (2004): 39–72.

López-Beltrán, Carlos. "The Medical Origins of Heredity." In *Heredity Produced*, ed. Müller-Wille and Rheinberger, pp. 105–32.

Louandre, Charles. *Les arts somptuaires: Histoire du costume et de l'ameublement et des arts et industries qui s'y rattachent.* 2 vols. Paris: Hangard-Maugé, 1857–1858.

Lucas, Prosper. *Traité physiologique et philosophique de l'hérédité naturelle.* Paris: J.-B. Baillière, 1847.

Luchaire, Achille. *Les premiers Capétiens (987–1137).* In *Histoire de France depuis les origines jusqu'à la Révolution*, ed. Ernest Lavisse, vol. 2, part 2. Paris: Hachette, 1911.

Maccioni Ruju, P. Alessandra, and Marco Mostert. *The Life and Times of Guglielmo Libri, 1802–1869.* Hilversum: Verloren Publishers, 1995.

MacLeod, Roy, and Milton Lewis, eds. *Disease, Medicine, and Empire: Perspectives on Western Medicine and the Experience of European Expansion.* New York: Routledge, 1988.

Magnan, Valentin, and Paul-Maurice Legrain. *Les dégénérés, état mental et syndromes épisodiques.* Paris: Rueff, 1895.

Mallette, Karla. *European Modernity and the Arab Mediterranean: Toward a New Philology and a Counter-Orientalism.* Philadelphia: University of Pennsylvania Press, 2010.

Mantegazza, Paolo. *L'amour dans l'humanité: Essai d'une ethnologie de l'amour.* Trans. Emilien Chesneau. Paris: F. Fetscherin et Chuit, 1886 (1885).

Martin, Henri. *Histoire de France.* 17 vols. Paris: Furne, 1855–1860 (1838–1853).

Martin, Henry. *Catalogue des manuscrits de la Bibliothèque de l'Arsenal. Histoire de la Bibliothèque de l'Arsenal*, vol. 8. Paris: Librairie Plon, 1899.

Maury, Alfred. "De l'hallucination envisagée au point de vue philosophique et historique, ou Examen critique de l'opinion émise par M. Brierre de Boismont, touchant les caractères auxquels on doit reconnaître l'hallucination chez certains personnages célèbres de l'histoire." In *Annales médico-psychologiques. Journal de l'anatomie, de la physiologie et de la pathologie du système nerveux* 5 (mai 1845): 317–38.

Mauss, Marcel. "Body Techniques." In *Sociology and Psychology.* Trans. Ben Brewster. London: Routledge & Kegan Paul, 1979, pp. 95–123. ("Les techniques du corps." In *Sociologie et anthropologie.* Paris: Presses universitaires de France, 1950 [1936], pp. 365–86.)

McBride, Theresa. "Divorce and the Republican Family." In *Gender and the Politics of Social Reform in France, 1870–1914*, ed. Elinor A. Accampo, Rachel G. Fuchs, and Mary Lynn Stewart. Baltimore: Johns Hopkins University Press, 1995, pp. 59–81.

McCracken, Peggy. *The Curse of Eve, the Wound of the Hero.* Philadelphia: University of Pennsylvania Press, 2003.

Mellinkoff, Ruth. *Averting Demons: The Protective Power of Medieval Visual Motifs and Themes.* 2 vols. Los Angeles: Ruth Mellinkoff Publications, 2004.

Meltzer, Françoise. *For Fear of the Fire: Joan of Arc and the Limits of Subjectivity*. Chicago: University of Chicago Press, 2001.

Méon, Dominique Martin, ed. *Nouveau recueil de fabliaux et contes inédits, des poètes français des XIIe, XIIIe, XIVe et XVe siècles*. 2 vols. Paris: Chasseriau, 1823.

Méray, Antony. *La vie au temps des cours d'amour*. Paris: Claudin, 1876.

Mérimée, Prosper. *Notes d'un voyage dans l'Ouest de la France*. Paris: Librairie de Fournier, 1836.

Meurger, Michel. *Gilles de Rais et la littérature*. Rennes: Terre de Brume, 2003.

Michéa, Claude-François. *Des hallucinations. Thèse à la Faculté de médecine de Paris*, no. 159. Paris: Righoux, 1837.

Michéa, Claude-François. "Des hallucinations, de leurs causes, et des maladies qu'elles caractérisent." *Mémoires de l'Académie royale de médecine*, vol. 12. Paris: J.-B. Baillière, 1846, pp. 241–71.

Michel, Francisque. *Recherches sur le commerce, la fabrication et l'usage des étoffes de soie, d'or et d'argent et autres tissus précieux en Occident, principalement en France, pendant le Moyen Âge*. 2 vols. Paris: Crapelet, 1852–1854.

Michel, Francisque, and Édouard Fournier. *Histoire des hôtelleries, cabarets, hôtels garnis, restaurants et cafés, et des anciennes communautés et confreries d'hôteliers, de marchands de vins, de restaurateurs, de limonadiers, etc.* Paris: Seré, 1851.

Michelet, Jules. *Histoire de France*. In *Œuvres complètes*, vol. 8. Paris: Flammarion, 1893–1898 (1855).

Michelet, Jules. *Histoire du Moyen Âge*, in *Œuvres complètes*, vols. 1–6. Paris: Flammarion, 1893–1898 (1833–1844).

Michelet, Jules. *Joan of Arc*. Trans. Albert Guérard. Ann Arbor: Ann Arbor Paperbacks, University of Michigan Press, 1967.

Michelet, Jules. *La sorcière*. Paris: Flammarion, 1966 (1862).

Michelet, Jules. "Préface de 1869." *Œuvres complètes*, vol. 1. Paris: Flammarion, 1893–1898, pp. i–xlv.

Michelet, Victor-Émile. *Le secret de la Chevalerie*. Paris: Ch. Bosse, 1928.

Mignard, Prosper. *Éclaircissements sur les pratiques occultes de Templiers*. Dijon: Douillier, 1851.

Mignard, Prosper. *Monographie du coffret de M. le Duc de Blacas*. Paris: Dumoulin, 1852.

Milo, Daniel. "Street Names." In *Realms of Memory: Rethinking the French Past, Volume 2: Traditions*, ed. Pierre Nora. Trans. Arthur Goldhammer. New York: Columbia University Press, 1992, pp. 363–90. ("Le nom des rues." In *Les lieux de mémoire*, ed. Nora, vol. 2, pp. 1887–918.)

Mirecourt, Eugène de. *Bibliophile Jacob, Paul Lacroix*. Coll. Histoire contemporaine, no. 10. Paris: Achille Faure, 1867.

Mirecourt, Eugène de. *Michelet*. Coll. Les contemporains. Paris: Gustave Havard, 1857.

Le modèle familial européen: Normes, déviances, contrôle du pouvoir. Actes des séminaires organisés par l'École française de Rome et l'Università di Roma (1984). Rome: École française de Rome, 1986.

Moll, Albert. *Die conträre Sexualempfindung*. Berlin: H. Kornfeld, 1891. (*The Perversions of the Sex Instinct: A Study of Sexual Inversion*. Trans. Maurice Popkin. Newark, N.J.: Julian Press, 1931.)

Monod, Gabriel. *Les maîtres de l'histoire: Renan, Taine, Michelet*. 2nd ed. Paris: Calman Lévy, 1894 (1875).

Monod, Gabriel. "Opinion de G. Monod." In Reinach, "Gilles de Rais." *Cultes, mythes, et religions*, ed. Duchêne, pp. 1045–46.

Montagnac, Elizé de. *Histoire des Chevaliers Templiers et de leurs prétendus successeurs, suivie de l'histoire des ordres du Christ et de Montesa*. Paris: Auguste Aubry, 1864.

Montaiglon, Anatole de, and Gaston Raynaud, eds. *Recueil général et complet des fabliaux des XIIIe et XIVe siècles*, 6 vols. Paris: Librairie des bibliophiles, 1872–1890.

Morache, Dr. Georges. *Le mariage: Étude de socio-biologie et de médecine légale*. Paris: Felix Alcan, 1902.

Moreau de Tours, Jacques-Joseph. *La psychologie morbide dans ses rapports avec la philosophie de l'histoire ou De l'influence des névropathies sur le dynamisme intellectuel*. Paris: Victor Masson, 1859.

Morel, B.-A. "Folie héréditaire. Condamnation pour outrage public aux mœurs, d'un individu qui comptait des aliénés dans son ascendance paternelle et maternelle. Considérations préliminaires sur les actes immoraux et pervers, commis par des individus non aliénés." *Gazette hebdomadaire de médecine et de chirurgie* 7:46 (15 novembre 1861): 734–37; 8:48 (29 novembre 1861): 768–70.

Morel, Benedict-Augustin. *Traité des dégénérescences physiques, intellectuelles et morales de l'espèce humaine*. Paris: J. B. Baillière, 1857.

Morel, Benedict-Augustin. *Traité des maladies mentales*. Paris: V. Masson, 1860.

Mucchielli, Laurent. "Criminologie, hygiénisme et eugénisme en France, 1870–1914: Débats médicaux sur l'élimination des criminels réputés 'incorrigibles.'" *Revue d'histoire des sciences humaines* 3 (2000): 57–88.

Murat, Laure. *La loi du genre: Une histoire culturelle du 'troisième sexe.'* Paris: Fayard, 2006.

Murat, Laure. *L'homme qui se prenait pour Napoléon: Pour une histoire politique de la folie*. Paris: Gallimard, 2011.

Murray, Jacqueline, and Konrad Eisenbichler, eds. *Desire and Discipline: Sex and Sexuality in the Premodern West*. Toronto: University of Toronto Press, 1996.

Müller-Wille, Staffan, and Hans-Jörg Rheinberger. "Heredity—The Formation of an Epistemic Space." In *Heredity Produced*, ed. Müller-Wille and Rheinberger, pp. 3–34.

Müller-Wille, Staffan, and Hans-Jörg Rheinberger, eds. *Heredity Produced: At the Crossroads of Biology, Politics, and Culture, 1500–1870*. Cambridge, Mass.: MIT Press, 2007.

Naquet, Alfred. *Le divorce*. 2nd ed. Paris: E. Dentu, 1881 (1877).

Naquet, Alfred. *Le divorce, Chambre des Députés, séance du 27 mai 1879, Extrait du Journal officiel*. Paris: Bureaux de l'École laïque, 1879.

Newman, F. X., ed. *The Meaning of Courtly Love*. Albany: State University of New York Press, 1968.

Nichols, Stephen G. "Introduction: Philology in a Manuscript Culture." In *The New Philology*, ed. Nichols, pp. 1–10.

Nichols, Stephen G. "The New Medievalism: Tradition and Discontinuity in Medieval Culture." In *The New Medievalism*, ed. Marina Brownlee, Kevin Brownlee, and Stephen G. Nichols. Baltimore: Johns Hopkins University Press, 1991, pp. 1–26.

Nichols, Stephen G., ed. *The New Philology. Speculum* 65:1 (1990).

Nirenberg, David. "Mass Conversion and Genealogical Mentalities: Jews and Christians in Fifteenth-Century Spain." *Past and Present* 174:1 (2002): 3–41.

Nodier, Charles. "Introduction." In La Curne de Sainte-Palaye, *Mémoires sur l'ancienne chevalerie*, ed. Nodier, vol. 1, pp. i–xxv.

Nodier, Ch., P. Lepeintre, L. Thiessé, and A. Lesourd, eds. *Bibliothèque dramatique, ou Répertoire universel du théâtre français*. Paris: Dabo-Butschert, 1824.

Nora, Pierre, ed. *Les lieux de mémoire*. 3 vols. Paris: Gallimard/Quarto, 1997 (1986).

Nye, Robert A. *Crime, Madness, and Politics in Modern France: The Medical Concept of National Decline*. Princeton, N.J.: Princeton University Press, 1984.

Nye, Robert A. "Heredity or Milieu: The Foundations of Modern European Criminological Theory." *Isis* 67:3 (1976): 334–55.

Nye, Robert A. "Honor, Impotence, and Male Sexuality in Nineteenth-Century French Medicine." *French Historical Studies* 16:1 (1989): 48–71.

Ory, Pascal. *Histoire culturelle*. Paris: Presses universitaires de France, 2004.

Ory, Pascal. "Le 'Grand Dictionnnaire' de Pierre Larousse. Alphabet de la République." In *Les lieux de mémoire*, ed. Nora, vol. 1, pp. 227–38.

Oulebsir, Nabila. *Les usages du patrimoine: Monuments, musées et politique coloniale en Algérie (1830–1930)*. Paris: Éditions de la Maison des sciences de l'homme, 2004.

Parchappe, Maximien. *Études sur le sang dans l'état physiologique et l'état pathologique*. Paris: J.-B. Baillière, 1856.

Parent-Aubert. *Almanach des mystères de l'amour conjugal et de l'hygiène du mariage*. Paris, 1852.

Parent-Duchâtelet, Alexandre-Jean-Baptiste. *De la prostitution dans la ville de Paris, considérée sous le rapport de l'hygiène publique, de la morale et de l'administration: Ouvrage appuyé de documens statistiques puisés dans les archives de la Préfecture de police*. 2 vols. Paris: J.-B. Baillière, 1836.

Paris, Gaston. "Études sur les romans de la table ronde." *Romania* 10 (1881): 465–96.

Paris, Gaston. "Études sur les romans de la table ronde: *Lancelot du Lac*." *Romania* 12 (1883): 459–534.

Paris, Gaston. "*La Chanson de Roland* et la nationalité française. Leçon d'ouverture au Collège de France, lc 8 décembre 1870." In *La poésie du Moyen Âge*, vol. 1, pp. 87–118.

Paris, Gaston. *La poésie du Moyen Âge*. 2 vols. Paris: Librairie Hachette, 1885.

Paris, Gaston. "La poésie du Moyen Âge." In *La poésie du Moyen Âge*, vol. 1, pp. 1–39.

Paris, Gaston. *Les contes orientaux dans la littérature française du Moyen Âge*. Paris: Librairie A. Franck, 1875.

Paris, Gaston. "Les cours d'amour du Moyen Âge." In *Mélanges de littérature française du Moyen Âge*, ed. Mario Roques. Paris: Honoré Champion, 1966 (1888), pp. 473–97.

Paris, Gaston. "Paulin Paris et la littérature du Moyen Âge." In *La poésie du Moyen Âge*, vol. 1, pp. 211–54.

Paris, Gaston. "Préface." In *La poésie du Moyen Âge*, vol. 1, pp. v–xiv.

Partner, Peter. *The Murdered Magicians: The Templars and Their Myth*. Oxford: Oxford University Press, 1982.

Pastoureau, Michel. "Ceci est mon sang. Le christianisme médiéval et la couleur rouge." In *Le pressoir mystique*, ed. D. Alexandre-Bidon. Paris: Le Cerf, 1990, pp. 43–56.

Pauphilet, Albert. "Le mythe du Moyen Âge." In *Le legs du Moyen Âge. Études de littérature médiévale*. Melun: Librairie d'Argences, 1950, pp. 23–63.

Pedersen, Elisabeth. *Legislating the French Family: Feminism, Theater, and Republican Politics, 1870–1920*. New Brunswick, N.J.: Rutgers University Press, 2003.

Peeters, Kris. "La découverte littéraire du Fabliau au XVIIIe siècle: le Comte de Caylus dans l'histoire d'un genre médiéval." *Revue d'histoire littéraire de la France* 4:106 (2006): 827–42.

Périer, Jean-Alexandre-Napoléon. *Essai sur les croisements ethniques*. Paris: Martinet, 1865.

Petit de Julleville, Louis. *Histoire du théâtre en France. La comédie et les mœurs en France au Moyen Âge*. Paris: L. Cerf, 1886.

Petit-Dutaillis, Charles. *Charles VII, Louis XI et les premières années de Charles VIII, 1422–1492*. In *Histoire de France depuis les origines jusqu'à la Révolution*, ed. Ernest Lavisse, vol. 4, part 2. Paris: Hachette, 1902.

Pick, Daniel. *Faces of Degeneration: A European Disorder, c. 1848–c. 1918*. Cambridge: Cambridge University Press, 1989.

Pinel, Philippe. *Nosographie philosophique*. 3 vols. 6th ed. Paris: J. A. Brosson, 1818 (1797).

Pître-Chevalier. *La Bretagne ancienne et moderne*. Paris: W. Coquebert, 1844.

Plane, J. M. *Apologie des Templiers et des Francs-Maçons*. Meudon, 1797.

Pomian, Krzysztof. "Franks and Gauls." In *Realms of Memory: Rethinking the French Past, Volume 1: Conflicts and Divisions*, ed. Pierre Nora. Trans. Arthur Goldhammer. New York: Columbia University Press, 1992, 25–76. ("Francs et Gaulois." In *Les lieux de mémoire*, ed. Nora, vol. 2, pp. 2245–300.)

Postel, Jacques, and Claude Quetel, eds. *Nouvelle histoire de la psychiatrie*. Paris: Dunod, 1994.

Poulot, Dominique. "Alexandre Lenoir et les musées des monuments français." In *Les lieux de mémoire*, ed. Nora, vol. 1, pp. 1515–43.

Poupin, Victor. "Préface." In Armand Hayem, *Le mariage*, 2 vols. Paris: Librairie de la bibliothèque démocratique, 1876, pp. 6–7.

Prevost, Marie-Laure. "Vrain Lucas, le Balzac du faux." *Revue de la Bibliothèque nationale de France* 13 (2003): 59–69.

Prouteau, Gilbert. *Gilles de Rais ou La gueule du loup*. Monaco: Éd. du Rocher, 1992.

Quinet, Edgar. "Études sur le Moyen Âge." In *La pensée d'Edgar Quinet: Étude sur la formation de ses idées avec essais de jeunesse et documents inédits*, ed. Willy Aeschimann. Paris: Éditions Anthropos; Genève: Georg, 1986, pp. 469–509.

Rabello, F. E. "Les origines de la syphilis." *Nouvelle presse médicale* 2:20 (1973): 1376–80.

Raffalovich, Marc André. *Uranisme et unisexualité: Étude sur différentes manifestations de l'instinct sexuel.* Lyon: A. Storck; Paris: Masson, 1896.

Raynouard, François Just Marie. *Choix des poésies originales des troubadours.* 6 vols. Paris: F. Didot, 1816–1821.

Raynouard, François Just Marie. "*Fundgruben des Orients*" (Review). *Journal des sçavans* (mars 1819): 151–61; (avril 1819): 221–29.

Raynouard, François Just Marie. *Les Templiers, tragédie. Précédée d'un précis historique sur les Templiers.* Paris: Giguet et Michaud, 1805.

Raynouard, François Just Marie. *Monumens historiques relatifs à la condamnation des Chevaliers du Temple, et l'abolition de leur ordre.* Paris: A. Égron, 1813.

Regnault, Dr. Jules. *La sorcellerie, ses rapports avec les sciences biologiques.* Bordeaux: Imprimerie du Midi, P. Cassignol, 1896.

Reinach, Salomon. "Bas-reliefs énigmatiques." *Revue africaine* 52:269 (1908): 155–70.

Reinach, Salomon. "Gilles de Rais." In *Cultes, mythes, et religions*, ed. Hervé Duchêne. Paris: Robert Laffont, 1996 (1904), pp. 1026–49.

Reinach, Salomon. "La tête magique des Templiers." In *Cultes, mythes et religions*, ed. Hervé Duchêne. Paris: Robert Laffont, 2000 (1911), pp. 1015–25.

Rétif de la Bretonne, Nicolas Edme. *Le pornographe ou Idées d'un honnête homme sur un projet de règlement pour les prostituées, propre à prévenir les malheurs qu'occasionne le publicisme des femmes.* London; Paris, 1769.

Rétif de la Bretonne, Nicolas Edme. *Les Parisiennes, ou XL caractères généraux pris dans les mœurs actuelles, propres à servir à l'instruction des personnes du sexe.* Paris, 1787.

Rétif de la Bretonne, Nicolas Edme. *Tableaux de la vie, ou les Mœurs du dix-huitième siècle.* Strasbourg, 1767.

"Review of Auguste Brachet." In *Revue des cours littéraires de la France et de l'Étranger*, 5e année, no. 9 (1er février, 1868): 152.

Rey, Alain. "Les trésors de la langue." *Les lieux de mémoire*, ed. Nora, vol. 2, pp. 2189–205.

Ribot, Théodule. *L'hérédité, étude psychologique.* 5th corrected and rev. ed. Paris: Felix Alcan, 1894.

Ribot, Théodule. *L'hérédité, étude psychologique sur ses phénomènes, ses lois, ses causes, ses conséquences.* Paris: Ladrange, 1873.

Richer, Jean. "Deux lettres de Victor Hugo à Paul Lacroix: P. Lacroix, Vigny, Hugo. Une source probable du *Roi s'amuse.*" *Revue des sciences humaines* 98–99 (1960): 199–201.

Ridoux, Charles. *Évolution des études médiévales en France de 1860 à 1914.* Paris: Honoré Champion, 2001.

Dr. Riolan. *Accouchement. Nouvelle collection exclusive d'hygiène et de médecine*, vol. 7. Paris: F. Pierre, 1909.

Dr. Riolan. *Chasteté, virginité et célibat. Nouvelle collection exclusive d'hygiène et de médecine*, vol. 3. Paris: F. Pierre, 1909.

Dr. Riolan. *Les maladies honteuses. Nouvelle collection exclusive d'hygiène et de médecine*, vol. 12. Paris: F. Pierre, 1909.

Dr. Riolan. *Pédérastie et homosexualité. Nouvelle collection exclusive d'hygiène et de médecine*, vol. 10. Paris: F. Pierre, 1909.

Rollet, Joseph. "Syphilis (Synonymie; Historique)." In *Dictionnaire encyclopédique des sciences médicales*, ed. Dechambre, 3e série, t. 14 (1884), pp. 256–90.

Ronsin, Francis. *La grève des ventres: Propagande néo-malthusienne et baisse de la natalité française, XIXe–XXe siècles*. Paris: Aubier Montaigne, 1980.

Ronsin, Francis. *Le contrat sentimental: Débats sur le mariage, l'amour, le divorce, de l'Ancien Régime à la Restauration*. Paris: Aubier, 1990.

Ronsin, Francis. *Les divorciaires: Affrontements politiques et conceptions du mariage dans la France du XIXe siècle*. Paris: Aubier, 1992.

Rossiaud, Jacques. *La prostitution médiévale*. Paris: Flammarion, 1988.

Roubaud, Félix. *Le livre des époux. Guide pour la guérison de l'impuissance, de la stérilité, et de toutes les maladies des organes génitaux*. Paris: L'auteur, 1852.

Roudinesco, Élisabeth. *La part obscure de nous-mêmes: Une histoire des pervers*. Paris: Albin Michel, 2007.

Rougemont, Denis de. *L'amour et l'Occident*. Paris: Plon, 1939.

Routledge, Michael. "The Later Troubadours." In *The Troubadours: An Introduction*, ed. Simon Gaunt and Sarah Kay. Cambridge: Cambridge University Press, 1999, pp. 99–112.

Roy, Just-Jean-Etienne. *Les Templiers. Histoire et procès*. Paris: Jean de Bonnot, 1995 (1848).

Sabean, David Warren. "From Clan to Kindred: Kinship and the Circulation of Property in Premodern and Modern Europe." In *Heredity Produced,* ed. Müller-Wille and Rheinberger, pp. 37–59.

Sabean, David Warren. "Kinship and Class Dynamics in Nineteenth-Century Europe." In *Kinship in Europe: Approaches to Long-Term Development (1300–1900)*, ed. David Warren Sabean, Simon Teuscher, and Jon Mathieu. New York: Berhahn Books, 2010 (2007), pp. 301–13.

Sabean, David Warren, and Simon Teuscher. "Kinship in Europe: A New Approach to Long Term Development." In *Kinship in Europe: Approaches to Long-Term Development (1300–1900)*, ed. David Warren Sabean, Simon Teuscher, and Jon Mathieu. New York: Berhahn Books, 2010 (2007), pp. 1–32.

Sacy, Silvestre de. "Livres divers" (Review of Hammer-Purgstall). *Magasin encyclopédique* 6 (1810): 145–75.

Said, Edward W. *Orientalism*. New York: Vintage Books, 2003 (1978).

Saint-Jorre, John de. *Fernand Fleuret et ses amis*. Coutances: Imprimerie P. Bellée, n.d. (1945?).

Samuels, Maurice. *The Spectacular Past: Popular History and the Novel in Nineteenth-Century France*. Ithaca, N.Y.: Cornell University Press, 2004.

Schneider, William. "Toward the Improvement of the Human Race: The History of Eugenics in France." *Journal of Modern History* 54:2 (1982): 268–91.

Schultz, James A. *Courtly Love, the Love of Courtliness, and the History of Sexuality.* Chicago: University of Chicago Press, 2006.

Sedgwick, Eve Kosofsky. *Epistemology of the Closet.* Berkeley: University of California Press, 1990.

Dr. Serrurier. *Du mariage considéré dans ses rapports physiques et moraux. Inconvénients des unions entre des individus entachés d'un principe constitutif héréditaire.* Paris: J.-B. Baillière, 1845.

Simon, Walter Michael. *European Positivism in the Nineteenth Century: An Essay in Intellectual History.* Ithaca, N.Y.: Cornell University Press, 1963.

Sismondi, Jean Charles Léonard Simonde de. *De la littérature du Midi de l'Europe.* Paris: Treuttel et Würtz, 1813.

Soueix, Robert. *Gilles de Raiz devant les médecins, essai médico-psychologique.* Bordeaux: Imprimerie de l'Académie et des Facultés, 1931, no. 83.

Sourdeval, Charles Mourain de. "Les Sires de Retz et le château de Machecoul." *Mémoires de la Société archéologique de Touraine* 1 (1842): 27–70.

Spiegel, Gabrielle M. "Foucault and the Problem of Genealogy." *Medieval History Journal* 4:1 (2001): 1–14.

Spiegel, Gabrielle M. "Genealogy: Form and Function in Medieval Historiography." In *The Past as Text: The Theory and Practice of Medieval Historiography.* Baltimore: Johns Hopkins University Press, 1997 (1983), pp. 99–110.

Mme de Staël. *De l'Allemagne.* Paris: Garnier Flammarion, 1968 (1813).

Stahuljak, Zrinka. "'Archéologie pathologique': Le Moyen Âge et l'histoire de la médecine." In *Fantasmagories du Moyen Âge, Senefiance* 56, ed. Élodie Burle-Errecade and Valérie Naudet. Aix-en-Provence: Publications de l'université de Provence, CUER MA, 2010, pp. 209–16.

Stahuljak, Zrinka. *Bloodless Genealogies of the French Middle Ages:* Translatio, *Kinship, and Metaphor.* Gainesville: University Press of Florida, 2005.

Stendhal. *De l'amour.* Paris: Gallimard/Folio, 1980 (1822).

Stendhal. *Mémoires d'un touriste.* 2 vols. Paris: A. Dupont, 1838.

Stockhorst, Stefanie. "Passionate Pilgrims: Secular Lead Badges as Precursors of *Emblemata Amatoria.*" In *Profane Images in Marginal Arts of the Middle Ages*, ed. Elaine C. Block, with Frédéric Billiet, Sylvie Bethmont-Gallerand, and Paul Hardwick. Turnhout: Belgium: Brepols, 2009, pp. 157–81.

Stocking, George W. *Race, Culture, and Evolution: Essays in the History of Anthropology.* New York: Free Press, 1968.

Stocking, George W. *Victorian Anthropology.* New York: Free Press, 1987.

Stoler, Ann Laura. *Carnal Knowledge and Imperial Power: Race and the Intimate in Colonial Rule.* Berkeley: University of California Press, 2002.

Stoler, Ann Laura. *Race and the Education of Desire: Foucault's History of Sexuality and the Colonial Order of Things.* Durham, N.C.: Duke University Press, 1995.

Strauss, Jonathan. *Medicine, Death, and Desire in Nineteenth-Century Paris.* New York: Fordham University Press, 2011.

Strezova, Mila Simeonova. *Gilles de Rais. Étude médico-légale et psychiatrique. Thèse présentée pour le doctorat en médecine.* Faculté de Médecine de Strasbourg, 1934.

Sullivan, Karen. *The Interrogation of Joan of Arc.* Minneapolis: University of Minnesota Press, 1999.

Sunderland, Luke. "Entre résistance et révolte: Les enjeux du territoire." *Revue des langues romanes* 112 (2008): 465–90.

Dr. Surbled. *L'amour sain.* Paris: A. Maloine, 1903.

Sutphen, Mary P., and Bridie Andrews, eds. *Medicine and Colonial Identity.* New York: Routledge, 2003.

Taburet-Delahaye, Elisabeth, ed., with Jean-Christophe Ton-That and Marie-Alice Virlouvet. *Thermes et hôtel de Cluny, Musée national du Moyen Âge: Œuvres nouvelles, 1995–2005.* Paris: Réunion des musées nationaux, 2006.

Tarde, Gabriel. "Introduction." In Jacoby, *Études sur la sélection chez l'homme*, pp. i–iii.

Tardieu, Ambroise. *Étude médico-légale sur la folie.* Paris: J.-B. Baillière, 1872.

Tarsot, M. L. *Fabliaux et contes du Moyen Âge.* Paris: Renouard, 1913.

Tatar, Maria. *Secrets beyond the Door: The Story of Bluebeard and His Wives.* Princeton, N.J.: Princeton University Press, 2004.

Theis, Laurent. "Guizot et les institutions de mémoire." In *Les lieux de mémoire*, ed. Nora, vol. 1, pp. 1575–97.

Thoinot, Léon-Henri. *Attentats aux mœurs et perversions du sens génital.* Paris: Octave Doin, 1898.

Tourdes, G. "Mariage (Médecine légale)." In *Dictionnaire encyclopédique des sciences médicales*, ed. Dechambre, 2e série, vol. 5 (1874), pp. 83–104.

Tournier, Michel. *Gilles & Jeanne.* Trans. Alan Sheridan. London: Minerva, 1987. (*Gilles & Jeanne.* Paris: Gallimard/Folio, 1983.)

Trautmann, Thomas R. "Dravidian Kinship and the Anthropology of Lewis Henry Morgan." *Journal of the American Oriental Society* 108:1 (1988): 93–97.

Trautmann, Thomas R. *Lewis Henry Morgan and the Invention of Kinship.* Berkeley: University of California Press, 1987.

Trélat, Ulysse. *La folie lucide étudiée et considérée au point de vue de la famille et de la société.* Paris: A. Delahaye, 1861.

Trigger, Bruce G. *A History of Archaeological Thought.* 2nd ed. Cambridge: Cambridge University Press, 2006 (1989).

Uzanne, Octave. *Parisiennes de ce temps en leurs divers milieux, états et conditions. Études pour servir à l'histoire des femmes, de la société, de la galanterie française, des mœurs contemporaines et de l'égoïsme masculin.* Paris: Mercure de France, 1910.

Vialleton, Louis-Marius. *Principales théories de l'hérédité.* Lyon: A. Storck; Paris: G. Masson, 1893.

Villemont, Dr. Michel. "Cours d'amour." In *Dictionnaire historique et scientifique de l'Amour et du mariage.* Paris: A. Lévy et Cie, 1886, pp. 112–16.

Villemont, Dr. Michel. *Hygiène de l'amour et du mariage*. Paris: Librairie des publications nouvelles, n.d. (1883).

Villemont, Dr. Michel. *Histoire des maladies vénériennes*. Paris: Librairie des publications nouvelles, n.d. (1882).

Villemont, Dr. Michel. *L'amour conjugal*. Paris: A. Lévy, 1885.

Viollet-le-Duc, Eugène. "Restauration." In *Dictionnaire raisonné de l'architecture française du XIe au XVIe siècle*, vol. 8. Paris: B. Bance, 1866, pp. 14–34.

Virey, J. J. *De la femme sous ses rapports physiologique, moral et littéraire*. Paris: Crochard, 1823.

Viriville, Auguste Vallet de. *Histoire de Charles VII, roi de France, et de son époque, 1403–1461*. 3 vols. Paris: Renouard, 1862–1865.

Viriville, Auguste Vallet de. "Rais ou Retz, Gilles de." In *Nouvelle biographie générale*, ed. Dr. Hoefer, vol. 41. Paris: Firmin Didot Frères, 1862, pp. 498–501.

Visker, Rudi. *Michel Foucault: Genealogy as Critique*. Trans. Chris Turner. London: Verso, 1995.

Voisenet, Jacques. "Le tabou du sang dans les pénitentiels du haut Moyen Âge." In *Le sang au Moyen Âge*, Les Cahiers du C.R.I.S.I.M.A. 4. Montpellier: L'Université de Paul-Valéry Montpellier III, 1999, pp. 111–25.

Voltaire. *Essai sur les mœurs et l'esprit des nations, et sur les principaux faits de l'histoire, depuis Charlemagne jusqu'à Louis XIII*. In *Œuvres*, ed. M. Beuchot, vols. 15–18. Paris: Werdet et Lequien, 1829 (1756–1763).

Wallon, Henri. "Notice sur la vie et les œuvres de M. Paulin Paris." *Académie des Inscriptions et Belles-Lettres*, 4e série, t. 10 (1882): 330–88.

Warren, Michelle R. *Creole Medievalism: Colonial France and Joseph Bédier's Middle Ages*. Minneapolis: University of Minnesota Press, 2011.

Weismann, August. *Die Continuität des Keimplasmas als Grundlage einer Theorie der Vererbung*. Jena: G. Fischer, 1885.

White, Owen. *Children of the French Empire: Miscegenation and Colonial Society in French West Africa*. Oxford: Oxford University Press, 1999.

Williams, Elizabeth A. *The Physical and the Moral: Anthropology, Physiology, and Philosophical Medicine in France, 1750–1850*. Cambridge: Cambridge University Press, 1994.

Williams, Harry F. "French Fabliau Scholarship." *South Atlantic Review* 46:1 (1981): 76–82.

Wilson, Geoffrey. *A Medievalist in the Eighteenth Century: Le Grand d'Aussy and the Fabliaux ou Contes*. The Hague: Nijhoff, 1975.

Wohnlich-Despaigne, Isabelle. "Introduction." In *Les historiens français de la médecine au XIXe siècle*. Paris: J. Vrin, 1987, pp. 11–33.

Zorzi, Diego. "Un frammento provenzale della *Doctrina Pueril* di Raimondo Lull." *Aevum* 28:4 (1954): 345–49.

Zumthor, Paul. *Speaking of the Middle Ages*. Trans. Sarah White. Lincoln: University of Nebraska Press, 1986. (*Parler du Moyen Âge*. Paris: Éditions de Minuit, 1980.)

WEB

Crimino Corpus. Dossier "Autour des *Archives d'Anthropologie Criminelle*." http://crimino-corpus.revues.org/161. Accessed November 6, 2011.

"Francis I." *Encyclopaedia Britannica Online*. http://www.britannica.com/EBchecked/topic/216656/Francis-I. Accessed May 26, 2011.

"Georges Hérelle (1848–1935)." http://www.garae.fr/spip.php?article220. Accessed October 11, 2011.

"Lamartine, orateur lyrique. Discours parlementaires (1834–1851)." http://www.assemblee-nationale.fr/histoire/lamartine/discours.asp#titre1. Accessed February 29, 2012.

Index

Acknowledgments

The book's first chapters were written at the Institute for Advanced Study, Princeton. I thank Joan W. Scott for her generous intellectual support and a fabulous year away (2005–2006), and the amazing staff of the IAS, especially Gabriella Hanson and Linda Geracci, and my IAS colleague, Michael Peletz, with whom I was fortunate enough to spend the same year there and who understood this project long before I did. The rest of the book would still not have seen the light of day had it not been for Dominic Thomas, chair of the Department of French and Francophone Studies at University of California at Los Angeles from 2005 to 2011, and the funding generously provided by the UCLA Academic Senate and Dean of the Humanities that supported the research at the Bibliothèque nationale de France, Bibliothèque de l'Arsenal, and the Musée national du Moyen Âge-Thermes de Cluny from 2006 to 2011. I also thank the UCLA Center for Medieval and Renaissance Studies for additional funds for image reproductions and permissions.

Enthusiasm, precious advice, and access to rare materials carried this book forward. I thank Monique Nemer, who was the source of great references and philological debates; Pierre Lartigue, who knew all the nineteenth-century forgeries but sadly didn't live to see this artifact; Emmanuel Pierrat, who opened his personal library of rare books of nineteenth- and twentieth-century *curiosa*; Sylvain Bellenger and Jean-Loup Champion, who guided me through art historical issues; and the staff of the Musée national du Moyen Âge-Thermes de Cluny, especially Marie-Alice Virlouvet, Jeannine Mercier, and Jean-Christophe Ton-That.

I am deeply indebted to Peggy McCracken for her enduring support and guidance in all aspects of intellectual and professional life, Virginie Greene for her great generosity, James D. Schultz for intellectual companionship and epistolary exchange, and my writing companions Christine Chism, Caroline Ford, Barbara Fuchs, Hannah Landecker, and Sarah Stein. I have also benefited greatly from the suggestions and support of Bill Burgwinkle, Howard Bloch, Sarah Kay, and Helen Solterer.

I especially thank colleagues who gave me an opportunity to test my ideas: Jonathan Strauss, and the faculty and graduate students at Miami University in Oxford, Ohio; Noah D. Guynn at the University of California, Davis; Deborah McGrady and the graduate students at the University of Virginia; Cynthia J. Brown, Jody Enders, and Carol Pasternack at the University of California, Santa Barbara; Maurice Samuels at the University of Pennsylvania; and the faculty of the Department of French and Romance Philology at Columbia University.

I am most grateful to Alisa Belanger for wonderful translations from modern French and her precious editorial help; to Katherine McLoone, my research assistant of many years who must have seen nearly all the references go through her most able and professional hands; and to Michael Weinberg for his bibliographic assistance.

The one that comes last comes first in this *Pornographic Archaeology*: I thank Laure Murat, whose curiosity and encouragement transformed a passing thought into a research project and whose love and patience nurtured it until it became a book.